Introduction to
ASTRONOMY

ASTRONOMY, *from Giotto's Campanile in Florence, sculptured about 1275 by Andrea Pisano. He is observing, with a quadrant, an instrument for measuring angles. Three signs of the Zodiac (Capricornus, Aquarius, and Pisces) are seen in the background (Photograph by Alinari).*

Introduction to
ASTRONOMY

Cecilia Payne-Gaposchkin

PH.D., D.SC., SC.D., PHILLIPS ASTRONOMER
HARVARD UNIVERSITY

Prentice-Hall, Inc. ENGLEWOOD CLIFFS, N. J.

PRENTICE-HALL PHYSICS SERIES

Donald H. Menzel, Editor

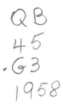

First printing ....June, 1954
Second printing ....May, 1955
Third printing ....September, 1958

47809

To Peggy Brown, Miles Davis, Fred Franklin,
Elizabeth Hollander, Jean Lockman,
Jean McCollum and John Olney
for their help and encouragement

PREFACE

This book is intended to introduce the elements of astronomy to the student and to the general reader who may have little background in mathematics or physics.

My chief reason for writing the book was a desire to give emphasis to stars and stellar systems as well as to the solar system, which usually occupies the major part of an elementary book. The greater part of modern astronomy is concerned with stars, and even an introductory text should reflect this emphasis.

I hope that this book will serve the purpose, among other things, of introducing astronomy to the liberal arts student who is "fulfilling a science requirement." With this in mind, I have been at pains to point out associations with the fields of language, literature, and history. Astronomy has played no small part in the cultural development of the human race.

Many students are unable to devote more than one semester to the study of astronomy; for these the book has been organized so that some chapters could be omitted altogether without impairing the understanding of the remainder. The reader without historical interests could, without loss, omit Chapters I and VII. Chapter IV, which discusses technical matters, could largely be omitted in a short course. So also could the two long chapters that deal in detail with stars (Chapters XII and XIII). Chapter XI, which briefly surveys stellar astronomy, would provide a sufficient introduction to the exciting studies that are today being made of systems of stars and of the remoter regions of our universe (Chapters XIV to XVIII).

My especial thanks are due to Dr. E. C. Slipher, of the Lowell Observatory, for preparing the beautiful planetary photographs that illustrate Chapter VIII. I am greatly indebted to friends and colleagues who have given the manuscript careful reading and valuable criticism: Dr. Frank K. Edmondson, Dr. Helen Sawyer Hogg, Dr. C. M. Huffer, Dr. R. W. Long, and Dr. D. H. Menzel, editor of the series of which this book forms a part. Valuable advice on special topics has been given by Mr. Wilbur Cheever and Dr. Taylor Starck. I should be ungrateful if I did not acknowledge the help given to me by the members of the small class in astronomy on whom I first inflicted the material of this book, and to whom I now dedicate it. My thanks are also due to Miss Sybil Chubb for her careful work in the typing of these chapters.

I acknowledge with thanks the permission given by various publishers for the use of some of the quotations in the text: Burns, Oates and Washbourne for *In No Strange Land* by Francis Thompson; Constable and Company, Ltd., for a quotation from *Lucifer in Starlight* by George Meredith; Doubleday and Co., Inc., for an excerpt from *Passage to India* by Walt Whitman; Houghton Mifflin Company for a verse from the Bayard Taylor translation of Goethe's *Faust*; The Oxford University Press, Inc., for permission to quote from *Theoretical Astrophysics* by Svein Rosseland, and from *The Stars* by Gerard Manley Hopkins; and to the Society of Authors and Mrs. Laurence Binyon for permission to use a number of extracts from Laurence Binyon's translation of the *Divine Comedy*.

CECILIA PAYNE-GAPOSCHKIN

CONTENTS

Introduction to

ASTRONOMY

INTRODUCTION:
THE DAWN OF ASTRONOMY

> When Science from Creation's face
> Enchantment's veil withdraws,
> What lovely visions yield their place
> To cold material laws!
>
> THOMAS CAMPBELL

> Who has not experienced the mysterious thrill of springtime in a forest, with sunbeams flickering through the foliage, and the low humming of insect life? It is the feeling of unity with nature, which is the counterpart of the attitude of the scientist, analysing the sunbeams into light quanta and the soft rustlings of the dragon-fly into condensations and rarefactions of the air. But what is lost in fleeting sentiment is more than regained in the feeling of intellectual security afforded by the scientific attitude, which may grow into a trusting devotion, challenging the peace of the religious mystic. For in the majestic growth of science, analytical in its experimental groping for detail, synthetic in its sweeping generalizations, we are watching at least one aspect of the human mind, which may be believed to have a future of dizzy heights and nearly unlimited perfectibility.
>
> SVEIN ROSSELAND
> *Theoretical Astrophysics* (Introduction)

Astronomy is certainly the oldest, yet perhaps the simplest, of the sciences. In a way it contains the broad generalization of all physical science, for it spans conditions wider than can be produced on earth. It has

> . . . such large discourse,
> Looking before and after . . .

that the astronomical scale of time is reckoned, not in centuries or millenniums (thousands of years), but in hundreds of millions of years.

Science is not a form of magic, popular opinion to the contrary. Magic was, perhaps, its ancestor, in the days when man personified the phenomena of nature as gods to be feared and placated. But as knowledge grew, the world was found to be predictable. A calendar could be drawn up to foretell the rhythm of the seasons. A still further step revealed that the world can be analyzed, and some phenomena controlled: *science* and *technology* were born.

Neither is science a system of mere technical description. Most of its concepts can be expressed in simple, everyday language. But such statements, if they are to be both correct and comprehensible, are apt to be lengthy. Science uses technical words and symbols to save time and to be definite. An idea can often be expressed more briefly and more clearly in symbols than in words.

For example, contrast the statement that "the sum of two numbers, when multiplied by itself, is equal to the sum of the first number multiplied by itself, the second number multiplied by itself, and twice the number obtained by multiplying the two numbers together" with the statement,

$$(a + b)^2 = a^2 + 2ab + b^2.$$

The advantage of the shorthand symbolic expression should be obvious. Again, consider how much time is saved by expressing "the point in a planet's orbit where it is furthest from the sun" by the single word "aphelion." Our everyday language is actually permeated by shorthand technical terms (fuse, radio tube, accelerator). Science uses a technical vocabulary to save both time and confusion. But this vocabulary is a means, not an end. We should never make the mistake of supposing that a technical expression is an explanation; it is nothing but a label.

Even astronomy, remote as it appears to be from everyday life, has practical applications to navigation, surveying, and the firing of projectiles. But from the human point of view, its greatest importance lies in its cultural influence. Astronomy first revealed to man the existence of natural law, and development of the science has steadily broadened our horizons.

To primitive man, the earth was the center of the universe. Gradually our planet emerged as but one of many that circle the sun; the sun assumed the central role. Further widening of the horizon revealed that the sun is but one star among many, and that other stars may rival or excel our own sun in size and brilliance. At first the system of stars was pictured as arranged around the sun, but as knowledge accumulated, the sun itself, with many other stars, was found to be moving in a huge orbit about a distant center. Our system of stars is itself a gigantic, rotating body, isolated in space. A further expansion of our horizon revealed that our stellar system (a *galaxy*) does not fill the whole universe. Millions of essentially similar, widely separated stellar systems populate space. True, our own is one of the largest, but it is in no way unique. Nor does it appear to occupy a favored position in the universe. The advance of astronomical knowledge has successively dethroned the earth, the sun, and the stellar system from their supposed unique and central stations.

1. THE CONSTELLATIONS

I know of nightly star-groups the assemblage
And those that bring to men winter and summer.
Bright dynasts, as they pride them in the aether—
Stars, when they wither, and the uprisings of them.

AESCHYLUS, *Agamemnon,* 4–5

The *constellations* carry us back to the dawn of astronomy. They have been called the fossil remains of primitive stellar religion, and as such they have extraordinary interest.

The erratic arrangement of the naked-eye stars has remained essentially unchanged since the time of the first records. The earliest extant list of stars was compiled in about 120 B.C. by the Greek astronomer Hipparchus, and all the stars that he described can be found, with the same brightness and practically at the same place, in the skies of today. The apparent immutability of the "eternal stars" is an illustration of the extreme slowness of cosmic processes, for (as we shall see) they are all in motion and must all be undergoing change. But there has been no gross alteration in the constellations since the days of Hipparchus, or indeed for millenniums before his time.

The whole sky has been arbitrarily divided into eighty-eight areas, which differ greatly in size and shape. Each area embraces a "constellation," or group of stars, and is known by a mythical or semi-mythical name. Rather more than half the constellations were recognized and mentioned by Hipparchus (and by Ptolemy, who edited the star catalogue, which has come down to us through later Moslem scholars under the name "Almagest"). The remaining constellations, which Hipparchus did not observe, lie in the Southern Hemisphere and were not named until the sixteenth and seventeenth centuries. (A list of the constellations is given in Appendix II.)

Few of the groups of stars that form the constellations look like the object for which they have been named. The most aptly titled is the Scorpion, whose "tail" curls up over the Milky Way. The Crab, the Dog, the Lion, and the Whale call for more imagination. Probably the mythological figures were not planned to fit the stars; rather, the constellations were chosen to fit the primitive gods. Thus the dawn of astronomy is bound up with the history of primitive astrology, a subject still rich in problems and possibilities.

The stars form the background on which the sun, moon, and planets move, and the *ecliptic* (the sun's apparent path in the heavens, so called because eclipses occur when the moon crosses it) supplied the most primitive star figures. The sun makes its circuit of the stars once a year; the moon once in about thirty days. Thus the year contains rather more than twelve lunar months (new moon to new moon), and the sun's path against the stars is divided into twelve sections or constellations, the so-called "signs of the zodiac." All but one of these zodiacal constellations bear the names of living creatures, and the word "zodiac" is derived from the Greek *zōon* (a living thing); sometimes the zodiac is called the "Circle of the Beasts." Perhaps Hercules (sometimes identified with a sun-god) typified the sun's progress through the twelve signs by his twelve "labors."

The jingle of Isaac Watts is a convenient mnemonic for the order of the zodiacal constellations.

		Latin Names
The Ram, the Bull, the Heavenly Twins,		Aries, Taurus, Gemini
And next the Crab, the Lion shines,		Cancer, Leo
The Virgin and the Scales;		Virgo, Libra
The Scorpion, Archer, and Sea-goat		Scorpio, Sagittarius, Capricornus
The Damsel with the Watering-pot		Aquarius
The Fish with glittering tails.		Pisces

Each of these constellations has the shorthand symbol shown in the table. Like the other Ptolemaic constellations (those contained in the "Almagest"), they were already known by the Greek equivalents of their present Latin names more than two thousand years ago. But they were not invented by the Greeks; they came to Greece from the earlier civilizations of the Euphrates valley, and their names are found in Euphratean tablets of about 600 B.C., which embody ideas of an even remoter age.

Constellation	*Symbol*	*Euphratean Name*
Aries	♈	Ram, Messenger
Taurus	♉	Bull of Heaven, Bull in Front
Gemini	♊	Great Twins
Cancer	♋	Workman of the River Bed
Leo	♌	Lion
Virgo	♍	Proclaimer of Rain
Libra	♎	Life-maker of Heaven
Scorpio	♏	Scorpion of Heaven
Sagittarius	♐	Winged Fire-head, Star of the Bow
Capricornus	♑	Goat-fish
Aquarius	♒	Urn
Pisces	♓	Cord-place (joining the Fish)

The beginning of the year was reckoned in primitive days from the start of spring or *vernal equinox,* the day when the sun crosses the celestial equator from south to north in its path around the ecliptic, and day and night are of equal length (p. 36). The constellation through which the sun is passing at the time of vernal equinox changes slowly with the centuries, and therefore the stars associated with the season of spring also change slowly. This gradual drift of the *constellations* with respect to the *seasons* is a result of the *precession of the equinoxes,* to be described and explained in Chapter II (p. 37). In the time of Hipparchus the sun was in Aries at the time of vernal equinox; today it is in Pisces. That the Euphratean constellations are much older than Hipparchus is shown by the fact that they began with Taurus, the "Bull in Front." If the sun was in

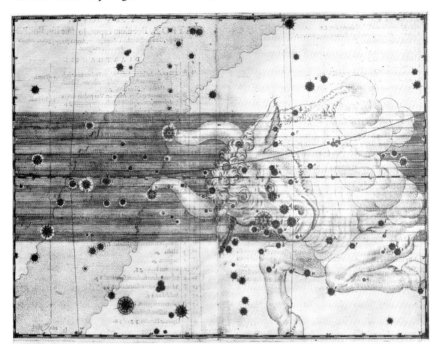

Fig. 1.1. The constellation Taurus, the Bull, from Johann Bayer's *Uranometria,* 1603.

Taurus at vernal equinox when the constellation was named, the date would have been about 2450 B.C. Virgil was echoing this tradition when he wrote:

> The gleaming Bull opens the year with golden horns, and the Dog sinks low, his star (Sirius) averted,

but he was already more than two millenniums out of date.

> ARIES bears a symbol that denotes the horns of a butting animal; the Greeks may have associated it with the ram whose fleece of gold was the goal of the Argonauts.

> TAURUS, the Bull, also bears a horned symbol. Perhaps he typifies the Cretan bull, or half-bull (all the drawings of the constellation show only his forequarters). The transit of the sun through the sign may have been the Bosporos [bull-carrying] of the Greeks.

> GEMINI, the Twins, are denoted by a Roman II. They have been associated with many mythical twins: Sun and Moon (in Babylonia), Horus and Harpocrates (in Egypt), Castor and Pollux (Greece and Rome).

> The sign for CANCER may denote the claws of the Crab, or the two "asses" that flank the "manger" [*Praesepe*]. The Babylonians sometimes

showed it as a Tortoise, and in Egypt it appears as the Scarabaeus or Sacred Beetle.

The symbol for LEO is probably a cursive capital *lambda,* the Greek initial letter of its name, though some have seen in it a crude picture of the Lion's tail. The constellation contains the bright star Regulus [little king], who was said by Copernicus to "rule the heavens."

VIRGO, the Virgin, has an ancient history. The symbol has been said to be a contraction of *Maria Virgo* [the Virgin Mary], but the sign is far older than Christianity. Various peoples have identified the Virgin with Erigone [the Early-born], Persephone, Ishtar, or Isis. Probably the symbol is a contraction of the Greek letters ΠAP [initial letters of *parthenos,* a virgin]. She bears the bright star Spica [the Ear of Corn], and Protrugater [the Fruit-plucking Herald, used in primitive Greece to date the harvest] is one of her fainter stars.

LIBRA is a conventionalized balance. In very early times this constellation was the Claws of the Scorpion [*chelae*]. Rather before the beginning of the Christian era the *autumnal equinox* (six months later than the vernal equinox, when again day and night are of equal length) was marked by the passage of the sun through the Claws. The constellation was renamed "the Scales" in recognition of the "balance" of day and night.

When Libra has made equal the hours of day and slumber, and divides the sphere half in light and half in shadow.
VIRGIL, *Georgics,* iv, 208–9

The symbol of SCORPIO is a conventional picture of the animal.

The Archer, SAGITTARIUS, is always represented as half-man, half-beast. Usually he is represented as a centaur, and Manilius (beginning of the Christian era) called him *mixtus equo* (combined with a horse). Eratosthenes, the Greek-Egyptian, called him a satyr. There was an Asiatic Indian constellation, "the Horseman," perhaps three thousand years ago.

CAPRICORNUS, half-fish, half-goat, is a very ancient symbol, often found on Babylonian gems. Perhaps the sign is a crude picture.

The symbol of AQUARIUS is the Egyptian hieroglyph for "water." This may be the oldest of all the signs if, as some believe, the sun entered it at the time of the Nile flood when it was named. It would then be fifteen thousand years old; but no other zodiacal constellation is of Egyptian origin.

The Fishes, PISCES, are joined in the symbol by a line (the "cord" of the Euphratean list?). Perhaps they represent the Fish-god, Dagon.

Besides the twelve zodiacal signs, the Babylonian constellations included twelve figures north of the ecliptic and twelve south of it. They are closely related to constellations that we still recognize.

The Egyptian constellations were not the same as those of Babylonia and Greece. The most famous Egyptian star map is the Zodiac of Denderah, which is dated near the beginning of the Christian era. It may well show the ancient traditional constellations of Egypt. We can recognize the signs of our zodiac, but the rest of the constellations differ from ours, and the stars seem differently distributed among the figures. Ursa Major appears as a female crocodile and a bull's thigh. Ursa Minor is a small dog. Orion

NORTHERN		SOUTHERN	
Latin	*Euphratean*	*Latin*	*Euphratean*
Cassiopeia	The Fertilizer	Eridanus (River)	Strong One of the Plain
Auriga (Charioteer)	The Chariot	Orion	Son of Life (Tammuz)
Cepheus (the King)	Numerous Flock	Canis Major (Dog)	Dog of the Sun
		Canis Minor (Little Dog)	Crossing-the-Water Dog
Ursa Minor (Little Dipper)	Small Chariot	Argo Navis (Ship)	Ship of the Canal of Heaven
Ursa Major (Big Dipper)	Long Chariot	Hydra (Watersnake)	Great Snake
Boötes (Herdsman)	Shepherd Spirit of Heaven	Corvus (Crow)	Great Storm Bird
Ophiuchus (Snake Bearer)	Prince of the Serpent	Centaurus (Centaur)	Horned Bull
Hercules	King (Gilgamesh)	Lupus (Wolf)	Beast of Death
Lyra (Lyre)	Vulture	Ara (Altar)	Ancient Altar Below
Aquila (Eagle)	Eagle	?	?
Pegasus	Horse	Piscis Austrinus (Southern Fish)	Fish of the Canal
Andromeda	Pregnant Woman	Cetus (Whale)	Great Dragon

is represented by Horus, followed by a hawk, and Canis Major appears as a cow in a boat, a star between her horns (Sirius, called Sothis by the Egyptians, and symbolical of the soul of Isis).

Greek literature reflects the vital impact of the stars on the life of an agricultural and seafaring people. Homer (9th century B.C.) notes the Bears, Boötes, the Pleiades and Hyades, and the star Sirius, all by the names we use today.

> [Odysseus is leaving the island of Calypso.] Odysseus . . . sat by the stern-oar steering like a seaman. No sleep fell on his eyes; but he watched the Pleiades and the late-setting Wagoner, and the Bear, or the Wagon as some call it, which wheels round and round where it is, watching Orion, and alone of them all never takes a bath in the ocean. Calypso had warned him to keep the Bear on his left hand as he sailed over the sea.
>
> *Odyssey,* v, 270–76

The Bear (Ursa Major) typifies the great Mother Goddess; Aristotle suggested that the constellation was named because the bear is the only creature that dares to invade the frozen north. The Greeks of Homer's time used Ursa Major to steer by. The name of "the Wagon" recalls the Euphratean "Long Chariot"; in England the constellation is still called "Charles' Wain," as it was in Shakespeare's day:

Fig. 1.2. The Zodiac of Denderah. Notice the hippopotamus, the thigh, and the dog, which represent Ursa Major and Ursa Minor. The twelve zodiacal constellations can be traced.

> An't be not four by
> The day I'll be hanged; Charles' Wain is over
> The new chimney and yet our horse not pack'd.
> *King Henry IV,*
> Part I, Act ii, Scene 1

The "Charles" may be Charlemagne, but more likely it is a corruption of "Churl's." Homer calls Boötes the "late-setting" because the kite-shaped constellation sinks sidewise towards the horizon, whereas it rises vertically.

> [Achilles rushes out to the fight.] His armor shone on his breast, like the star of harvest whose rays are most brilliant among many stars in the murky night; they call it Orion's dog. Most brilliant is that star, but he is a sign of trouble, and brings many fevers for unhappy mankind.
>
> *Iliad*, xxii, 25–30

This is the first of many references in Greek literature to Sirius, the "Dog Star," in the constellation that the Babylonians called the "Dog of the Sun."

Hesiod (9th century B.C.) makes very practical use of the stars in his delightful farmer's calendar, *Works and Days*. They are the signs by which farmer and sailor may time their operations.

> When the Pleiades, daughters of Atlas, are rising (early May), begin your harvest, and your ploughing when they are going to set (November). Forty nights and days they are hidden, and appear again as the year moves round, when first you sharpen your sickle.
>
> But when the house-carrier, the snail, leaves the ground and crawls upon the plants, fleeing from the Pleiades, then the early harvest must be begun.
>
> When the Pleiades, the Hyades and the strength of Orion set, then be mindful of timely ploughing. . . . When the Pleiades, fleeing from the mighty strength of Orion, fall into the murky sea, the sailing season is over.
>
> But when Orion and Sirius are come into mid-heaven and the rosy-fingered dawn sees Arcturus, then . . . cut off all the grape clusters and bring them home.
>
> But when the force and sultry heat of the sun abate, and almighty Zeus sends the autumn rain (October) . . . the star Sirius passes over the heads of men . . . only a little while by day and takes greater share of the night.
>
> *Works and Days,*
> 383–7; 571–2; 615–7; 619–20; 609–11; 414–9

Sirius, the Dog Star (its Greek name means "the scorcher"), recurs in a more convivial strain. Alcaeus (7th–6th century B.C.) sings:

> Soak your throttle in wine; for the star is coming round again, the season is hard to bear with the world athirst because of the heat; the cricket sounds sweetly from the leaves of the tree-top, and lo! the artichoke is blowing; now are women at their sauciest, but men lean and weak, because Sirius parches head and knees.

Theognis (6th century) echoes the same note:

> Foolish are men and senseless, who drink not wine when the Dog-star rules.

The Pleiades, daughters of Atlas, were the most beloved of all constellations. They had been turned into a flock of doves as they fled from Orion (so the legend said), and were placed in the sky. Pindar wrote of them (5th century B.C.):

> And meet it is that Orion should not move far behind the seven mountain Pleiades.
>
> *Nemean Odes,* ii, 17–18

Simonides (5th century B.C.) wrote of:

> Mountain Maia of the glancing eye, who was the fairest of all Atlas' violet tressed daughters dear, that are called the heavenly Peleiades.

The gem among poetic references to the stars is perhaps Sappho's exquisite verse (7th century B.C.):

> The moon is gone
> And the Pleiads set,
>   Midnight is nigh:
> Time passes on
> And passes; yet
>   Alone I lie.

The first systematic account of the stars in literature is contained in the *Phenomena* of Aratus. Perhaps this poet of the third century B.C. received more attention from the Christian world because Saint Paul quoted one of the early lines of the *Phenomena* in his sermon on the Unknown God: "As certain also of your own poets have said: for we also are his offspring." (*Acts* 17:28)

Aratus put into verse the early fruits of Greek observation. Cicero (*De republica,* i, 14) recorded that:

> The first Hellenic globe of the sky was made by Thales of Miletus (636–546 B.C.), having fallen into a ditch or well while star-gazing. Afterwards Eudoxus of Cnidus (408–355 B.C.) traced on its surface the stars that appear in the sky; and . . . many years after, borrowing from Eudoxus this beautiful design and representation, Aratus had illustrated it in his verses, not by any science of astronomy, but by the ornament of poetical description.

Aratus described forty-three constellations: nineteen in the north, the twelve of the zodiac, and twelve in the south. He also named five individual stars: Arcturus (the Bear-watcher); Stachys (the Ear of Corn, now called Spica); Protrugater (the Fruit-plucking Herald, now called Vendemiatrix, in Virgo); Sirius(the Scorcher); and Procyon (the Dog's Forerunner).

Each constellation is described. Here, for example, is the story of the Lost Pleiad:

> Near [Perseus'] left knee all the Pleiades are gathered together. Seven in number are placed on record, but to our sight only six appear. It is not known for certain when the missing star disappeared. Nevertheless they each have their respective names. Alcyone, Merope, Celaeno, Electra, Sterope, Taygeta, and the queenly Maia.

The account of the polar stars throws an interesting light on the state of astronomical understanding in the third century B.C.

> The axis is always firm, although it may appear to shift a little, while the earth maintains its equilibrium in the center, and around it the sky turns itself. Also the two poles terminate at their extremity; one indeed is not visible, but the other to the northward rises high above the ocean. Surrounding it, the two Bears lie circularly, which are usually called the Wains. . . . They called one of these Bears Cynosura, and the other Helice. The Greeks placed faith in Helice in respect of their naval affairs, and the direction of shipping. The Phoenicians have confidence in Cynosura during

their voyages. Helice is clear and readily observed, shining brightly at the commencement of the night. The other, Cynosura, is comparatively obscure, but nevertheless more useful to the sailor, because it revolves in a lesser circle. By this latter the Sidonians navigate their ships with great accuracy.

Clearly Aratus recognized the earth as at the center of the heavens, which turn about two poles, one visible, one hidden. The slight shifting of the axis suggests a suspicion of precession, which was not actually discovered until a century later. Cynosura [the Dog's Tail] is Ursa Minor; Helice [the Twister] is Ursa Major. We note the advance in navigation represented by the use of Ursa Minor (i.e., our polestar, Polaris) instead of Ursa Major (as Odysseus did). The Greeks were indebted to the Phoenicians for knowledge both in arithmetic and in astronomy.

A hundred years after Aratus came the great Hipparchus (160–125 B.C.?). Tradition has it that he was attracted to astronomy by seeing a "new star," which was probably a bright comet. He set himself to compile the star catalogue that has come down to us as the "Almagest." Astronomy emerges from the mists of tradition and becomes an observational science.

2. THE SOLAR SYSTEM

The constellations form the background for the motions of our near neighbors, the members of the solar system. Today we recognize the sun as its central body, with the planets (Mercury, Venus, Earth, Mars, Jupiter, Saturn, Uranus, Neptune, Pluto) moving around it, and also a host of smaller bodies (asteroids, comets, meteors, and interplanetary dust). Many of the planets are accompanied by moons of their own. These facts are commonplaces to us, but they have been realized slowly. One of the themes of the first section of this book will be the gradual growth of our knowledge of the solar system.

Sun and moon seem to move around the earth in a definite path among the stars, and this path, as we have seen, was recognized in very ancient times and divided into twelve "signs of the zodiac." That sun and moon look so nearly the same size is one of the strangest astronomical coincidences, for the sun is by far the largest body in our solar system, and the moon (though by no means the smallest) is smaller than any of the principal planets. The same path among the stars (known as the *ecliptic* because eclipses can occur only when the sun and moon are in or very near it) is rather closely followed by the naked-eye planets, and by other members of the solar system that can be studied only with the telescope.

Only the five planets readily visible to the naked eye were known in ancient times. Venus, the brightest of them, sometimes bright enough to cast a shadow, was associated with the sun and moon in a great triad (Shamash, Ishtar, and Sin) in Euphratean lands. The four other planets were in a separate class.

Nearly three thousand years before the Christian era, the Babylonians were already recording the *heliacal risings* (simultaneous with sunrise) of Venus. Their interest in astronomy was closely bound up with astrology. In the sixth century B.C., they were making extensive calculations of the positions of the sun and moon, and of approaches of sun, moon, and planets to one another in the sky. Babylonian astrologo-astronomy was the wellspring of Western astronomy. From it we inherit our seven-day week, our day of twelve (double) hours, our method of dividing the circle into 360 degrees, which has survived to the present day in a world that carries out other calculations by the decimal system.

An ancient inscription gives the Sumero-Akkadian (Euphratean) names of the planets:

> The god-the-Moon and the god-the-Sun, the god-the-Messenger-of-the-Rising-Sun [Mercury], the star-the-Ancient-Proclaimer [Venus], the star-the-Old-Sheep [Saturn], the god-the-Old-Sheep-of-the- Furrow-of-Heaven [Jupiter], the star-of-Death [Mars], the seven old sheep-stars are they.

The only planet mentioned by Homer is Venus, in her two roles of morning and evening star. She was called Eosphoros [Dawn-bearer] at morning, Hesperos at evening.

> But when Eosphoros came, proclaiming light on earth, and after him the saffron mantle of Dawn spread o'er the sea, at that hour did the flame (of the funeral pyre) die down.
>
> *Iliad*, xxiii, 226
>
> The great spear, which gleamed like the finest of all the stars of heaven, Hesperos brilliant in the dark night. . . .
>
> *Iliad*, xxii, 317

Sappho sings only of Venus among the planets:

> Evening star, that bringest back all that lightsome Dawn has scattered far, thou bringest the sheep, thou bringest the goat, thou bringest the child to the mother.

Probably Homer did not know that Eosphoros and Hesperos are one. The discovery (or at least its popularization) took place in about the sixth century B.C., and has been ascribed to the great Pythagoras.

PLANETARY NAMES AND ASSOCIATED DEITIES

MODERN NAME:	*Mercury*	*Venus*	*Mars*	*Jupiter*	*Saturn*
EUPHRATEAN GOD:	Nabou (wisdom)	Ishtar (love)	Nergal (war)	Marduk (lord)	Ninib (strife)
GREEK GOD:	Hermes (messenger)	Aphrodite (love)	Ares (war)	Zeus (lord)	Chronos (time)
DESCRIPTIVE NAME:	Stilbon (twinkler)	Phosphoros (light-bearer)	Pyroeis (fiery)	Phaethon (radiant)	Phainon (clear-shining)

Planets other than Venus do not seem to have had special names in ancient Greece. They were spoken of in association with certain gods. The Greek astronomers at Alexandria in the second century B.C. gave each planet a name that described its appearance.

Not only do we derive our seven-day week from the Babylonians; even the names of the days can be traced back to the planetary gods.

NAMES OF THE DAYS OF THE WEEK

PLANET:	*Moon (Luna)*	*Mars*	*Mercury*	*Jupiter*
SYMBOL:	☾	♂	☿	♃
ITALIAN:	Lunedi	Martedi	Mercoledi	Giovedi
FRENCH:	Lundi	Mardi	Mercredi	Jeudi
ANGLO-SAXON:	Monandæg	Tiwesdæg (Tiw)	Wodneesdæg (Woden)	Thuresdæg (Thor)
ENGLISH:	Monday	Tuesday	Wednesday	Thursday
GERMAN:	Montag	Dienstag	Mittwoch	Donnerstag

PLANET:	*Venus*	*Saturn*	Sun
SYMBOL:	♀	♄	☉
ITALIAN:	Venerdi	Sabato	Domenica
FRENCH:	Vendredi	Samedi	Dimanche
ANGLO-SAXON:	Frigedæg (Frigga)	Sæternesdæg	Sunnandæg
ENGLISH:	Friday	Saturday	Sunday
GERMAN:	Freitag	Sonnabend	Sonntag

Monday is always clearly associated with the moon. The Italian and French names for *Tuesday* point to Mars, the god of war, and the Anglo-Saxon and English words relate to his Germanic counterpart, Tiw; the German *Dienstag* refers to the *Ding* [*Thing*], or deliberative assembly: Tuesday was the "day of deliberation." *Wednesday* is Mercury's day in Italian and French; in Anglo-Saxon and English it is associated with his Germanic counterpart Woden (Odin); the Germans call the day "midweek." *Thursday* goes with Jove (Jupiter) in Romance tongues; in Anglo-Saxon and English it takes its name from the corresponding Thor, god of thunder, and the Germans simply call it "thunder-day." *Friday* is associated with Venus and her counterpart, Frigga. *Saturday* goes with Saturn in French, Anglo-Saxon, and English; the Italian name is a contraction of *Sabbati dies* (the Sabbath Day), and the German name means "Sunday-eve." *Sunday* is dedicated to the sun in Anglo-Saxon, English, and German; the Italian and French words are contractions of *Dies Dominica* (the Day of the Lord).

The planetary symbols represent respectively: the caduceus, a staff wreathed with serpents (Mercury); the handled mirror (Venus); the shield and spearhead (Mars); the thunderbolt (Jupiter); and the scythe (Saturn). Solar and lunar symbols are stylized pictures.

The order of the days of the week, as well as their names, comes to us from ancient Babylonia. Successive hours of the twenty-four-hour day were supposed to be governed in turn by the planets. The Babylonians did not know the actual distances of the seven heavenly bodies, but they did know how long each requires to make a complete circuit of the heavens. Saturn takes the longest time; then, in order, come Jupiter, Mars, the sun, Venus, Mercury, and the moon. The planet that controlled the first hour was, to the Babylonians, the ruling planet and gave the day its name.

If Saturn controlled the first hour, the day (in our language) would be Saturday. Jupiter (under the control of Saturn for the day) would control the second hour, Mars, the third, and so on. If the successive hours are numbered, and assigned in rotation to the seven planets, the ruling planet for the following day will be found to be the sun, for the next day the moon, and so on. A little table illustrates the relationship. Two complete cycles are shown in full. For later cycles only the first hour is indicated throughout a complete week; the planet that controls this hour gives the name of the day.

DAY	Saturn	Jupiter	Mars	Sun	Venus	Mercury	Moon
Saturday:	1	2	3	4	5	6	7
	8	9	10	11	12	13	14
	15	16	17	18	19	20	21
	22	23	24				
Sunday:				1	2	3	4
	5	6	7	8	9	10	11
	12	13	14	15	16	17	18
	19	20	21	22	23	24	
Monday:							1
Tuesday:			1				
Wednesday:						1	
Thursday:		1					
Friday:					1		
Saturday:	1						

The nine planets (Mercury, Venus, Earth, Mars, Jupiter, Saturn, Uranus, Neptune, and Pluto) all go around the sun in the same direction and in orbits that are nearly circular. The primitive observation that all the naked-eye planets move in or near the ecliptic is evidence of the *flatness* of the solar system, one of its most remarkable properties, which was already appreciated two thousand years ago. The fact that the sun is the central body, containing most of the matter in the system, did not win general recognition until the sixteenth century, and part of the theme of the first half of this book will be the gradual steps by which this recognition was attained. The cardinal facts in the study of the arrangement and history of the solar system are its flatness; the tendency of all its members to revolve about the sun in the same direction; and the concentration of the greater part of its material in the sun itself.

THE EARTH

And swift, and swift beyond conceiving,
The splendor of the world goes round,
Day's Edenbrightness still relieving
The awful Night's intense profound:
The ocean-tides in foam are breaking,
Against the rocks' deep bases hurled,
And both, the spheric race partaking,
Eternal, swift, are onward whirled!

GOETHE, *Faust* (Prologue)
Translated by Bayard Taylor

The earth is one of the heavenly bodies, a planet that goes around the sun once a year in a nearly circular path. Among the planets the earth is fifth in diameter, fifth in mass, and third in order outward from the sun. It is very like several of the other planets. Perhaps its only uniqueness is in the suitability of its surface to life like our own.

Our deepest mines and wells barely prick the earth's outer skin. We have gone down only about three-thousandths of its total diameter, and our knowledge of what lies below is entirely indirect.

The mass of the earth is 5.975×10^{24} kilograms* (see p. 27); its average radius is 6371.23 kilometers (see p. 26). Its mean density (the average mass contained in unit volume) is therefore 5517 kilograms per cubic meter, so it has 5.517 times the density of water (a cubic meter of water under standard conditions weighs 1000 kilograms). The rocks at the earth's surface have a mean density of only about 2700 kilograms per cubic meter, so our planet must be far denser inside than it is at the surface.

1. THE EARTH'S INTERIOR

The probable structure of the earth is shown in Figure 2.1. The rocky crust (*lithosphere*), mostly of igneous rocks with a small surface layer of

* The shorthand way of expressing 1,000,000,000,000,000,000,000,000, or 1 followed by 24 zeros, is to write it as *a power of 10,* in the form 10^{24}. This practice is followed for expressing many astronomical quantities that involve large numbers.

The units in which various quantities are expressed are defined in Appendix I, which should be consulted whenever a new unit is encountered.

sedimentary rock, is about 30 miles thick. About 60% of the crust is silicon dioxide; about 15%, aluminum oxide; the remainder (in decreasing order) consists of oxides of iron, magnesium, calcium, sodium, potassium, hydrogen (water), and other elements. Most of these elements are common throughout the universe: the earth is a representative sample of cosmic stuff, though it contains less hydrogen and less of the rare gases than the stars.

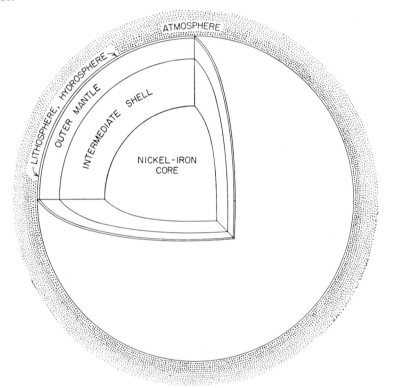

Fig. 2.1. The structure of the earth.

The lower layers of the lithosphere are igneous, granite-like rocks, which are mixtures of quartz (silica, or silicon dioxide), micas, felspars, and tourmalines (all complex silicates of aluminum). As their name implies, igneous rocks bear evidence that they were once at high temperature.

The thin surface layer of *sedimentary rocks* (usually laid down by water) consists of clays and limestones. *Clays* have been formed from granitic rocks by the weathering action of water with acids in solution, or of volcanic gases. *Limestones,* principally calcium carbonate, are mainly of organic origin, the debris of small creatures that populate ocean bottoms. The microscopic shells of these creatures can be seen in the structure of *chalk.*

The very thin fertile layer of the outer crust, which sustains plant life and, indirectly, animal life, owes its fertility to the organic content of the *humus,* replete with decaying animal and vegetable matter. The humus teems with bacteria, which capture the free nitrogen of the atmosphere and enrich the soil with natural fertilizer. The fertile layer is at most a few feet thick; the sedimentary layer ranges from a few thousand feet to little or nothing.

The lithosphere, about 30 miles in thickness, floats on a basaltic layer (complex silicates of aluminum, magnesium, and other elements), crystalline above and glassy below. The chemistry and physics of the earth's crust, so important to organic life, is a very complex subject, the province of the geologist and petrologist. From the astronomical point of view the lithosphere is almost negligible.

The oceans, lakes, and rivers (the *hydrosphere*) cover more than half the earth's surface, but they are not as deep or as dense as the lithosphere and therefore weigh much less. The hydrosphere is principally water, with chlorides (of sodium, magnesium, calcium, and so on), bromides, and sulfates in solution. These substances have been washed out of the rocks by rivers. If we measure how much of them the ocean contains and also how fast the rivers are bringing them down, we can estimate how old the oceans are (p. 465). Although the oceans contribute very little to the earth's mass and density, their tides produce surprisingly large effects on its rotation (p. 147).

Our most powerful means for studying the conditions in the interior of the earth is *seismology* (the science of earthquakes). The earth's crust is continually subject to internal stresses, and from time to time a fracture or slipping takes place as a consequence of these stresses. Geologists consider that an earthquake is the complex vibration that arises when such a fracture occurs. Some parts of the crust are under greater stress than others, and form the "earthquake belts" that run from the Mediterranean through the Himalayas to the East Indies, through Japan and Alaska, and along the western coast of the Americas.

Waves travel through matter at a speed that is determined by the properties of the substance through which they pass. Sound waves in air, or some other medium, are examples of *compressional* or *longitudinal* waves, which are transmitted by to-and-fro motions of the medium *in the direction of the wave's motion*. The waves in a vibrating string, on the other hand, are transmitted by motions of the medium *perpendicular to the direction of the wave's motion;* they are known as *distortional* or *transverse* waves. All substances resist compression to some extent, so all substances can transmit compressional waves. Solids resist distortion, and therefore transmit transverse waves; even liquids resist distortion to some extent (as the inexpert diver well knows), but under ordinary conditions they transmit transverse waves feebly. We can therefore make the broad distinction that a solid allows waves of both types to pass, but a liquid (or gas) carries compressional waves.

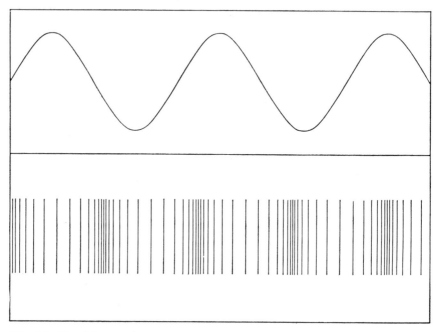

Fig. 2.2. The difference between transverse or distortional waves (above) and longitudinal or compressional waves (below).

The speeds of both sorts of waves are determined by the resistance to compression or distortion: the greater the resistance, the greater the speed; they also vary with the local density: the greater the density, the smaller the speed. If both compressional and distortional waves pass through the same substance, the compressional wave always travels faster.

When an earthquake occurs, the disturbance sets both kinds of waves going through the interior of the earth, and they travel in all directions. The longitudinal wave, with a greater speed, outruns the transverse wave and arrives at a given spot first. At the earth's surface these waves travel at about 8 kilometers a second, but those that pass nearer to the earth's center go almost twice as fast. When we recall that the density of the earth is greater in the interior than at the surface, we realize that this increase of speed must be a result of a very great increase of resistance to compression and distortion—in fact, when allowance is made for the probable difference of density, the deeper portions of the earth are seen to behave as though they were *more rigid than steel*!

The longitudinal waves sent out by a disturbance of the crust arrive, after an appropriate interval (which depends on the distance and on the conditions in the interior), at all points of the earth's surface. The transverse waves, which travel more slowly, arrive some time later, and (by the same kind of calculation that enables one to estimate the distance of a thunderstorm by measuring the interval between the lightning flash and the sound of the thunderclap) an observer can form an estimate of the distance of the earthquake by noting the length of time that elapses between the successive arrivals of the two kinds of waves. But at some points on the earth's surface, the transverse waves never arrive at all. They seem to be stopped by a region in the earth's interior, which, while it appears to exhibit a rigidity greater than that of steel, thus displays the apparently contradictory property of behaving like a fluid and transmits transverse waves feebly, if at all.

Many attempts have been made to express these observed facts—the varying speed of the compressional waves, and the complete stoppage of the transverse waves by a certain part of the earth's core—in terms of known materials under probable conditions. It is generally thought that the center of the earth consists of a mass of metal of great density and under high pressure, probably principally nickel-iron. The suggestion that it contains a core of highly condensed gases such as hydrogen and helium has also been made, but it is not as yet generally accepted by geophysicists.

Earthquakes are detected by an instrument known as the *seismograph* —a heavy pendulum adjusted to record automatically the vibrations of the earth beneath its point of support. A typical seismogram shows first a record of the compressional waves, and then, after an interval, the transverse waves may appear; both have traveled through the interior of the earth. Finally a long train of waves, which have skirted round the crust of the earth, complete the record; these are known (after their discoverers) as *Rayleigh waves* and *Love waves*. Some earthquakes originate near the earth's surface; others have been recorded that came from disturbances as deep as 550 kilometers. Deep earthquakes, such as this, are recognized because they produce relatively feeble surface waves.

A few earthquakes have been produced by observable causes. Near the end of World War I, at an ammunition factory in Oppau, Germany, 4500 tons of a certain high explosive blew up and produced 6×10^{19} ergs of energy. Most of this went into the air, and the noise was heard in many distant parts of Europe. But the measured energy of the actual earthquake produced was only 5×10^{16} ergs, less than one-thousandth of that of the explosion.

A larger earthquake was caused when the top of a mountain in the Pamirs broke off and crashed down into a valley. Here, most of the energy went into producing an actual earthquake, and such energy was about 10^{21} ergs or ten thousand times as much as that of the Oppau "earthquake."

Even the Pamirs earthquake was a small one, as earthquakes go. The California earthquake of 1906—naturally produced by nonobservable causes—had a thousand times as much energy, about 10^{24} ergs, and others even more powerful have been recorded, such as the shock that destroyed Lisbon in 1755.

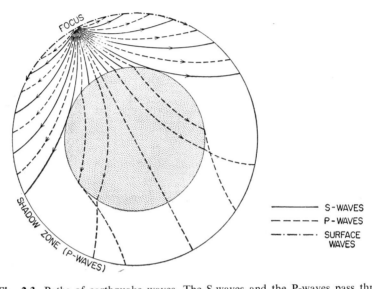

Fig. 2.3. Paths of earthquake waves. The S-waves and the P-waves pass through the earth's interior; the surface waves travel along or near the surface. The S-waves are stopped at the discontinuity between the solid and quasi-liquid zones; the P-waves are transmitted by the core, but are bent (refracted) at the discontinuity, so that shadow zones are produced at certain parts of the surface.

The Pamirs earthquake involved about as much energy as the uranium bomb.* A really destructive earthquake, such as those that ravaged Lisbon and San Francisco, is therefore much more effective than any man-made device that has been used as yet.

That the earth behaves like a magnet, is shown by the behavior of the magnetic compass. The magnetic poles of the earth are not at the terrestrial poles and are continually shifting. The motion of the magnetic poles shows that the earth's magnetic field is probably not caused by a solid magnetic body within. Terrestrial magnetism, although well known and widely used, is incompletely understood. The most satisfactory theory relies on electric currents that circulate within the core, and produce magnetic fields of the kind familiar in physics.

---

* The complete fission (see p. 345) of 1000 grams of uranium 235 would liberate about 8×10^{20} ergs; the mass involved in the uranium bomb is probably of this order.

2. THE ATMOSPHERE

The gaseous envelope of the earth is of great importance to us, as living creatures, as human beings, and as astronomers. Both animals and plants depend for existence on it: most animals breathe the oxygen it contains, and its carbon dioxide is essential to plants. But to astronomers the atmosphere spells almost nothing but trouble.

The approximate volume composition of the atmosphere is: nitrogen, about 78%; oxygen, about 21%; argon, 1%; carbon dioxide, 0.03%; hydrogen, 0.01%; neon, 0.0012%; helium, 0.0004%; and the other rare gases in even smaller quantities.

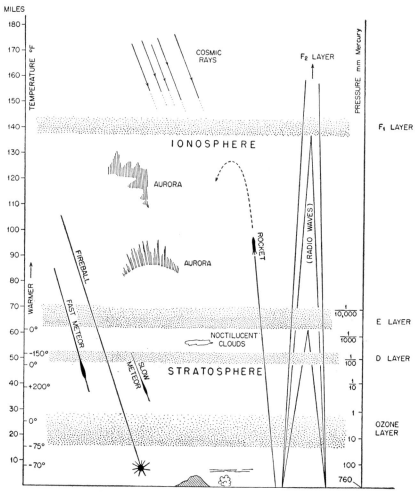

Fig. 2.4. The structure of the earth's atmosphere.

The average pressure of the atmosphere falls off rapidly with height above the earth. The lower levels are accessible by balloons: explorers have not yet penetrated above a few kilometers, but sounding balloons can be sent up to about 30 kilometers. Higher levels (up to about 300 kilometers) can be explored by rockets loaded with instruments. Nor are we limited to man-made instruments in exploring the atmosphere. From 50 to 110 kilometers up, the bright streaks left by meteors (p. 247) tell of the atmospheric conditions that heat these small bodies in their flight from outer space. Higher levels, up to perhaps 800 kilometers, can be explored by means of the Aurora Borealis, which gives evidence of smaller bodies, electrons and atomic nuclei, ensnared by the earth's magnetic field and spiraling into the tenuous outer atmosphere.

Atmospheric conditions are a very complex subject and outside the scope of an elementary book. One of the most striking facts is the variation of temperature (see Fig. 2.4), which does not fall off the higher we go, as might have been expected; after falling to 75° below zero (Fahrenheit) at 26 kilometers, it rises, just below 60 kilometers, to very nearly the boiling point of water. At about 80 kilometers it has fallen to $-30°F$; at 100 kilometers it is zero and is warmer above. This surprising conclusion has been reached from the study of meteors. The temperature inversion near 40 kilometers is probably associated with the ozone layer; the one at about 70 kilometers may be similarly connected with a layer of molecular oxygen.

From the human standpoint, the atmosphere is important because it carries the weather. However, all weather phenomena are very low lying; few clouds are found more than 10 miles up. The ionization layers make radio transmission for large distances possible. And most important of all is the ozone layer.

Ozone, the triple oxygen molecule, occupies an atmospheric layer from 10 to 30 miles up. Its importance lies in the fact that it absorbs ultraviolet light almost completely. Astronomically speaking, the ozone layer is regrettable, for it cuts off the ultraviolet light of all astronomical objects, and, as we shall see, much of the most interesting light of stars lies in the ultraviolet region. We should know far more about astrophysics if the atmosphere contained no ozone. Even a rocket that goes some way up through the ozone can furnish a revelation.

Because the atmosphere lies between us and all astronomical objects, it cuts off some of their light. The nearer they are to the horizon, the thicker is the atmospheric layer through which they shine, and the more are they dimmed. This is common experience: we can look at the rising or setting sun with the naked eye, but the sun overhead is unbearably bright. One can be blinded by looking directly at the noonday sun. All observations of the light of sun, stars, and planets must be corrected for this dimming effect (*atmospheric extinction*), a very difficult matter when great accuracy is desired, especially because the effects vary with the condition of the air, its temperature, humidity, and so forth.

The molecules in our atmosphere scatter light. The bluer the light, the more of it is scattered out of the direction of the beam. Thus the air molecules rob the sun of its bluer light and confer on it a reddish hue. The setting sun, which we see through a thicker layer of air than the high sun, is redder than the sun at noon. Strictly speaking, the setting sun is not redder than the high sun, but *less blue*. On the other hand, the light that is scattered away from the direct beam of the sun can reach us only if it is rescattered from air molecules in other parts of the sky. Because blue light is scattered more than red light, the light rescattered in our direction is even bluer than the original scattered light. So it comes about that the sky far from the sun becomes progressively bluer than that near the sun's edge.

The blue of the sky seen from near sea level is pale in comparison with the intense purplish-blue seen from a mountain peak. The lower atmosphere is full of small particles that reflect sunlight as well as scatter it, and the reflected light is yellower than the scattered blue that dominates at levels above the layer of dust.

Not only do the air molecules scatter light, but the particles in interstellar space do also, and, as we shall see, distant stars seen through thick layers of interstellar particles appear reddened, much as the setting sun is reddened.

The atmosphere has another effect that is important in astronomy. Because of the variation of density with height, the atmosphere acts like a prism and bends the rays slightly, especially those rays passing near the horizon. This effect (*atmospheric refraction*) causes objects to seem higher above the horizon than they would seem to be if the earth had no atmosphere. It must be allowed for in all accurate measurements of the positions of the heavenly bodies.

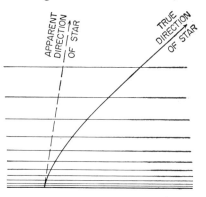

Fig. 2.5. The atmosphere refracts (bends) light rays. Therefore the apparent direction of a star corresponds to a greater altitude than its true direction.

3.   THE EARTH AS A PLANET

The earth is a typical planet. Its density increases greatly toward the center. Although earthquakes cannot be studied on the other planets, sources of information show that there, also, density increases inward. The earth has an extensive atmosphere, and many (though not all) of the other planets share this property, though all planetary atmospheres are not of the same chemical composition. An observer outside the earth would see it as a bluish planet with rather a bright surface. At some phases of the moon we can actually see the light of the earth's surface reflected from the dark face of the moon ("the old moon in the new moon's arms"). Here we see reflected sunlight that has passed twice through the earth's atmosphere, once on the way in, once on the way out, and has been reflected back to us from the brownish surface of the moon and passed a third time through our atmosphere to reach us. This "earthlight" is our only means of measuring the reflecting ability of the earth's surface (its *albedo*). As a reflector, the earth is intermediate between Venus and Mars (p. 180).

An observer on a nearby planet would certainly see the weather zones of the earth, variable cloud bands that run essentially parallel to the equator, and produce an effect rather like that of the surface of Jupiter. The continental masses would be visible as dim, permanent markings. And at certain angles the oceans would act as mirrors and would reflect a brilliant point of sunlight.

4. SHAPE AND SIZE OF THE EARTH

The "roundness" of the earth is a commonplace to most of us, although a small group of fanatics maintains even today that the earth is flat. At a time when trips around the world are an everyday occurrence, the rotundity of our planet is easy to visualize, but it cannot have been so when the explored areas were small and isolated.

Pythagoras (6th century B.C.) is recorded as the first to guess the sphericity of the earth, very likely because the sphere was considered the perfect figure. The disappearance of ships below the horizon is an elementary piece of evidence. More sophisticated is the recognition that the earth can cast a circular shadow on the moon; Anaximenes (585–528 B.C.) correctly understood eclipses of the moon as being caused by the shadow of the earth cast by the sun. Strictly speaking, however, the shape of the shadow shows only that the earth is roughly circular in section.

A better argument for a rounded earth is the fact that the sun and stars are seen to rise to different altitudes from different parts of it. Since 1940 the curved horizon of the earth has actually been photographed from rockets and stratospheric balloons. We know today that although the earth is round, it is not perfectly spherical, but bulges at the equator.

The diameter of the earth was measured by the Greek astronomer Eratosthenes (276–195 B.C.?), who worked in Egypt. He noticed that at midsummer the sun's rays fell vertically down a well at Syene (the modern Assuan, at the First Cataract of the Nile), while at Alexandria, almost due north of Syene, they were about 7° from the vertical. By an extraordinary flight of imagination he realized that if the earth is spherical, and the sun so far away that its rays may be considered parallel, a knowledge of this angle and the distance form Alexandria to Syene would suffice to determine the earth's diameter. He expressed the circumference of the earth in stades; if (as is generally believed) one stade equals 517 feet, the result is 24,500 miles, within about 50 miles of the modern value! If Eratosthenes had worked at three stations instead of two he would have obtained *evidence* that the sun is very distant, as the student can easily verify.

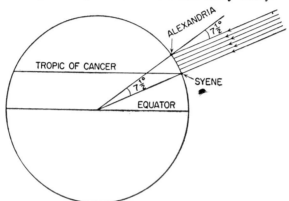

Fig. 2.6. Principle by which Eratosthenes measured the earth's diameter. The sun's rays were vertical at Syene, and they made an angle of 7½° with the vertical at Alexandria, due north of Syene. This angle is the angle at the center of the earth that corresponds to an arc of the surface equal in length to the distance from Alexandria to Syene. The base of the triangle and the angle of the vertex determine the height of the triangle, which is approximately equal to the radius of the earth. The angle is considerably exaggerated in the diagram.

Attempts to measure the size of the earth accurately were impeded, for almost two millenniums, by the difficulty of measuring large distances on the surface. In 1671 the French scientist Picard devised a wheel with a mechanical counter (rather like the modern device, often miscalled a "speedometer," that records the mileage of a car, correctly called an "odometer"), measured a large distance accurately for the first time, and made possible the determination of the earth's size.

The earth bulges at the equator: it is *oblate,* like many, if not all the other planets. A body that is longer through the equator than through the poles, like a pumpkin, is oblate; a body that is longer through the poles than through the equator is *prolate,* like a football. A numerical measure of oblateness is the ratio of the difference of equatorial (assumed constant) and polar diameters to the mean equatorial diameter. If a is the diameter measured through the equator, and b that measured through the poles, the *oblateness* is expressed by

$$\frac{(a - b)}{a}.$$

The earth's oblateness is 1/297: the equatorial diameter is 12756.78 kilometers (7926.68 statute miles) and the polar diameter, 12713.82 kilometers (7899.98 miles), so the distance between opposite points on the equator is about 43 kilometers (nearly 27 miles) greater than the distance between the poles; both, of course, are measured through the earth's center.

These very precise figures have been obtained by means of a network of accurately determined distances on the earth's surface—a simple geometrical problem, difficult only in practical execution. For example, the length of a degree of latitude (fixed by astronomical measures of the altitudes of stars) is different at different latitudes: 110.6 kilometers at the equator against 111.7 kilometers at the poles. The network of distances measured on the earth's surface determines both shape and size.

Another way of determining the shape of the earth (but not the size) depends on the fact that the force of gravity varies with distance from the center. Because the poles are nearer to the center of the earth than a point on the equator, gravity must be stronger at the former, and, if the earth were not spinning, the ratio of the gravity at poles and equator would give the relative diameters immediately. But the earth also produces a force that acts in the opposite direction to gravity and is clearly greatest at the equator, which travels fastest, and zero at the poles. The measured difference of gravity results from both the oblateness and spin of the earth; the oblateness can be determined if an allowance is made for the effect of spin.

The acceleration of gravity is measured by the period (time of swing, T) of a pendulum of known length, l. The value of gravity, g, is deduced from the relation

$$T = 2\pi \sqrt{l/g}. \qquad g = \frac{4\pi^2 l}{T^2}$$

This formula applies to the "simple pendulum"; in practice a "compound" (or heavy, rigid) pendulum is used, and the theory is appropriately modified. If the time of swing of a given pendulum is measured at a number of places on the earth's surface, the shape of the earth can be calculated.

The great French mathematician Clairaut proved that

$$W = \tfrac{5}{2}\phi - \varepsilon$$

where W is the excess of the polar, over equatorial, gravity, ε is the oblateness of the earth, and ϕ is the ratio of centrifugal force to gravity at the equator. He also showed that if the earth were homogeneous (of the same density throughout),

$$\varepsilon = W = \tfrac{5}{4}\phi$$

would be true. The measures of W and ϕ show that this is not the case: the earth is not homogeneous, as we have already concluded from its density.

Purely astronomical methods, based on dynamics, can be used to determine the earth's oblateness: the effects of the equatorial bulge on the precession of the equinoxes (p. 38) and on the motion of the moon (p. 130) give an independent value of 1/294. The difference between 1/294

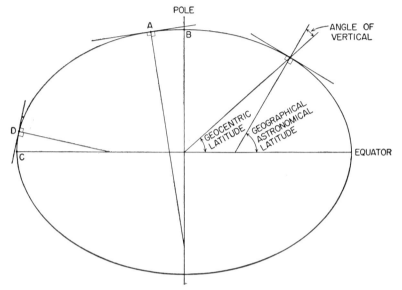

Fig. 2.7. Relation of geocentric latitude to geographical or astronomical latitude. The oblateness of the earth is greatly exaggerated.

and the directly determined value 1/297 is only 1%, well within the margin of uncertainty of the observations. Because the earth is not a sphere, a line perpendicular to its surface will not pass through the center, save for points on the equator and at the poles. The angle between the perpendicular at the surface and the line to the earth's center (the direction of gravity) is the difference between the actual (*geographical* or *astronomical*) latitude and the latitude that the given place would have if at the surface of a spherical earth (the *geocentric latitude*). The difference, known as the *angle of the vertical,* is quite appreciable at intermediate latitudes and amounts to 11′ at 45°.*

5.   MASS OF THE EARTH

The earth's mass is measured by a comparison of the force it exerts on a body of known mass with the force exerted by another body of known

* In circular measure, which is used for geographical and astronomical distances, 60″ (seconds of arc) = 1′ (minute of arc); 60′ = 1° (degree); 360° = the circumference of a circle.

mass. The earth is the only astronomical body whose mass can be measured in the laboratory, and our knowledge of all other masses depends directly upon it.

The law of gravitation, formulated by Newton (p. 170), states that the attractive force between two particles is proportional to the product of their masses and inversely proportional to the square of the distance between them:

$$F = G\,\frac{M_1 M_2}{d^2}$$

where F is the force; M_1, M_2, the masses; d, the distance; and G, a constant of proportionality, which depends on the units in which the other quantities are expressed, and which is known as the *gravitational constant*.

The experiment of "weighing the earth" depends on very delicate measurements with an accurate balance. This ambitious and fundamental experiment was first carried out by an eccentric scientific recluse, Henry Cavendish, in 1797–8. A more precise determination on the same principle, made by von Jolly in 1881, gave the earth's mass as 6.15×10^{24} kilograms (6.78×10^{21} short tons weight). The best modern determination gives 5.975×10^{24} kilograms. If the MKS system of units (based on meter, kilogram, second) is used, G, the constant of gravitation, has the value 6.670×10^{-6}.

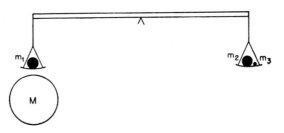

Fig. 2.8. Principle by which the earth is weighed. Two equal masses, m_1 and m_2, balance exactly. When the large mass M is placed below m_1, its attraction disturbs the balance, which is restored by the addition of the small mass m_3 to m_2. If the mass of M and its distance from m_1 are known, and also the distance to the earth's center (where its mass may be taken to be concentrated), the mass of the earth can be calculated in terms of the four known masses M, m_1, m_2, and m_3.

The actual experiment is very delicate, and the diagram only illustrates its principle.

The gravity balance gives the most accurate mass for the earth, but another method had been used even earlier. A mountain exerts a pull on the bob of a pendulum and deflects it from the vertical; the pendulum achieves a compromise between response to the earth and the mountain. The small deflection is measured, the mass of the mountain is estimated, and the mass of the earth is then obtained. The attraction of Chimborazo, in Ecuador, was used by Bouguer in 1740 to measure the earth's mass, and in 1772 a similar determination was made by the English Astronomer Royal, Maskelyne, with the peak of Schiehallion in Scotland. The results of plumb bob measures are not nearly as accurate as those from the gravity balance.

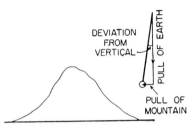

Fig. 2.9. The mass of the earth may be estimated, in terms of the mass of a mountain, by the deflection of a plumb bob from the vertical by the attraction of the mountain.

6. THE EARTH'S ORBIT ABOUT THE SUN

The sun appears to move eastward among the stars. Although the constellations are not visible in the day, those on the other side of the sky from the sun appear during the night, and change steadily with the seasons. Leo and Boötes are rising in March, followed by Virgo; summer sees the appearance of Ophiuchus, Libra, and Scorpio; Orion and Taurus shine in the winter sky. At midsummer the sun is in these latter constellations, and the heat of the "dog days" has been associated with Sirius "the Scorcher" since before the days of Hesiod.

Many of the constellations just named are in the zodiac (p. 4), and the sun appears to move eastward during the course of the year through the twelve zodiacal constellations. As mentioned earlier, the apparent path of the sun is known as the *ecliptic*.

The motion of the sun against the background of stars, taken by itself, could equally well be explained by motion of the sun round the earth or motion of the earth round the sun. But other facts have made it clear that the latter is the correct explanation.

When Copernicus suggested that the sun is the central body of the solar system, about which the planets move, the objection was raised that the earth's orbital motion ought to cause displacements of the stars in the course of the year. The conclusion was a perfectly correct one, but the objectors were mistaken in deciding that, because such displacements had not been observed, Copernicus was wrong. They had no idea how distant the stars are and, consequently, how small the displacement must be. Before the invention of telescopes the effect would have been undetectable, and in fact it was not measured until 1838 (p. 273). The apparent annual displacement of the stars by the effect of the earth's orbital motion (the *parallax*, p. 273) is today one of the best attested evidences of that motion.

A second effect of the earth's motion is the *aberration of light,* an apparent displacement of the position of a star, which results from a combination of the motion of the earth with the speed of light coming from the star. It was discovered in 1725 by Bradley, who got the idea from noticing the changes in direction of the vane on a sailboat as the boat tacked. A simple illustration of aberration is given by falling raindrops (which may be compared to the light falling from the star). A person standing still in a vertical rainstorm sees the drops falling straight down, but if he is moving, the drops seem to come to meet him; the faster he moves, the more horizontal is their approach. From a rapidly moving car the drops always seem to be coming almost straight towards one.

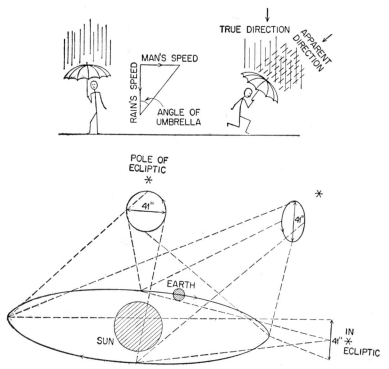

Fig. 2.10. Aberration of light illustrated (above) by the analogy with the apparent direction of falling raindrops for a stationary and a moving observer.

Below are shown the apparent displacement of the position of stars that result from aberration of light. The earth's orbit is shown in perspective, and the apparent changes of position are shown for three stars, one at the pole of the ecliptic (a circle), one in the plane of the ecliptic (a straight line), and one in an intermediate position (an ellipse). Notice that the diameter of the circle, the major axis of the ellipse, and the length of the straight line are all equal.

The largest aberrational displacement of the image of a star takes place when the earth is moving perpendicular to the star's light. This displacement is still very small, only 20″.5, but it is readily measured with precise instruments. A star that lies on the ecliptic seems to move back and forth along a line 41″ long (20″.5 each way). A star that lies as far as possible from the ecliptic seems to describe a little circle of radius 20".5, and stars in intermediate positions are ostensibly caused by aberration to describe little ellipses, with major axes (greatest diameters) 41" long. Accurate measures of star positions must be corrected for aberration.

A third way of demonstrating the annual motion of the earth around the sun is to measure the rate at which we approach or recede from a particular star as the year makes its cycle. The rate is measured by means of the spectroscopic "Doppler effect" (p. 290). The star Arcturus has been used for this test, which tells not only the fact that the earth is moving in an orbit, but also the size of this orbit. For if we knew the speed of the earth in its orbit (which we may for the moment assume to be circular) in kilometers a second, and multiply this number by the number of seconds in a year (the time taken to go around once), we obtain the circumference of the orbit in kilometers. The earth's mean orbital speed is 29.80 kilometers a second; there are 31,558,149.5 seconds in a (sidereal) year; *

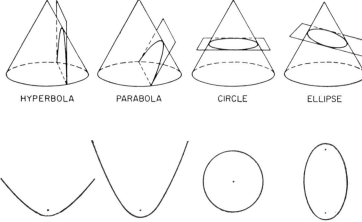

HYPERBOLA   PARABOLA   CIRCLE   ELLIPSE

Fig. 2.11. The hyperbola, the parabola, the circle, and the ellipse are all curves made by the intersection of a plane with a cone. For the *parabola,* the plane of intersection is parallel to the edge of the cone; its arms continually diverge towards the base of the cone, so it is an open curve. If the plane is more nearly parallel to the axis of the cone than to the edge, it cuts the cone in an *hyperbola,* also an open curve. If the plane is more inclined to the cone's axis than to one of the cone's edges, it cuts the cone in an ellipse, which is a closed curve. If the plane is perpendicular to the axis of the cone, the intersection is a circle, a special case of the ellipse.

The ellipse is the *conic section* with which we shall be most concerned. An important geometrical property of the ellipse is that for any point on it, the sum of the distances from two fixed points (the *foci*) is constant. A circle may be considered to be an ellipse with the two foci coincident.

The conic sections are important to us because (as we shall see later) bodies that move under gravitational attraction travel in conic sections; when they describe closed orbits, the attracting body is situated at one of the two foci.

* The time taken by one circuit of the earth in its orbit (see p. 36); it is measured relative to the stars, which provide an essentially fixed system of reference.

hence, the radius of the orbit, our distance from the sun, is about 150 million kilometers, or about 93 million miles.

Shape of the Earth's Orbit.—The earth does not move exactly in a circle around the sun; rather, its orbit is very nearly an ellipse. When Newton formulated the laws of motion, he showed that the motion of two bodies under gravity will be in one of the conic sections (ellipse, parabola, or hyperbola); the circle is a special case of the ellipse.

The sun lies at one *focus* of the elliptical orbits of the planets. The distance of the planet from the sun's center at any part of the orbit is called the *radius vector. Eccentricity* expresses the shape of an ellipse and is defined by the ratio of the distance between the foci to the longest diameter. All ellipses have eccentricities between zero and one: the eccentricity of a circle is zero, and a conic with eccentricity one or greater is not a closed curve, but a parabola or hyperbola. The point on the orbit at which the radius vector is smallest, and the planet is accordingly nearest to the sun, is called the *perihelion* [Greek: *peri,* near; *helios,* sun]; the point where the radius vector is greatest is the *aphelion* [Greek: *up,* away from]; finally, the line joining perihelion and aphelion, the longest axis of the ellipse, is the *line of apsides.* The position of the planet in its orbit is defined by the angle between the radius vector and the longest axis, measured from perihelion in the direction of motion, and known as the *anomaly.* The *mean distance* of the planet from the sun is half the sum of perihelion and aphelion distances, which is half the major axis of the ellipse.

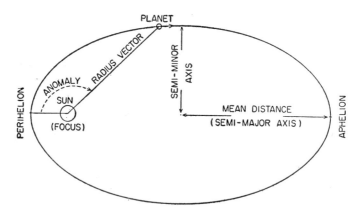

Fig. 2.12. The orbit of a planet, an ellipse with the sun at one focus.

The fact that the earth goes around the sun in an ellipse is shown by its variable speed in the orbit (see Kepler's Laws, p. 165). The interval from the beginning of spring in the Northern Hemisphere (*vernal equinox,* p. 37) to the beginning of autumn (*autumnal equinox,* p. 51) is 186

days, but from autumnal to vernal equinox it is only 179 days. This difference was discovered by Hipparchus in 120 B.C.; he believed that the sun
went around the earth and thought that the difference showed that the earth
was not at the center of the sun's circular orbit. However, we know today
(p. 165) that it is a necessary consequence of elliptical motion. The eccentricity of the earth's orbit is very small (0.01674) and the ratio of major to
minor axis is less than one in a thousand.

Since the earth's orbit is elliptical, the earth is nearer to the sun when
at perihelion (at midwinter in the Northern Hemisphere) than at aphelion,
and the sun must therefore show a larger apparent diameter then. The
difference, however, is but 3% in the sun's apparent size (although the
difference in distance is nearly three million miles), and so it is not
readily noticeable, though quite easily measurable.

7. THE ROTATION OF THE EARTH

The apparent motion of the sun against the background of stars is satisfactorily explained by the orbital motion of the earth. But the more
obvious changes of the sky involve an apparent turning of the whole sky
about the earth once a day (p. 43). In ancient times, the heavens were
actually supposed to turn about the earth, but already in the fourth century
B.C., Heracleides of Pontus recognized that it is actually the earth which
turns.

As soon as men realized that the stars are at immense distances, the
rotation of the whole heavens was seen to be fantastically unlikely. Also,
the other planets are seen to rotate, and the earth is a typical planet. Again,
the earth is oblate, and rotation would account for its oblateness; moreover the apparent gravitational pull at the equator is smaller than it should
be if the earth were not spinning, and the difference is of the right size to
be the effect of the earth's spin, if the earth were to rotate once in twenty-
four hours. For all these reasons, therefore, a daily rotation of the earth from
west to east appeals to common sense.

In 1851 the French physicist Foucault devised an experiment that
impressively illustrates the earth's rotation. He suspended a long pendulum
from the dome of the Panthéon in Paris, set it swinging, and marked the
line of swing along the floor. After a few hours the pendulum no longer
swung along the line; *the earth had turned under the pendulum,* which
maintained a fixed direction *in space.* At the poles the earth makes a complete turn under the point of support once in 23 hours 56 minutes (the
sidereal day, p. 56). At lower latitudes the angular motion is slower. At
New York, for example, the pendulum would appear to deviate 10° an
hour, or 240° a day, and would require one and a half days to come around
to the starting direction. At the equator the pendulum would not turn at all.

The pendulum deviates clockwise in the Northern Hemisphere, counter-clockwise in the Southern.*

A body that is dropped from a height retains the eastward motion that it had when it was dropped, except as this is modified by the friction of the air. In 1803 an experiment was made by dropping a weight from a 76-meter tower near Hamburg, Germany; it struck the ground about a centimeter *east* of the vertical. The explanation is that the top of the tower was moving faster eastward than its base (since it was 76 meters farther from the earth's center), and the weight started out with this component of velocity and kept it, in addition to the speed acquired by falling. The ground at the base of the tower, moving eastward more slowly, lost one centimeter on the weight during the fall, hence the latter struck the ground east of the vertical.

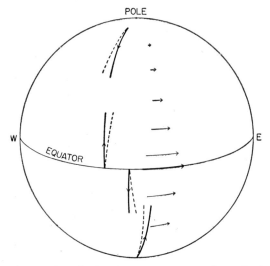

Fig. 2.13. Deviation of projectiles by the earth's rotation. The heavy lines show the direction of firing; the broken lines, the direction of flight as modified by the earth's rotation. Arrows indicate the difference of the earth's rotational velocity at various latitudes. Notice that (whether the projectile is fired north or south) it is always deviated to the right in the Northern Hemisphere, to the left in the Southern.

For similar reasons a projectile deviates to the right in the Northern Hemisphere, to the left in the Southern. Suppose a rocket is fired northward

* The detailed theory of the Foucault pendulum should take account of the curvature of the earth's orbit, which changes the orientation of a parallel of longitude by about one degree a day. The fundamental principle by which the behavior of the pendulum is interpreted depends on the assumption that the swing maintains a fixed direction *in space*. The conception of a fixed direction in space is philosophically a very difficult one, and one which is connected with the vexed problem of the nature of absolute rotation. A discussion of the question is outside the scope of an elementary book.

from a point on the equator. In addition to its initial, or muzzle, velocity, it starts out with the rotational speed that the earth has at the point of firing. It flies northward over parts of the earth that are turning with smaller linear speeds, and it pulls ahead of them in consequence of its greater east-ward "equatorial" component. Therefore, it falls to the earth to the east of the firing position and seems to deviate to the right. A rocket fired south-ward from a place near the pole moves from a place where the rotational motion is small and passes over land that has a greater eastward motion, hence it falls behind in the rotation and deviates to the west. The change in direction is again to the right, because the observer is now looking in the opposite direction to that of the man who fired northward from the equator. The student can verify the fact that the deviation in the Southern Hemis-phere will always be to the left. With modern artillery speeds and ranges, the effect is so large that allowance, sometimes of several miles, must be made for it in aiming at distant targets. The force that produces this deflec-tion relative to the earth is termed the *Coriolis force.*

The motion within weather cyclones, tornadoes, and hurricanes can be seen as a result of the earth's rotation, though the detailed theory is very complicated. Cyclones turn counterclockwise in the Northern, clock-wise in the Southern, Hemisphere. But the systematic air motions are verti-cal as well as horizontal, and cyclones are an illustration of the effects of the earth's rotation, rather than a simple proof of it.

8.   PRECESSION

The year is not marked only by the motion of the sun against the back-ground of the zodiac. Much more obvious is the variation of the midday altitude of the sun with the season. In the middle latitudes of the earth and in the Northern Hemisphere, the sun rises highest at midsummer and reaches the least altitude at midwinter. At midsummer, the days are longer than the nights; at midwinter, the nights are longer than the days.

These changes in the sun's altitude and the length of daylight are caused by the fact that the earth spins on an axis that is not perpendicular to the plane of its orbit, but is tipped at an angle of very nearly 23°.5 away from the perpendicular to the orbit plane.

The rotation of the earth on its axis causes the whole sky to appear to spin around, once in about twenty-four hours, about two points, one in the northern sky and one in the southern, that seem to be fixed. These are the *poles of the sky,* and the positions toward which the two ends of the earth's axis of rotation are directed. We may therefore picture the heavens as a hollow sphere that turns with the two poles fixed. The line that encircles the sky halfway between the poles can be regarded as the equator of the heavens (the *celestial equator*). If the axis on which the earth spins were exactly perpendicular to the orbit in which it moves, the ecliptic would coincide with the celestial equator. But the axis is tipped at an angle of about 23.5° to the perpendicular to the orbit. Therefore the sun seems to move against the stars in a path that is inclined to the equator by this amount (the *obliquity of the ecliptic*).

The ecliptic must cut the equator at two points on opposite sides of the celestial sphere. When the sun is at either of these two points, daylight and night will be of equal length (neglecting refraction), and therefore they are known as the *equinoxes* [Latin: *nox,* night]. Between equinoxes there are two points where equator and ecliptic are farthest apart; these are known as the *solstices.* At the solstices the midday altitude of the sun reaches its highest and lowest values; the increase or decrease of altitude stops and reverses [Latin: *sistere,* to cause to stand still]. The time of solstice can be determined by means of a *gnomon,* a vertical or slanting stick whose shadow is measured. At summer solstice (in the Northern Hemisphere) the shadow will be shortest, and at winter solstice, longest.

The year of the seasons is defined by the interval between summer or winter solstices, or between alternate equinoxes, and is known as the *tropical year* [Greek: *trope,* turning].

The rising of a given star at the same time as the sun (*heliacal rising*) defines a different kind of year, the *sidereal year* [Latin: *sidus,* a star]. A third kind of year is defined by the interval between two successive passages of the earth through perihelion, the point on its orbit where it is nearest to the sun. This is called the *anomalistic year* because it refers to the *anomaly* of the earth in its orbit. The three kinds of year differ slightly in length:

	d	h	m	s	
Tropical year (seasons):	365	5	48	46.0	= 365.24220 days
Sidereal year (stars):	365	6	9	9.5	= 365.25636
Anomalistic year (perihelion):	365	6	13	53.0	= 365.25964

The differences between the three types of year are very significant. To visualize their cause, we must realize that the earth turns on its axis in the same sense that it moves in its orbit, and that it rotates more than 365¼ times with respect to the stars during one trip around the sun. If the earth turned as the moon does and always kept the same face toward the sun, it would rotate on its axis *exactly* once a year with respect to the stars, and there would be no alternations of day and night. The sun would always seem to be in the same part of the sky, and the solar day would be infinitely long. However, as the earth went around its orbit (from west to east), the sun would appear to be displaced against the background of stars, and we should see it move *eastward* against the stellar background, exactly as we actually do, but not at the same rate.

Suppose that our imaginary earth, instead of rotating *exactly* once a year with respect to the stars, were to begin to spin faster, still from west to east (the same as the direction in the orbit). The sun would then no longer remain poised in the same part of the sky. As the earth turned toward the east, the sun would appear to move in the opposite direction, toward the west. This is the actual state of affairs: the earth spins more than 365 times on its axis while it goes once around its orbit *in the same direction*. Therefore, the sun appears to move *westward* across the sky once about every twenty-four hours, and *eastward* among the stars, completing the circuit once a year.

The tropical year is defined by successive passages of the sun from solstice to (similar) solstice, or from equinox to (similar) equinox. The sidereal year, however, is defined by the intervals between the sun's return to exactly the same point in the sky, *relative to the stars*. Already in 125 B.C., the great observer Hipparchus noticed that the two intervals are not the same; the tropical year is about twenty minutes shorter than the sidereal year. In other words, the position of the equinox (the point of intersection of equator and ecliptic) does not remain fixed among the stars. In the earliest days of the zodiac (see p. 5), Taurus, the "Bull-in-Front," opened the year. The *vernal* (spring) *equinox* was in the constellation of Taurus. In the days of Hipparchus it had moved into Aries. Today it is in Pisces, and it is moving steadily westward (backward) through the zodiacal constellations. By comparing his own with earlier observations of the bright star Spica, Hipparchus concluded that the equinoxes are moving backward by 40″ a year—a surprisingly good result, the modern value being 50″.3 a year.

The intersection of ecliptic and equator moves westward to meet the sun (which moves eastward among the stars), and the motion is known as *precession*. Precession is evidence of a slow *swinging motion* of the earth's axis of rotation, which always remains inclined at nearly the same angle to the earth's orbit, but which also swings around, describing a cone, and makes a complete turn in about 26,000 years.

The earth may be likened to a gigantic spinning top. The familiar peg top does not ordinarily spin in a vertical position; when tipped, it swings around in the same direction as the direction of spin, and its axis sweeps out a cone in space. For the top, this "precession" is caused by the interaction of an upward force, caused by the friction of the toe of the top on the surface on which it spins, and a downward force, exerted by the earth's gravitational pull. The former acts on the toe of the top, the latter, through the center of gravity of the top; hence they do not act in the same line but constitute a turning force or *couple*. If this couple were to act on a nonspinning top, it would, of course, cause it to fall down, as it does when the top ceases to spin. But while the top is spinning, the interaction of the spin and the couple produces a precessional force *at right angles to them both*.

A similar "precessional force" may be experienced by the student if he turns a sharp corner on a rapidly moving bicycle; he has undoubtedly learned to allow for the resulting tip of the vehicle, which is at right angles to the direction in which the wheel spins, and to the direction in which he has turned.

The earth is also caused to precess by a couple, but this couple is produced by different means from those operating on a peg top. The earth bulges at the equator, but (because of the inclination of the earth's axis to its orbit) the bulge does not lie in the plane of the orbit. The gravitational effect of the sun (which, of course, lies in the plane of the earth's orbit) tends to pull the equatorial bulge into the orbital plane. The moon contrib-

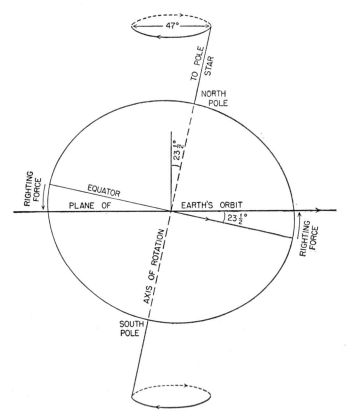

Fig. 2.14. The earth is caused to precess by the righting force, exerted by sun and moon on its equatorial bulge, which tends to pull the earth's equator into the plane of its orbit. The righting force produces a gyroscopic turning force that is at right angles both to the axis of rotation and to itself and causes the axis of the earth to sweep out a cone in space. The consequence is that the poles of the sky (the positions toward which the earth's axis points) move in a circle of radius 23½° (the obliquity of the ecliptic) in a period of about 26,000 years. The apparent paths of the poles in the sky are shown as circles drawn in perspective; the near side is shown by a continuous line, the far side by a broken line.

utes a similar effect. A couple is thus exerted on the spinning earth (see Fig. 2.14), and acts upon it as the couple acted upon the spinning top: it causes the axis of the earth to precess. The earth, however, differs in this respect from the top, for it precesses in the direction opposite to that of its spin, whereas the top precesses in the same direction as its spin. Notice, however, that the effects of the couple also are opposite in the two cases; for the earth, the couple would render the axis perpendicular to the orbit if the earth did not spin, whereas the couple that acts on the top would render the axis horizontal if the top were not spinning. A spinning top that is suspended from above becomes vertical, not horizontal, when it ceases to spin; it precesses as the earth does, in the opposite direction to its spin.

The axis of the precessing earth does not continue to lie in the same direction relative to the stars. The part of the sky toward which it points (the *pole* of the sky) is therefore changing continually, and this pole describes a circle of radius very nearly 23°.5 among the stars. The *inclination* of the earth's axis to its orbit remains very nearly the same, but the *direction* changes steadily in a cycle of about 26,000 years. The pole of the sky (round which the heavens seem to turn once a day) therefore shifts among the stars during the same cycle. Our polestar, Polaris, has not always been, and will not continue to be, near the celestial pole. In 3000 B.C. the "polestar" was α Draconis; in 7500 A.D., α Cephei will be near the pole, and in 14,000 A.D., Vega. In about 28,000 A.D. Polaris will be the polestar once more.

As the pole swings round among the stars, the visible constellations shift. The Southern Cross, now never visible from the latitude of New York, was visible there in 4000 B.C.

The moon exerts a "righting force" on the earth's equatorial bulge, just as the sun does; the moon varies in its distance from the ecliptic (p. 131), and the result is a small additional oscillation of the pole with a cycle of 19 years. The consequent motion of the pole of the equator in the sky (*nutation*) is a line that waves back and forth about the precessional circle in a rather complex way. This "nutation," or wobble, accounts for a slight shift of inclination such that the average is near 23°.5. The planets, too, have a small effect, about one-fortieth of the luni-solar precession.

One obvious effect of precession is a continual change of the positions of all the stars, when referred to the pole of the sky to which the earth's axis points. For this reason, accurate statements about a star's position must include the *epoch,* or the time to which the position refers, and, before using the recorded position of a star for some time in the past in order to locate the star now, corrections for precession and nutation are necessary.

The earth is not a perfect clock. Its rate of rotation relative to the stars varies by a very small amount. The day is steadily lengthening by about a thousandth of a second a century as a result of the action of the tides (p.

147). There are also very small erratic changes in the rate of rotation (about a ten thousandth of a second), which have only been detected in recent years as a result of new, very accurate ways of measuring time (p. 53). The earth may "run fast" by twenty seconds or so for a couple of decades, then become "slow." These small irregularities are incompletely understood. They are probably connected with changes in the distribution of our atmosphere, perhaps seasonal and perhaps of longer cycle. The atmosphere contains a very small fraction of the earth's substance, but enough to produce a perceptible effect.

Besides the motions of precession and nutation, which are swinging, nodding motions of the earth's axis, the axis itself executes some slight wobbles within the earth. These motions of the axis are probably caused by redistribution of the earth's mass as the polar ice melts and freezes. They cause the extremely small changes of the position of the pole *on the earth* and are known as the *variation of latitude*.

9. THE SEASONS

Our seasons are caused by the revolution of the earth in its orbit, and the resulting changes in the direction of its axis of rotation relative to the sun. The climate at a given place and time depends partly on the length of the day, partly on the altitude of the sun. When the sun is low, its rays fall obliquely on the earth's surface. The heating produced by a given beam of sunlight is in inverse ratio to the size of the surface illuminated.

The atmosphere acts like a blanket and holds back some of the heat that falls on the earth's surface. The temperature at any time represents a balance between heat received and heat lost. The hottest time of year in the Northern Hemisphere is not the time when days are longest and the sun's rays most vertical (near June 21), but lags behind by about a month because of the blanketing effect. The coldest time is nearer February 1 than December 21, near the winter solstice, for the same reason. Daily temperatures show the same effect: the hottest hour of a summer day is not noon but nearer 2 p.m.

Because the earth's orbit is elliptical, and perihelion occurs near the winter solstice, the earth *as a whole* receives 6% more heat from the sun at that time than at summer solstice. In the Northern Hemisphere this excess of heat is more than offset by the obliqueness of the sun's rays, but the December temperatures in the Southern Hemisphere are higher, on the average, than the June temperatures in the Northern because of the greater proximity of the sun. Although the Southern Hemisphere receives more heat from the sun at the winter solstice than the Northern Hemisphere does at the summer solstice, the earth moves faster at perihelion than at aphelion, and the total amount of heat received by the two hemispheres between alternate equinoxes is about the same.

At the summer solstice, the tipping of the earth's axis causes the north-

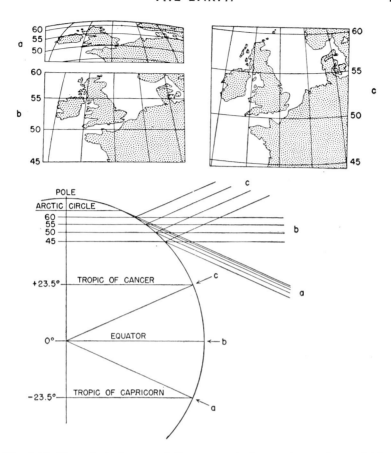

Fig. 2.15. The principal cause of the seasons is the varying angle at which the sun's rays fall on different parts of the earth's surface. Views a, b, and c show part of the coast of Europe, *as seen from the sun,* at winter solstice, vernal equinox, and summer solstice.

The lower part of the diagram shows the angle at which the rays of the sun fall on the corresponding latitudes for views a, b, and c. Notice the curvature of the upper edge of a, which defines the *Arctic Circle,* in shadow at winter solstice. Notice also that at winter solstice (a) the sun's rays fall vertically on the Tropic of Capricorn; at vernal equinox (b) they are vertical at the Equator; and at summer solstice (c) they are vertical at the Tropic of Cancer.

ern pole of the earth to be inclined toward the sun, which does not set at midsummer for latitudes north of the Arctic Circle, 66°.5 (the "midnight sun").* The North Pole itself is slightly inclined toward the sun for six months (from vernal to autumnal equinox), and the daylight

* If the earth had no atmosphere, the center of the sun would be on or above the horizon on midsummer day for 24 hours or more at all latitudes north of +66°.5. However the apparent altitude of the sun is increased by atmospheric refraction.

lasts six months. Between the Arctic Circle and the Pole, the "day" at mid-summer is shorter the farther the station is from the Pole, and south of the Arctic Circle the sun always sets for some period of time every twenty-

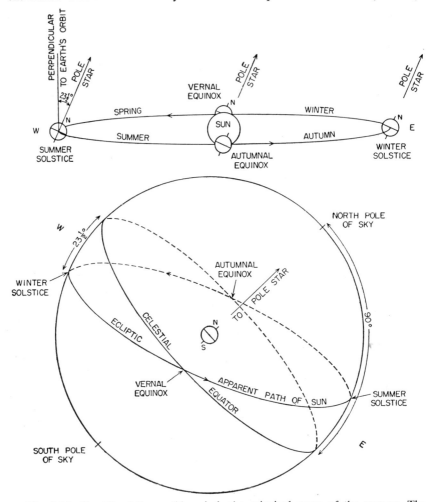

Fig. 2.16. The tilt of the earth's axis is the principal cause of the seasons. The upper diagram shows the earth going round its orbit (drawn in perspective) during one year. Notice that the axis always points in the same direction. The lower diagram shows the corresponding positions of the sun on the celestial sphere.

four hours. Within the Arctic Circle, the midnight sun circles the sky with varying altitude, never greater than $23°.5$. Even a little south of the Circle, daylight persists for twenty-four hours, though the sun does not remain above the horizon (the so-called "white nights"). Here one sees the sunset colors move around the sky from west through north to east between sunset and sunrise.

At winter solstice the southern pole of the earth is inclined toward the sun, and the southern polar regions experience conditions similar to those at the North Pole at summer solstice. Actually a square mile at the South

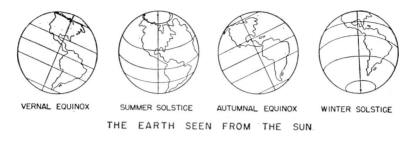

VERNAL EQUINOX   SUMMER SOLSTICE   AUTUMNAL EQUINOX   WINTER SOLSTICE

THE EARTH SEEN FROM THE SUN

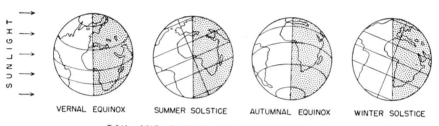

VERNAL EQUINOX   SUMMER SOLSTICE   AUTUMNAL EQUINOX   WINTER SOLSTICE

DAY AND NIGHT AT DIFFERENT SEASONS

Fig. 2.17. The different aspects of the earth at the solstices and the equinoxes. The upper row shows the earth as seen from the sun. Notice that the North Pole is tipped toward the sun at summer solstice; the South Pole, at winter solstice.

The lower row illustrates the relationship of day and night at the same seasons. The earth is seen illuminated from the left by the sun; the globe is shown, therefore, turned through about 90° relative to the views of the upper row. Length of day and night are determined by the places at which the shadow crosses the latitude considered. Notice that at the equinoxes, the shadow cuts each parallel of latitude in half: day and night are equal everywhere. At summer solstice, no shadow falls on the Arctic Circle, but the shadow covers the Antarctic Circle completely (midnight sun and 24-hour night, respectively). In northern latitudes at summer solstice, the zone of darkness covers less than half the circles of latitude: day is longer than night. Day and night are equal at the equator, and in southern latitudes, night is longer than day. The corresponding seasons are reversed at winter solstice.

Pole in twenty-four hours receives 25% more light than a square mile at the equator in the same interval (the daylight at the Pole is twice as long); if it were not for the cooling effect of the polar ice (which has accumulated during the long, dark winter), the South Pole at winter solstice would be the hottest spot on earth! The North Pole experiences a six months' night from autumnal equinox to vernal equinox; the South Pole is in darkness the other half of the year.

At the equator (neglecting refraction), daylight and night are always of equal length, and at the equinoxes the noonday sun passes exactly overhead. Between the equator and the tropics (up to latitude 23°.5) the sun is overhead at noon at some time between solstice and equinox. North and south of the tropics the sun can never pass directly overhead.

10. OUR CALENDAR

The rotation of the earth, and the motions of the sun and moon, are fundamental bases for reckoning time. However, an attempt to use more than one of them leads to complications. It would be an improbable coincidence if the earth executed an exact number of turns on its axis with respect to the sun while making one trip around the sun, or while the moon made one trip around the earth; nor would it be likely for the moon to make an exact number of circuits of its orbit during one orbital period of the earth. In fact, day, month, and year are not integral multiples one of another, and a system of time based on any one is bound to get out of step with both the others. The history of the calendar is a record of successive compromises between systems of time based on the earth, the sun, and the moon.

The most important interval in human affairs (particularly for agricultural peoples, who made the primitive calendars) is the year of the seasons (the *tropical year,* the interval between alternate equinoxes or solstices, of 365.24220 days). It is, and has been for millenniums, the basis of civil reckoning.

The phases of the moon also left their mark on early calendars: the interval from new moon to new moon, the *synodic* month (not the same as the time taken by the moon to describe its orbit with respect to the stars, p. 120), is about 29½ days. Twelve lunar months are 11¼ days less than a solar year. The Mohammedan calendar uses the lunar month, hence the months and the festivals connected with them move with respect to the seasons, and their calendar gains on ours by one year in thirty-three.

The Egyptians of two thousand years ago had effected a partial compromise between solar and lunar reckoning. Diodorus of Sicily (about 50 B.C.) recorded of them:

> They have made special arrangements concerning the months and the years. For they do not reckon days by the moon, but by the sun, putting thirty days in the months, and adding five days and a quarter to those of the twelve months, and thus they fill out the yearly circle.

The ancient Roman calendar was luni-solar; the year began in March and consisted of 354 days, and to keep in step with the seasons, the Romans added an extra or intercalary month every two years. But this system resulted in confusion and political abuses. Julius Caesar, in 45 B.C., took the advice of the astronomer, Sosigenes, for the reform of the calendar and abandoned the attempt to combine lunar and solar reckoning. His system was very near the one that we still use, with a common year of 365 days, and a 366-day year (leap year) every fourth year. Since the tropical year contains very nearly 365¼ days, this expedient accounted for the extra quarter of a day and kept the solar and terrestrial reckoning in step. Two extra months were put in to bring the current year into line with the seasons, and 46 B.C. (or 708 A.U.C., *anno urbis conditae,* the year as dated from the founding of Rome) was the last "year of confusion." The calendar devised by Julius Caesar is known as the *Julian calendar.*

The original Julian calendar assigned thirty-one days to January, March, May, Quintilis (July), September, and November—alternate months. April, June, Sextilis (August), October, and December received thirty each, and February, twenty-nine, or thirty in leap year. Quintilis ["fifth month"] was renamed "July" in honor of Julius Caesar. His successor, Augustus, called Sextilis [the "sixth month"] "August" after himself; he is said to have taken one day from February to add to August, so that his month should not be inferior to that of his predecessor. The fact that the Roman year began in March accounts for the fact that September, October, November, and December (the seventh, eighth, ninth, and tenth months of the Roman calendar) are our ninth, tenth, eleventh, and twelfth months, since our year now begins in January.

Although the Julian calendar removed much of the confusion of civil reckoning of time, and was adopted throughout the Roman Empire, the moon continued to cause complications. The feasts of the Jewish and Christian years were fixed by the moon; the Passover was set by the date of the full moon in the month of which the fourteenth day (from new moon) fell on or after the vernal equinox. The date of Easter in turn depended on the date of Passover: most Christians wanted Easter to be the Sunday following the fourteenth day of the moon, and those who placed it on the fourteenth day were regarded as heretics, and called Quartodecimans. The problem of the date of Easter was officially settled at the Council of Nicaea in 325 A.D. Easter Day was to be (and still is) the first Sunday after the fourteenth day of the moon that occurs on, or directly after, the vernal equinox. Thus the date of Easter is set by luni-solar reckoning, and its fluctuations by more than a month from year to year are a good illustration of the complications involved.

The Julian year was a great improvement over earlier calendars, but it was still longer than the tropical year by eleven minutes and fourteen seconds—a small interval, but one which amounts to one day in 128 years, or about three days in 400 years. Consequently the calendar slowly fell out of step with the seasons. In the year 730, the Venerable Bede, the medieval English scholar, showed that the deviation was more than three days; Roger Bacon, in 1200, found an error of seven or eight days. Dante, soon after 1300, was well aware of the need of calendar reform. He prophesied great changes in the world:

> . . . ere from winter January advance
> To spring, through the hundredth part which ye
>   neglect.
>
> *Paradiso,* Canto XXVII, 142–3

In 1474 Pope Sixtus IV invited the scientist, Regiomontanus, to revise the calendar and bring it into line; but the pope's death interrupted the plan.

It was not until 1582 that Pope Gregory XIII revised the calendar, dropped the ten days of accumulated error, and decided on the rule that only such centesimal years were to be leap years as were divisible by four hundred. This is the calendar that we still employ, called the *Gregorian calendar* after its author. It is more nearly satisfactory than the Julian calendar, but still not perfectly so, and after a few thousand years another slight revision (such as making only such millenniums into leap years as are divisible by four thousand) will be necessary. The problem is one that cannot be solved except approximately, since the periods of rotation and revolution of the earth are in no simple ratio.

The Julian calendar was promptly adopted throughout the Roman Empire, which comprised the whole civilized Western world. But the Gregorian calendar was not so fortunate, because it was formulated after the Reformation, and the pope's decision was not at once accepted everywhere. The Catholic nations adopted it at once, but the Protestant nations fell into line only gradually. Germany adopted the Gregorian calendar in 1700; England, in 1751. By this time the two calendars were out of step by eleven days, and there was rioting in England with the slogan: "Give us back our eleven days." In spite of the great efforts that were made to avert injustices in the payment of wages and so forth, people felt that they were being robbed of something! The year began in England on January 1, instead of March 25, for the first time in 1752, though the change had been made in Scotland in 1600. Russia adopted the Gregorian calendar only after the Revolution of 1918, so that many Russians now living were born under the "Old Style" (Julian calendar). The Russian Orthodox Church still uses the Julian calendar, which is why the Russian Easter may differ by one or two weeks from the date of Easter used by the rest of the Christian world.

There have been many proposals for the further reform of the calendar, principally in the direction of abandoning the vestiges of lunar reckoning and assigning such feasts as Easter to fixed date. There is the further complication that the seven-day week (a legacy from ancient Babylonia) does not divide the year exactly, and some proposals suggest that the calendar should be so adjusted that the same date always falls upon the same day of the week. To many people, however, the calendar as we know it today seems to contain interesting vestiges of human history which it would be a pity to destroy.

For scientific and chronological purposes, the expression of dates by year, month, and day is a clumsy expedient, and the interval between two dates involves calculations that are unnecessarily awkward. The Renaissance scholar Joseph Justus Scaliger suggested in 1582 that all dates be referred to an arbitrary initial date, January 1, 4713 B.C., which he chose in connection with his work on early chronology. The date thus reckoned is known as the *Julian day,* named by its inventor in honor of his father, Julius Scaliger, and not to be confused with the Julian calendar, with which it has no connection. Julian days are used for expressing the times of most astronomical observations. They are reckoned from noon, and parts of a day are expressed in decimals to the necessary degree of precision. On January 1, 1950, the Julian date was 2,432,282.

CHAPTER
III

TOOLS AND METHODS

Euclid alone
Has looked on Beauty bare. Fortunate they
Who, though once only and then but far away,
Have heard her massive sandal set on stone.
EDNA ST. VINCENT MILLAY

Science is essentially quantitative, and any treatment of astronomy that is not purely descriptive must continually make accurate statements in terms of standard quantities and concepts. The most fundamental ideas in the study of the universe are those of time, distance, mass, and temperature. These are the "dry bones" of astronomy, but without their support her form would be spineless and undistinguished. With their aid she appears in all her coherence and beauty, *simplex munditiis,* dignified and austere.

The previous chapters have already called upon some of the fundamental definitions, which must have aroused the student to the need for precise formulation. The present chapter is a means to this end.

1. CELESTIAL SURVEYING

Position in the sky is fixed on exactly the same principle as position on the earth's surface.

Any place on earth can be specified by its *latitude* and *longitude.* Latitude is measured from the equator (the circle, equidistant from the poles, that marks the earth's largest circumference). The whole circle contains 360°; the angular distance from pole to pole is 180°. The latitude at the equator is chosen as the zero quantity; latitudes north of the equator have positive values from 0 to +90°, those south of the equator, negative values from 0 to −90°.

Longitude is measured around the earth; there is a natural zero for latitude (the equator), but none for longitude, and an arbitrary choice is made. Longitude is measured from the great circle running through the poles and the position of Greenwich (England); it is counted through 180° east and west.

48

Longitude may also be expressed in hours, minutes, and seconds, rather than in degrees. The whole circle comprises 24 hours $= 360°$; thus 1 hour $= 15°$, and $1° = 4$ minutes of time. Note, therefore, that $4^m = 60'$, or one minute of time is equivalent to 15 minutes of arc. When the longitude of a place is expressed in time, it states how much earlier (east or negative) or how much later (west or positive) noon occurs there than at Greenwich.

The latitude and longitude of several places on the earth's surface are:

				h	m	s
Evanston, Illinois:	$+42°$	03′	27.2″	$+5$	40	41.84
Fort Davis, Texas:	$+30$	40	17.0	$+6$	56	05.36
Cambridge, England:	$+52$	12	53.3	-0	00	22.77
Bosque Alegre, Argentina:	-31	35	53	$+4$	18	11.2
Wellington, New Zealand:	-41	17	03.9	-11	39	03.69

These are commonplaces to everybody. Positions on the sky can be described in exactly the same way. The equatorial system most used in astronomy is thus defined. For purposes of fixing the position of a heavenly body, the sky is pictured as a sphere with the earth at the center—the *celestial sphere*. Points on the celestial sphere are located by means of circles that are arranged exactly in the same way as the circles of latitude and longitude on the surface of the earth.

There are several systems of celestial co-ordinates, a situation that the student sometimes finds confusing. Actually they exist to avoid confusion. Even on the earth we do not always use the same basis of reference. We locate the post office by saying that it is across the street from the railroad station; we say that Kankakee is fifty miles south of Chicago, and that Samoa is at longitude 173° west, and 14° south of the equator. We may describe a runner as being three quarters of the way around the track. For each place, we choose the most natural point of departure and specify distance and direction from there.

If (1) we are pointing out an object in the sky to a companion, we may find it convenient to say how far it is above the horizon, and how far around the sky from a fixed direction (say the south). If (2) we are describing the position of a planet, we may say where it is in the ecliptic, the circle in the sky near which sun, moon, and planets move. A star (3) is most conveniently located with reference to the celestial poles, about which the heavens seem to turn in their daily motion. When looking at an object (4) outside our flattened, pancake-shaped system of stars (the *galaxy*), we use the flat edge of the system to count from. The four kinds of celestial co-ordinates that we have mentioned are, respectively, the (1) *horizon system,* (2) the *ecliptic system,* (3) the *equatorial system,* and (4) the *galactic system.* Each one is used when most convenient. They are compactly illustrated in Figure 3.1.

The *Horizon System* refers positions to the *horizon.* The point directly overhead for the observer is the *zenith;* the point underfoot, opposite to the

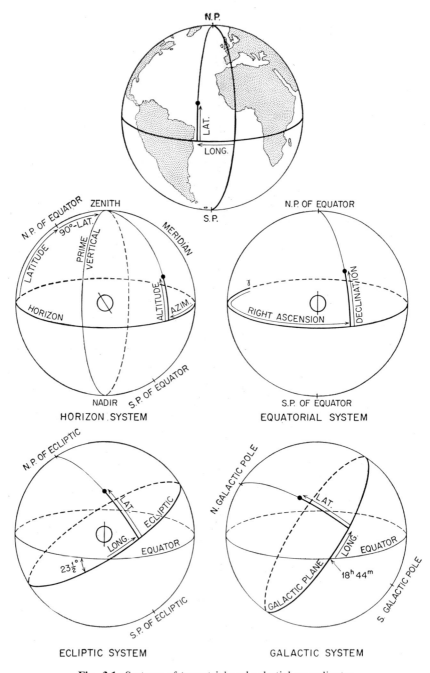

Fig. 3.1. Systems of terrestrial and celestial co-ordinates.

zenith, is the *nadir*. The north-south great circle, passing through the zenith, is the *meridian;* the east-west great circle, passing through the zenith 90° from the meridian, is the *prime vertical*. The position of an object is described by its *altitude,* measured in degrees up toward the zenith from the horizon (negative altitudes refer to points below the horizon), and its *azimuth,* measured along the horizon in degrees westward from the south point of the meridian (by astronomers). Surveyors and navigators measure an azimuth in the same way, except that they count from the north point on the horizon, through the east. The *zenith distance* of the object is 90° minus its altitude. On this system, the altitude of the north pole of the sky is equal to the latitude of the place at which the observation is made; the zenith distance of the pole is 90° minus the latitude. The horizon system of co-ordinates has the great disadvantage that in it the position of a heavenly body varies with the place from which it is observed, and also that time as well as position must be specified. It is therefore of no value in specifying positions that are to be used by observers elsewhere.

The *Ecliptic System* is defined by the *plane of the earth's orbit,* which fixes the position of the *ecliptic*. It is useful in describing the location of members of the solar system, most of which move in or near the ecliptic. As mentioned earlier (Chapter II), the ecliptic intersects the equator in two points (the *equinoxes*). The point in which the sun's path along the ecliptic crosses the equator from south to north is called the *vernal equinox;* it crosses the equator again from north to south six months later, at the *autumnal equinox*. The intersections of ecliptic and equator are used to fix the zero of *celestial longitude,* which is measured in degrees eastward from the vernal equinox. The vernal equinox is sometimes called the First Point of Aries, and denoted by the symbol ♈ ; however, because of precession the First Point of Aries is actually not now in Aries but in Pisces. *Celestial latitude* is measured north (positive) or south (negative) from the ecliptic, in degrees. As the ecliptic cuts the equator at an angle of nearly 23°.5, the pole of the ecliptic is nearly 23°.5 from the pole of the equator.

The *Equatorial System,* the one most generally used in astronomy, is defined by the *direction of the earth's axis*. The north and south poles of the sky are the points to which the ends of the axis are directed; around them the heavens seem to turn. The *celestial equator* is the great circle equidistant from these poles. The angular distance of a star from the equator toward the nearest pole is the *declination,* and is measured in degrees, north or south (positive or negative). The co-ordinate that corresponds to longitude is called the *right ascension,* and it is measured eastward in degrees or time along the equator from the vernal equinox. The 360° of right ascension are divided, like the 360° of longitude on the earth, into 24 hours, and hence 1 hour in right ascension corresponds to 15°, and 1° of arc to 4 minutes of time. The north (or south) polar distance of a star is ±90° minus its declination. Because of precession, which moves the vernal equinox backward (westward) along the ecliptic, the zero of right ascension is steadily changing relative to the stars, and both the right ascensions and declinations of all stars are altering continuously. Accurate positions are referred to a particular date or *epoch,* and positions for another date require a correction that is largest nearest the poles and increases with the elapsed interval.

The *Galactic System* of co-ordinates is exactly similar in principle; it is defined by the *plane of the galaxy* (very nearly, but not quite, the same as the plane of the sun's orbit round the galactic center). *Galactic latitudes* are measured (positive or negative) from the galactic plane toward the north or south galactic pole. *Galactic longitudes* are measured eastward from the point of intersection of the galactic plane and the celestial equator at 18^h44^m right ascension. It is rather unfortunate that this is not the direction of the center of the galaxy, which is in galactic latitude 0, galactic longitude 327°, on this system. To refer longitudes on the galactic system to the center of the galaxy, 33° must therefore be added to the conventional galactic longitude.

For some purposes, *galactocentric latitude and longitude* must be used. These treat the *center of our galaxy* as the central point, and are therefore fundamentally different from all the co-ordinate systems hitherto described which subdivide the celestial sphere. Galactocentric co-ordinates are used only in the three-dimensional study of our galactic system.

The co-ordinates in any one system are readily transformed into those in another by the appropriate formulas of spherical trigonometry.

2. TIME

The alternations of day and night, and the passage of the year, are the obvious crude means of measuring time. The section on the calendar has described some of the inevitable complexities.

For the measurement of shorter intervals on a uniform system, the rate of some dependable physical process must be used. One of the most ancient was the rate of flow of water from a vessel (the waterclock or clepsydra), or of sand (the sandclock). The rate, however, is not constant but depends on the "head of water," and such devices are not capable of accuracy. The rate of burning of a candle was another such ancient "clock," which depended for success on the making of uniform candles.

In medieval days, time was registered by simple weight-driven clocks; however, they were both ponderous and very poor timekeepers, because they were governed only by the rate of descent of the weight, which turned a drum.

In 1581, Galileo made the major scientific discovery that the time of swing (*period*) of a pendulum is essentially constant and depends on the length of the pendulum. Not until Huyghens perfected the *escapement,* however, were pendulums used to regulate the rate of clocks. Until thirty years ago, the accurate measurement of time intervals rested on improvements in the pendulum clock. The variable rate of the pendulum, for example, caused by expansion or contraction under changes of temperature, was corrected by several ingenious devices. None the less, the best pendulum clocks still had errors amounting to between one-hundredth and one-thousandth of a second a day, or about one part in ten million. To construct a perfectly periodic mechanism is next to impossible.

A great advance in accurate time-keeping was effected by using the periodic piezoelectric properties of certain crystals (such as those of quartz) to regulate the rate of a pendulum clock electrically. The accuracy so attained (one part in a hundred million) was great enough for the detection of the very small changes in the rate of the earth's rotation.

The most accurate timepiece hitherto devised employs the rate (natural period) of vibration of a molecule as a time-measuring device. It has a theoretical accuracy of one part in a billion, or perhaps even ten billion. The molecule that was first used for the purpose was the ammonia molecule (NH_3), and the time standard is one of the natural periods of vibration (a wave length in the microwave region) of the molecule. The accuracy attainable with this method depends on the "fuzziness" (width) of the spectrum line; the *inherent* line width is so small that an accuracy of one part in a billion billion would be attainable if it could be used in the absence of outside disturbances of the atoms. The latter always contribute something to the line width, and therefore an accuracy better than one part in a hundred million, or perhaps one part in a thousand million, cannot be expected.

To summarize: the accuracy of all timepieces is limited by the degree of regularity of the *frequency* (number of vibrations in unit time, or reciprocal of period). The simpler the mechanism, the less will conditions complicate the frequency. A pendulum clock is a large mechanism which must be protected from variations of temperature and operated in a vacuum. A vibrating crystal is a simpler mechanism, and a molecule is simpler still. For the greatest accuracy these, too, must be protected from outside influences. But the ultimate precision of a molecular clock is limited by the natural fuzziness of the molecule itself, a property of matter that cannot be circumvented.

Time and Sun. The sun is our natural timekeeper. Its passage across the meridian defines *noon*. As the earth turns from west to east, the sun comes to the meridian successively at places farther and farther west. Thus the moment of noon differs in different longitudes. The moment, at any place, where the sun crosses the meridian is *apparent local solar noon*: apparent, because it is the observed moment; local, because it is peculiar to the longitude of the place; solar, because it is fixed by the sun.

The *apparent local solar time,* at a given place and a given instant, is the interval since the sun passed through the meridian at that place. The sun crosses the meridian twice in twenty-four hours: at noon it crosses the upper meridian, and is highest in the sky at that time; at midnight it crosses the lower meridian, and is farthest below the horizon then. As we count our days from midnight, apparent local solar time is the interval since the sun passed the *lower* meridian. If the day is divided into two successive twelve-hour periods, this interval is numerically the same as that since the sun crossed the upper meridian; however, all astronomical time is reckoned on a twenty-four-hour basis and counted from midnight. The interval between successive passages of the sun across the lower meridian is the *solar day*.

The sundial records apparent local solar time. A sundial consists of a *gnomon,* a vertical or inclined rod that casts a shadow, and a scale marked with the hours, on which the shadow falls. Anyone who has used a sundial knows that it does not keep in step with the time shown by a clock; the difference (fast or slow) may amount to a quarter of an hour. In addition, a sundial keeps local time, a clock is usually kept on standard time.

The reason for the difference between sundial and clock is twofold: the earth's orbit is an ellipse, and the ecliptic is inclined to the equator.

The ellipticity of its orbit causes the earth to move round the sun with variable speed, faster near perihelion than near aphelion (p. 165); therefore the sun's apparent speed in the sky, against the background of the stars, varies in the same way. The consequence is that the solar day varies in length: it is fifty-one seconds longer, from solar noon to solar noon, on December 23 than on September 23. The *sidereal day* (p. 56), the interval between successive meridian passages of a star, is of course not variable in length because it depends only on the earth's rotational speed, which is virtually constant.

For illustration we compare the actual speed of the sun with the speed of an imaginary sun (the *mean sun*) that would go around the equator once a year at a uniform rate. We recall that the earth's rotation causes the sun to seem to go around the sky once a day *from east to west,* but the earth's orbital motion causes it to move round the ecliptic once a year *from west to east.* The true sun will make the circuit of the ecliptic, and the mean sun will make the circuit of the equator, in exactly a year; also, the true sun will take exactly half a year to go from perihelion to aphelion, and the same time to go from aphelion to perihelion. Thus the true sun and the mean sun may be considered to be in step at these two points.

If we compare the motion of mean sun and true sun during the year, starting from perihelion, we find that the true sun, which is then going fastest, begins to pull ahead of the mean sun and gets to the *east* of it. In the daily east-to-west motion the true sun will therefore begin to lag *behind* the mean sun, and to be *later* than the mean sun in reaching the meridian. Six months later, when the earth is at aphelion, the mean sun has caught up with the true sun, and they are in step again. Now the true sun is moving more slowly than the mean sun, and begins to fall behind (westward); therefore in the daily east-to-west motion it begins to reach the meridian *before* the mean sun. Gradually it loses its westward lead, and at perihelion the two suns are again in step and cross the meridian together. These changes will evidently repeat themselves once a year, the true sun running alternately ahead of and behind the mean sun by 7.75 minutes.

The effect of the obliquity of the ecliptic is less obvious. If the equator and ecliptic coincided, the effect of orbital eccentricity would be the only difference between true sun and mean sun. If the obliquity were acting by itself (i.e., if the earth's orbit were circular, but inclined), its effect would be zero at the equinoxes and solstices: from March 21 to June 21 it would cause the true sun to be west of the mean sun and, therefore, to cross the meridian early; between June 21 and September 21, meridian passage would be delayed; it would be early again from September 21 to December 21 and late from December 21 to March 21. The obliquity of the ecliptic has, in fact, an effect with a cycle of six months; its range is ten minutes each way.

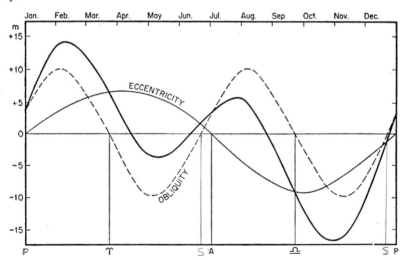

Fig. 3.2. The Equation of Time. The vertical co-ordinate shows by how many minutes the true sun is later than the mean sun, and therefore by how many minutes sundial time will be later than clock time. The light line shows the effect of the eccentricity of the earth's orbit throughout the year (the horizontal co-ordinate shows the months); it is zero at the times of perihelion and aphelion of the earth in the orbit (*P* and *A* in the figure). The light, broken line shows the effect of the obliquity of the ecliptic; it is zero at vernal (♈) and autumnal (♎) equinoxes, and at the solstices. The heavy line shows the result of summing the two curves: the Equation of Time.

The two causes mentioned add up to a fluctuating difference between sundial time and uniform time (*mean time,* because referred to the mean sun) that has two unequal maxima and minima in the course of the year. It is known as the *Equation of Time,* defined as apparent *minus* mean time. Thus mean time is derived from apparent, or sundial, time by subtracting the Equation of Time, and conversely.

Mean time is the time to which clocks are adjusted; it would be impractical to build clocks that ran at a nonuniform rate and kept pace with the true sun. True time is, in a sense, the time told by sun and sundial. When daylight saving time was introduced, there were some who objected to it on the ground that it was not "God's time." If any time deserves to be called "God's time," it must be apparent local solar time, which would be awkward and confusing to use. Clock time, even for the local meridian (*local mean (solar) time*), is seen to be quite artificial.

Local mean time, which differs with longitude, would also be most confusing. *Local civil time* is the same as local mean (*solar*) time and differs with longitude; it is counted from lower meridian passage of the mean sun at the given place.

Standard time on land, or *zone time* at sea, is arbitrarily fixed to keep the time the same over chosen ranges of longitude—the *time zones.* It is equal to local civil time at adopted central meridians, usually 15° apart, at about the middle of the time zones, and changes abruptly by a whole hour between one zone and the next. On opposite sides of the 180° meridian the time changes abruptly by a whole day (the *international date line*).

For astronomical purposes, time is often expressed as *Greenwich mean (solar) time,* now counted from the lower meridian passage of the sun at the longitude of Greenwich, the same as that from which our terrestrial longitudes are measured. This also defines *Greenwich civil time* and *universal time.* In earlier years, Greenwich mean time was reckoned from noon.

When an accurate statement is made of the time of an astronomical event (such as the minimum of a rapidly varying eclipsing-star) it may be necessary to allow for the time taken by light to cross the earth's orbit. At dates six months apart, the distance from star to earth may differ by the whole diameter of the orbit—184 million miles. Light takes over sixteen minutes to cross the earth's orbit. The times of observations made six months apart will be in error, relative to one another, by this amount in the plane of the orbit. The error will be less in higher latitudes, because of the effect of projection, and will be zero at the poles of the ecliptic. Such observations must be "corrected to the sun": the time of observation is calculated as it would be for a body in the center of the solar system. Time so expressed is *heliocentric mean time.* The kind of time in which an observation is made must always be stated.

Sidereal Time. The true *sidereal day* is the interval between successive meridian passages of a given *star.* In a sidereal year (equinox to equinox) there are 365.25636 mean solar days or 366.25636 sidereal days. The sun moves eastward among the stars and loses one lap a year on them.

The solar day is therefore about four minutes longer than the sidereal day.

Sidereal time is reckoned from upper meridian passage of the First Point of Aries (vernal equinox); at the autumnal equinox, therefore, solar and sidereal time (on the twenty-four-hour day) are in step. At any moment, the sidereal time that has elapsed since the meridian passage of a star is called the *hour angle* of the star. Because right ascension, like sidereal time, is counted from the First Point of Aries, it follows that sidereal time *minus* right ascension equals hour angle. Thus on the meridian (hour angle zero), sidereal time equals right ascension. Hour angle is counted east or west of the meridian and expressed in time: negative hour angles are east of the meridian (up to 12 hours); positive hour angles, west.

Almost all but the smallest portable telescopes are mounted equatorially (p. 88) so that they can be "set" by declination and hour angle. To point a telescope to a star of known position (right ascension and declination), it is necessary to know the sidereal time, which is obtained from a *sidereal clock* that is adjusted to run at the proper rate. The hour angle is calculated from the right ascension and the current sidereal time; the telescope is set by means of the circles, graduated in hour angle and declination; and the star can then be located.

3. LIGHT

O supreme light, who dost thy glory assert
High over our imagining, lend again
Memory a little of what to me thou wert.
Vouchsafe unto my tongue such power to attain
That but one sparkle it may leave behind
Of thy magnificence to future men.
DANTE,
Paradiso, Canto XXXIII, 67–72

Almost all information about planets and stars comes to us from the light that they send us. By means of their light we study their position, motion, and brightness. The qualities of such light reveal whether they are gaseous or solid, self-luminous or mere reflectors. The light of a planet is a clue to the nature, roughness, and temperature of its surface, to the presence and composition of its atmosphere, and to its speed of rotation. The light of a star yields the receding or approaching speed of the star in space and its rate of rotation, as well as the motion, chemical composition, density, and temperature of its atmosphere.

Light may be described as consisting of *electromagnetic waves:* periodic fluctuations in space of electric and magnetic fields in directions perpendicular to each other and to the direction of propagation. According to this view it may be pictured as a *transverse wave motion* (see Fig. 2.2, p. 18). When no matter is present, the speed of light waves is 3×10^{10} cm/sec, a speed which is the same for light of all colors, and which cannot be exceeded by that of any material object, according to the theory of relativity. When passing through matter, light waves are slower by an amount that depends on the properties of the substance and on the wave length.

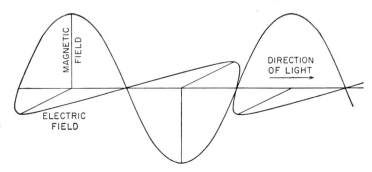

Fig. 3.3. Electromagnetic waves may be regarded as tranverse displacements of electric and magnetic fields. This three-dimensional diagram shows that the electric and magnetic displacements are *at right angles to one another and to the direction of travel of the light.* In ordinary (unpolarized) light there are electric and magnetic displacements in all directions about the line in which the light is traveling. When the light is polarized, they are confined to certain directions. For plane-polarized light, the "plane of polarization" is parallel to the displacement of the magnetic field. More complicated types of orientation also occur.

The picture of light as a wave motion suffices to describe most of its gross behavior. Many of the properties of light, however, can be described and understood only in terms of the idea that light acts like a stream of particles. Such, for example, are the facts that radiating atoms emit only discrete colors, and that the color of the light given out by a radiating surface varies with temperature. The wave and particle theories of light were in conflict since the days of Newton, but the modern *quantum theory* (p. 61) has shown that they are compatible. Light pulses combine the properties of wave and particle. As a particle (the *photon*), light is regarded as one of the fundamental constituents of matter. The properties of photons in large quantities approximate those of waves. This is but one of many instances in which detailed processes require the quantum theory for their description, but gross behavior fits the formulation of classical physics. We shall begin by speaking of light in terms of wave motion.

The *color* of light depends on the number of wave troughs or crests that meet the observer in a given interval. The number that arrive per second is called the *frequency*. For visible light the frequency is enormous: about five hundred million million (5×10^{14}) waves of red light strike the eye in a second. The distance between successive crests or troughs is called the *wave length*. It is the ratio of the speed of light to the frequency of that light:

$$v = c/\lambda \qquad \text{or} \qquad \lambda = c/v$$

where v is the frequency, λ the wave length, c the velocity of light (about 3×10^{10} cm/sec), which, as mentioned above, is constant and the same for all colors in a vacuum. The wave length that corresponds to a fre-

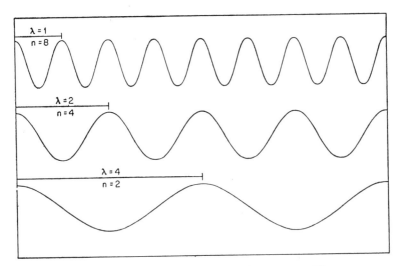

Fig. 3.4. Relation between wave length λ and frequency n. The waves are supposed to pass from left to right, across the figure, in unit time. The number of waves included within the figure is therefore the frequency n. Notice that wave length is inversely proportional to frequency, so that the product $n\lambda$ is a constant.

quency of 5×10^{14} is therefore $3 \times 10^{10} \div 5 \times 10^{14}$, or 6×10^{-5} centimeters—less than a thousandth of a millimeter. It is awkward to express such small wave lengths in centimeters; they are usually given in Angstrom units (one A $= 10^{-8}$ cm as used by astronomers), named in honor of a Swedish physicist (1817–1874) who made great contributions to the study of the solar spectrum. The wave length of the red light considered is thus 6000 Angstroms.

Our eyes are sensitive to a very small range of wave lengths. Violet light has a wave length of about 4000 Angstroms; blue, 4200 Angstroms; green, 5000 Angstroms; yellow, 5600 Angstroms; and red, 6800 Angstroms. Waves shorter than 3900 Angstroms or longer than 8000 Angstroms are invisible to most human eyes, though individuals differ somewhat. Other animals have not the same color range. Bees, for example, can perceive ultraviolet light of wave length 3900 Angstroms, but cannot see red light.

Our color perception is restricted, but radiating bodies give out waves of a variety of wave lengths, from about a mile to a ten-millionth of an Angstrom. These waves are all of the same kind, and travel with the same speed through space, as visible light. We describe them by a variety of names, but this is an accident that depends on the limitations of our methods of detecting them.

TYPES OF ELECTROMAGNETIC WAVES

Long-wave radio:	Over 1500 m
Aircraft control band:	750–1500 m
Radio broadcast band:	180–750 m
Short-wave radio:	15–180 m
Ultrahigh frequency and television:	2.5–15 m
Microwaves (radar):	0.1–250 cm
Extreme infrared light:	0.002–0.1 cm
Near infrared light:	15,000–200,000 A
Photographic infrared light:	7500–15,000 A
Visible light:	3900–7500 A
Ultraviolet light:	100–3900 A
X rays:	0.001–100 A
Gamma rays:	0.001–0.0000007 A

From long-wave radio to microwaves, detection is by induced oscillations of the appropriate electromagnetic circuits such as are used in the radio receiver. Infrared light is essentially heat, and is detectable by its

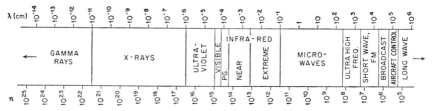

Fig. 3.5. The whole spectral range runs from gamma rays to long electromagnetic waves. Wave lengths (in centimeters) are marked above, frequencies (in cycles per second), below. Notice the narrowness of the band of visible radiations.

heating effects, for example with the thermocouple (electric current produced by heating one junction of certain metals, while the other junction is kept at some reference temperature). Waves from the photographic infrared through the X-ray region are observed through their chemical action on the photographic plate, often stimulated through absorption by appropriate dyes in the emulsion (*sensitized plates*). The sensitization of photographic emulsions has been an important contribution of applied chemistry and has steadily pushed the investigation of the spectrum into the infrared region. The photoelectric cell is also used for the detection and measurement of radiations near the visual region. The shortest known waves, the so-called gamma rays, are pulses of radiation that are given out by disintegrating atomic cores, or nuclei, such as those of the radium atom (*radioactivity*). They can be detected by the effects that they produce when they interact with matter.

In its early days, astronomy was concerned only with the small span of visible wave lengths. The universe was viewed, so to speak, through a narrow chink. The chink has been steadily widened during the past half-century. Heat waves from planets and stars are studied with the thermocouple, and the region from infrared to ultraviolet, with the photographic plate. Our astronomical survey of the ultraviolet is limited not by methods of detection, but by the ozone layer of the atmosphere, which cuts off light of wave length less than 2900 Angstroms. Even this barrier is now being broken down by rocket-borne spectrographs shot up through the layer of ozone.

Microwave astronomy has recently begun to expand the observed range toward longer wave lengths. This relatively new field is rich in possibilities. Already radar is reflecting waves from the moon and detecting outbursts of energy on the sun. The microwaves penetrate the galactic dust and haze better than radiations of shorter wave length, and are observed to come from distant parts of our galaxy, especially from its otherwise inaccessible center. Microwave radiations have even been detected from the Andromeda galaxy and other distant stellar systems. The possibilities of this new technique, and the range of new phenomena that it can reveal, are enormous and may well change our whole cosmic picture during coming decades.

The Laws of Radiation. A radiating body emits light of a large range of wave lengths. A simple picture of the process of the emission of electromagnetic waves from a surface depicted a large number of vibrating sources, with a variety of possible frequencies. Each vibrator was imagined to be radiating light, of wave length c/n, where c is the speed of light, n the frequency of vibration. The sum total of all the emitted light waves was supposed to add up to the total radiation from the surface of the body. The idea that a radiating body is equivalent to a continuous series of vibrators of all possible frequencies was developed at the end of the nineteenth century by the great theorist Lord Rayleigh. But it was found to predict two results that were not compatible with observation:

1. The shorter the wave length (or the greater the frequency) the more energy should there be in that wave length, so that the energy would all be crowded into the short wave lengths, whatever the temperature.

2. At all temperatures, the total energy emitted per unit of radiating surface should be infinite.

Neither of these predictions is fulfilled; the second is manifestly untrue, and the first contradicts the known fact that the hotter a surface is, the "bluer" is the radiation it emits (i.e., the shorter is the wave length at which the radiation is most intense).

The difficulty was resolved by means of the *quantum theory,* developed by the German physicist Max Planck about 1900. He suggested that energy is emitted only in parcels of definite size (quanta) such that:

$$E = h\nu,$$

where E is the energy, h is a universal constant, and ν the number of waves passing a given point per second (the *frequency*). Planck's suggestion led to results that were in full agreement with the observation of radiation from hot bodies. It was backed up by certain other information, for example the capacity of solids to absorb heat (worked out by Einstein) and the *photoelectric effect* (the ejection of electrons from the surfaces of metals by radiation that falls on them).

Planck developed a theory which predicted how the energy from a radiating body would be distributed among various wave lengths (or frequencies) at different temperatures. The theory was worked out (for simplicity) for a surface that is a perfect absorber and a perfect radiator (technically known as a *black body*). The theory is beyond the scope of the present book, but the shapes of the curves that result from it are important

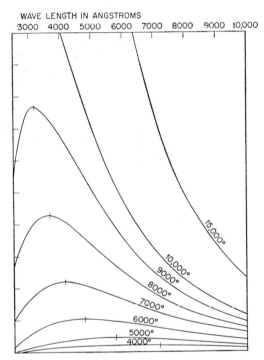

Fig. 3.6. The Planck Law of black-body radiation.

and easily memorized. Figure 3.6 shows the distribution of the energy among various wave lengths, for several temperatures.* Note especially

* Temperature in this book will usually be expressed on the *Kelvin scale,* which, like the Centigrade scale, covers 100° between the boiling and freezing points of water under standard conditions. It differs from the Centigrade scale by being reckoned from *absolute zero,* very nearly —273°C, so that a temperature on the

that *at all wave lengths* the curve for a given temperature lies wholly above the curve for lower temperature. Moreover, the wave length at which most energy is given out (the maximum of the curve) is smaller for higher temperatures.

If we can measure the amount of energy given out at a number of different wave lengths by a distant body (such as a star), we can then look among the Planck curves; *if the star is radiating like a black body* (i.e., if it is a perfect radiator), we can pick out one that exactly fits the observations in shape, and that will tell at once the temperature of the star.

Actually the stars are not black bodies, and the fit is never perfect. Also the part of the star (depth within it) that we see at different wave lengths may not be quite the same. Nevertheless, the correspondence is usually fairly close for the stars, and quite good approximations to stellar temperatures are obtained by means of the Planck formula.

For the sun, in red and violet light, the fit with the Planck curves leads to a temperature of 5800°; in the infrared, to 5600°; in the green, to about 6150°. The chief difficulty in using this method for the sun and stars is that of allowing for absorption in the earth's atmosphere, which not only cuts down the light but also distorts the energy curve.

There is a simpler way to determine temperature from observed distribution of energy among the various wave lengths (or frequencies). The *maximum* of energy is at shorter wave lengths for higher temperatures. Actually the wave length of maximum energy is inversely proportional to the temperature (Wien's Law). If the wave length at which the intensity is a maximum can be determined, the temperature follows at once, if the surface radiates like a black body. Note that Wien's Law, unlike Planck's Law, gives a single, unambiguous result if the energy curve is highest at one point. However, we cannot tell from the observation of maximum intensity whether we are justified in using the Wien relation, whereas the conformity of the observed energy-curve to the Planck Law is an immediate measure of applicability. Wien's Law leads to a temperature of 6150° for the sun's radiation.

Another application of the Planck formula uses the total area under the curve to measure the temperature. As the curves for higher temperature always lie wholly above those for lower temperature, the total area goes up with temperature, and can be shown to vary as its fourth power (the Stefan-Boltzmann Law).

Kelvin scale is the Centigrade temperature $+273°$ approximately. The Fahrenheit scale, in general everyday use, is related to the Centigrade and Kelvin scales as follows:

	Centigrade	Kelvin	Fahrenheit
Absolute zero:	$-273°$	$0°$	$-460°$
Water freezes:	0	$+273$	$+ 32$
Water boils:	$+100$	$+373$	$+212$

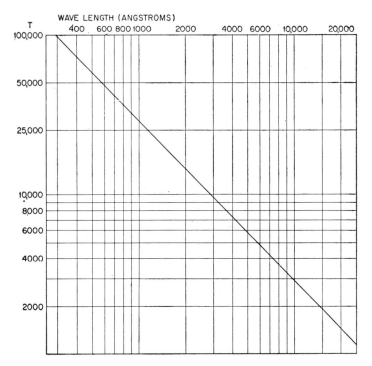

Fig. 3.7. The Wien Law of black-body radiation. The wave length at which the intensity is a maximum (compare Fig. 3.6) is plotted against temperature. In order to represent a large range of temperature and wave length, the diagram is plotted logarithmically.

The relation between wave length of maximum energy, λ_{max}, and temperature T, is $\lambda_{max}\,T = 2.885$. If the quantities were plotted linearly, the resulting curve would be a rectangular hyperbola. When the law is expressed logarithmically, log $\lambda_{max} + $ log $T = 0.460$, the relation is a straight line, shown in the figure.

Notice that for the coolest stars (temperature about 2000°), the maximum of energy is in the infrared; for stars of highest temperature (over 50,000°), it is in the ultraviolet with wave length less than 1000 Angstroms. Notice also that the maximum of energy lies within the range ordinarily observed (3000 to 7000 Angstroms) for stars of temperature between 4000° and 10,000°; for stars of about the sun's temperature (about 6000°), it is near 5000 Angstroms, the wave length to which our eyes are most sensitive.

Whereas the Planck and Wien laws refer to the shape and maximum of the energy-curve, the Stefan-Boltzmann Law involves the total energy that is radiated *by unit area of the surface*. A determination of temperature by this method, therefore, requires a knowledge of the total energy-output of a body and the area of its surface. We possess this information for the sun, and the resulting value of the solar temperature is 5750°.

The distribution of energy among the radiations of many stars can be used (as we shall see later) to determine their temperatures, by modifications of the Planck Law and of the Wien and Stefan-Boltzmann laws, which can be derived from it.

Compare the solar temperature determined in the three ways:

Planck Law:	(*shape* of the energy curve)	5600–6150°
Wien Law:	(*maximum* of the energy curve)	6150°
Stefan-Boltzmann Law:	(*area under* the energy curve)	5750°

The differences are a measure of the departure of the sun's radiation from that of a black body.

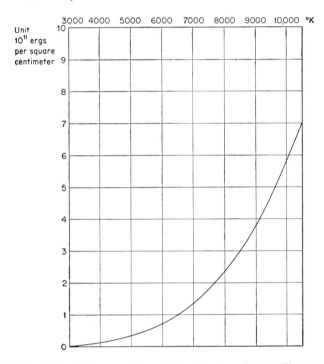

Fig. 3.8. The Stefan-Boltzmann Law of black-body radiation. The energy emitted by one square centimeter of surface is plotted against the temperature. Notice that the energy rises rapidly, as the fourth power of the temperature. Thus if the temperature is doubled, the energy radiated per square centimeter of surface is multiplied by 2^4, or 16, as may be verified from the diagram.

Radiation by Atoms. The laws of radiation of the preceding section apply to continuous radiating surfaces, such as those of solids and liquids. Even gases, under some circumstances, emit continuous radiation; otherwise we should not be justified in applying the Planck formula and its derivatives to the sun and stars, which are wholly gaseous. Stars are surrounded by semi-transparent atmospheres, but at a certain distance down toward their interiors (which may differ greatly for stars of different kinds) their substance becomes quite opaque, and the corresponding layer radiates like a solid or liquid.

An important feature of the laws of continuous radiation is that they are valid, *whatever the surface is made of* (provided, of course, that it complies with the definition of a black body, i.e., it absorbs and radiates perfectly in all wave lengths). The radiations of individual atoms, however, have no such simplicity. Not only do atoms of different kinds radiate differently; equally great variety is shown by the radiations of one sort of atom under different conditions.

When considering the atom, we leave the realm of familiar expressions. Many of the ideas involved in the study of atoms require a point of view that has no counterpart in the study of matter *en bloc*. They can be strictly expressed in rather recondite mathematical terms, which are beyond the scope of an elementary text. However, it is possible to describe the behavior of atoms by means of *analogies* with ideas already familiar, and this procedure will be followed. But when we speak of particles, orbits, etc., it must be remembered that though the phenomena are being described in terms of particles and orbits, we are not dealing with hard, round particles, or actual orbits like those of planets; we are merely using familiar concepts to build up a schematic picture.

The modern theory of matter recognizes that about a hundred kinds of atoms are sufficient to account for all known materials, and their motions and interactions are responsible for most physical and chemical phenomena. When the atoms were first named they were thought to be indivisible [Greek: *a,* not; and *tomos,* split: the "unsplittables"], but modern studies have shown that atoms are themselves complex.

Atoms consist normally of two parts: a *nucleus* (constituted principally of *neutrons* and *protons,* which will be described in a later chapter), and a number of *electrons,* which were once pictured as going round the nucleus in orbits (like a miniature solar system) but are now known rather to be distributed about it in a sort of haze. Both proton and neutron have masses of 1.67×10^{-24} grams; the mass of the electron is 9.11×10^{-28} grams. Every nucleus contains as many, or more, protons and neutrons combined as there are circumambient electrons, and therefore the nucleus carries most of the atom's mass, for an electron has less than one-thousandth of the mass of a proton. Each proton carries a unit positive electric charge; each electron, a unit negative one. Neutrons carry no charge.

The atom is held in its form by electric forces that determine how many electrons a given nucleus can carry. The positive charge on the nucleus, in the normal state, is equal to the negative charge on the haze of electrons, hence the whole atom is electrically neutral. Since each proton carries a single positive charge and each electron a single negative charge, the number of electrons (in a normal state) will be equal to the number of protons contained in the nucleus. It is the electrons that confer on a substance its chemical properties and most of its physical properties. Atoms interact and combine chemically by the interaction of the outermost electrons. Atoms also radiate light by the motions and interactions of their external electrons.

It is found that when the atoms are arranged in series, according to the number of external electrons, they have also been arranged in the order of their chemical and spectroscopic properties. But these properties do not change progressively as the number of electrons increases. At intervals there are repetitions of the type of behavior, a group of successive atoms shows properties similar to those of a previous group of successive atoms; the behavior differs quantitatively, but not qualitatively. These similar groups occur *periodically* when the atoms are arranged in order of number of electrons (called the *atomic number*), and this arrangement is known as the *periodic table*. It was discovered during the last century by Mendeleev, who even predicted "missing elements" by its means, and these missing elements, whose properties he accurately forecast, have since been found.

The common forms of the first nine atoms of the periodic table are enumerated below.

Atom	Mass of Nucleus*		Protons		Neutrons	Number of Electrons
Hydrogen	1	=	1	+	0	1
Helium	4		2		2	2
Lithium	7		3		4	3
Beryllium	9		4		5	4
Boron	10		5		5	5
Carbon	12		6		6	6
Nitrogen	14		7		7	7
Oxygen	16		8		8	8
Fluorine	19		9		10	9

* In units of the mass of the proton; the masses are given in round numbers. The proton has a mass of 1.67248×10^{-24} grams; the neutron, the slightly greater mass of 1.6848×10^{-24} grams.

Notice that the number of electrons is always equal to the number of protons and that the mass of the nucleus goes up as the number of electrons goes up, but more rapidly. The periodic table (as known at present) contains a complete series of atomic species, with numbers of electrons from one (hydrogen) to ninety-eight (californium). We may even regard the neutron as an atom of atomic number zero.

Electrons are distributed around the nucleus according to definite rules. The relation of any one electron to the nucleus is described (in terms of the quantum theory, which we have met in connection with continuous radiation) by numbers that specify essentially its distance from the nucleus, its energy, and its own directional properties.*

* The outmoded orbital model described the electron in terms of the shape and size of the orbit, the angular momentum in the orbit, and the spin of the electron itself. Quantum mechanics replaces some of these schematic conceptions by functional expressions that cannot readily be visualized.

The essential feature of the quantum picture of the atom is that the positions and motions of the electrons are not completely unrestricted; they are limited by the fact that *no two electrons can have wholly identical properties* (the *Pauli exclusion principle*). This principle, which is found to fit the facts in detail, has not been deduced theoretically. Its importance for our present purpose is that it limits the number of electrons that stand in any given relation to the nucleus. Two electrons, and two only, can fill the "inner shell" nearest to the nucleus. The next shell can contain up to two, and the next up to eight, in subshells of six and two, and so forth. The building of these successive shells leads to the periodic recurrence of physical and chemical properties that are characteristic of the series of atoms. Closed shells, which contain the maximum possible number of electrons, produce chemically inert atoms, like those of neon and helium. A shell with only one electron in it is associated with great chemical activity (hydrogen, lithium, etc.).

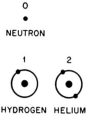

Fig. 3.9. The electron shells for the ten lightest elements. The atomic numbers are given above each illustration; the names of the atoms, below.

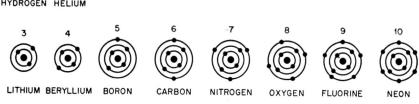

The shells for the first ten elements of the periodic table, and the neutron, are shown schematically in Figure 3.9. Those for atoms with one and two outer electrons are shown in Figure 3.10.

The alkali metals, lithium, sodium, potassium, rubidium, cesium, and francium, have similar chemical properties, and their atoms radiate light in a similar way because they all have one outer electron. The two outer electrons of beryllium, magnesium, calcium, strontium, barium, and radium are responsible for their chemical and spectroscopic similarity. Other similar series, for the same reasons, are arsenic, antimony, and bismuth; copper, silver, and gold; germanium, tin, and lead. These are but examples of regularities that run through the whole list of chemical elements.

Most atomic nuclei are very stable, but some of them (the radioactive nuclei) go to pieces spontaneously. Others can be caused to react or to subdivide under extreme conditions, such as exist in the interiors of stars, and can even be produced artificially on earth. These reactions, however, will not concern us now. They are responsible for the energy radiated by the stars, and are utilized in the manufacture and operation of atomic (more

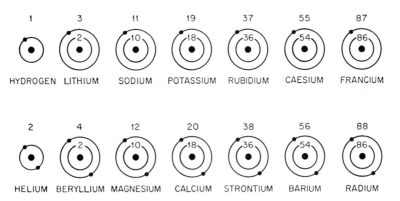

Fig. 3.10. The electron shells for atoms with one external electron (above) and two external electrons (below). The number inserted in the inner circle denotes the total number of electrons in *all* the inner shells.

properly, nuclear) bombs. We defer them to a later chapter and restrict ourselves to the comparatively mild conditions that affect the haze of electrons that surrounds the nucleus.

The external electrons are subject to many disturbances, both by interaction with other unattached electrons and atoms, and by radiation. They also affect one another to some extent. Theoretical spectroscopy is concerned with the "motions" and interactions of electrons. We shall now discuss the production of line spectra in very simple terms.

The quantum theory has already been mentioned in connection with continuous radiation. Its application to radiation from atoms is equally important. The relation $E = h\nu$ applies to the electrons around atomic nuclei and expresses the restriction that they do not move in just *any* way; they must move either with the absorption of a single quantum of energy, or with the emission of a single quantum of energy. If their energy were conceived as being associated with motion in an orbit (as for planets and double stars), this would mean that only certain orbits were possible; the electrons would be imagined as being able to pass from one orbit to another orbit that differs in energy by a fixed amount, governed by the quantum condition. We repeat that this is only a schematic picture. The essential point is to associate given, discrete amounts of energy with the electrons. Such definite amounts of energy can admittedly be associated with motion in definite orbits, but this simple picture in connection with an atom is obsolete; instead, we must associate discrete amounts of energy with various distributions of electron haze about the nucleus.

When an atom emits energy, and an electron passes from a state of higher, to a state of lower, energy, the difference in energy is given out, with frequency (and therefore wave length) specified by the quantum equation, $E = h\nu$. The atom then radiates energy of a given frequency

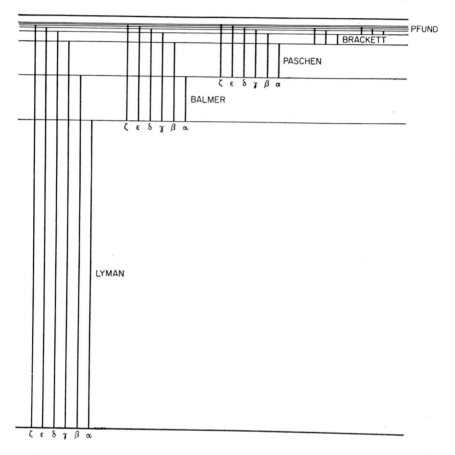

Fig. 3.11. The energy levels for the atom of hydrogen. The upper levels continue to crowd together toward the limiting line. Vertical lines denote the transitions between various levels that give rise to the line spectrum of atomic hydrogen. The Lyman series arises from the first level; the Balmer series, from the next lowest, and so on. Only the first few lines of each series are represented; they are conventionally denoted by Greek letters.

(wave length, color). Similarly, when an atom absorbs energy, and an electron passes from a state of lower energy to a state of higher energy, the wave length of the light absorbed follows the same rule. In the first case the atom will give out light of a fixed wave length; in the second it will absorb light of the same wave length, if the same change of energy is involved.

The electrons associated with the nuclei of all atoms are capable of assuming a large number of discrete energy states, and a transition from one of these states to another * involves the emission of light of a single wave length (or color) if the final state is of lower energy than the first. If the final state is of higher energy, energy of the same wave length is taken up. These are the processes by which atoms radiate and absorb light, respectively. The great variety of colors radiated (or absorbed) springs from the variety of possible energy states, which vary from atom to atom and are governed by the number of electrons associated with the nucleus, their arrangement, and the properties of the nucleus itself.

The simplest of all atoms is that of hydrogen, with but one electron. Even this single electron can assume an enormous number (in theory, if

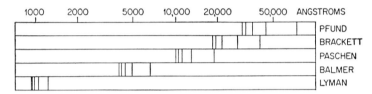

Fig. 3.12. The first five spectral series of hydrogen (compare Fig. 3.10). Wave lengths are drawn on a logarithmic scale.

isolated, an infinite number) of energy states. Transitions between these states can therefore lead to radiation or absorption of an infinite number of colors. The arrangement of a few of the energy states of the atom of hydrogen is shown in Figure 3.11, and some of the wave lengths that can be radiated or absorbed (the *spectrum* of the hydrogen atom) in Figure 3.12. The "series" in the spectrum of hydrogen are known (counting from short to long wave length) as the Lyman, Balmer, Paschen, Brackett, and Pfund series, after their discoverers.

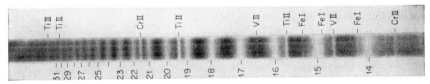

Fig. 3.13. The Balmer series of hydrogen in the spectrum of the supergiant star α Cygni. It is traceable as far as the twenty-ninth line (No. 31, as numbered in accordance with the quantum levels). No laboratory spectrum of hydrogen has shown so many lines. The enormous extent and small density of the atmosphere of α Cygni make possible the observation of a large number of lines.

The lines of a few other elements (neutral iron, ionized titanium, ionized chromium, ionized vanadium) are marked on the upper margin.

Photograph by Dr. Otto Struve, with the 100-inch telescope at Mount Wilson Observatory.

* The tendency of given states to combine with one another differs greatly and is subject to complex rules.

Each of the five series shown in the diagram is evidently tending toward a limit as the lines crowd more and more closely together toward the violet. The significance of this limit will be clear from Figure 3.11, which shows the relative energies of the various possible states of the electron. The bounding line at the top of the figure represents the largest amount of energy that the electron can take up and still remain attached by electrostatic attraction to the nucleus. If more energy than this amount is absorbed, the electron will be detached from the atom—the atom will become *ionized*. Nucleus and electron are now separate; the nucleus is a *positive ion* and the electron a *negative ion*. While the electron was part of the atom, the whole was electrically neutral, because the negative and positive charges of electron and nucleus compensated one another exactly. An electron that has taken up more than enough energy to detach it from the atom is no longer subject to the quantum conditions after it has escaped, and can absorb light of any wave length shorter than the limit of the series from which it started out. Beyond the limit of a series, an escaping electron can perform *continuous absorption*. Conversely, an unattached electron in the process of being captured by an ionized nucleus can execute continuous emission beyond the appropriate series limit.

Even with one electron, under actual conditions, an enormous number of transitions is possible. With two outer electrons, as for neutral calcium, the number of possibilities is enormously increased, and for such an atom as that of iodine, with five electrons in the outer shell, there are many thousand possible transitions. Each transition corresponds to light of given wave length and color.

The resulting spectrum is studied by means of a spectroscope or spectrograph (p. 89), which spreads the colors into a band that resembles a rainbow (but may, of course, contain colors to which the eye is not sensitive). A spectrum is made up of a series of images of the narrow slit of the spectroscope, which is so arranged that these images are perpendicular to the width of the spectrum. Each atomic transition produces a narrow range of color, and therefore each produces a narrow image of the slit at the appropriate part of the spectrum, which thus appears to be crossed by a number of bright lines perpendicular to its length, as in Figure 3.15. These are known as *spectrum lines*. An emitting atom gives a series of *bright lines*, but, as we shall see, the spectrum lines observed for most stars are *dark lines*, projected against the greater brilliance of the stellar surface.

The problem of disentangling the transitions between various states of energy—the "analysis" of an atomic spectrum—is often a very complicated one, and is still incomplete for many of the richer spectra. It has provided the basis of analyzing the atmospheres of stars, and many astrophysicists have busied themselves with such research during the past two decades.

When an atom of hydrogen has lost an electron and become ionized, it cannot produce its characteristic spectrum until it acquires another electron. Atoms with several electrons can be ionized several times, once for each electron removed. A singly ionized atom has one less electron than the "neutral" (i.e., un-ionized) atom of the same nucleus. Because the nature of the spectrum emitted or absorbed depends upon the number of external electrons, an ionized atom will clearly give a totally different spectrum from a neutral atom. A singly ionized atom *will behave qualitatively like the neutral atom preceding it in the periodic table.* Thus ionized helium will give a spectrum resembling that of hydrogen; ionized calcium, a spectrum resembling that of sodium, and so on. The correspondence between the spectra of hydrogen and ionized helium is illustrated in Figure 3.14.

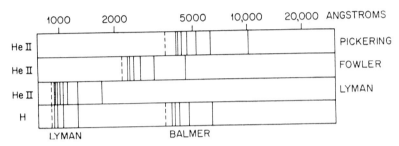

Fig. 3.14. Three of the spectra of ionized helium. Two hydrogen series are shown below for comparison. Note that alternate Pickering lines practically coincide with Balmer lines, and alternate Lyman lines of ionized helium practically coincide with Lyman lines of hydrogen. A second Lyman series of ionized helium lies further in the ultraviolet. Series limits are shown by dotted lines. The scale is logarithmic.

The state of *ionization* is concerned with the *number* of orbital electrons. When the atom is in its normal state, the number of such electrons is equal to the number of protons in the nucleus, or the nuclear charge. Thus an oxygen atom in its normal state has eight electrons, once-ionized oxygen has seven, twice-ionized oxygen, six, and so on. The corresponding spectra are known as the spectra of O I, O II, O III, . . . (the chemical symbol for the atom being used for brevity). The amount of energy that will just remove an electron from an atom is known as the *ionization potential* and is measured in electron volts.* Note that there is an ionization potential, not only for every atom, but for every stage of ionization of that atom. For instance, the ionization potentials for successive stages of the atom of oxygen are:

O	I	13.56 ev	O	IV	77.08 ev
O	II	35.00	O	V	113.38
O	III	54.71	O	VI	137.52

Successive electrons are always more and more difficult to detach.

* One electron volt $= 1.6020 \times 10^{-12}$ erg.

The state of *excitation* of an atom is concerned with the *arrangement* of the orbital electrons that are present. Here again there will be a normal state, of minimum energy, which is known as the *ground state*. Often, however, one or more of the electrons will have received more than the mini-

SPECTRA OF POTASSIUM
SINGLE LINE SPECTRUM

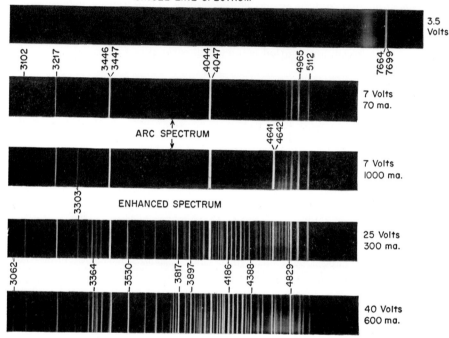

Fig. 3.15. Spectra of the atom of potassium under different conditions. These are emission spectra obtained in the laboratory, and are produced as the single outer electron moves from states of higher to lower energy. The degree of excitation (shown on the right in electron volts) determined how many atoms have their electrons excited to the higher energy states. For the first spectrum only the most easily excited lines (the ultimate lines at 7669 and 7664 A) are produced. As the excitation rises, lines that correspond to higher energy states appear, such as those of wave length 3217 and 5112 A. Under still higher stimulation, one electron is removed ("enhanced spectrum") and the spectrum of potassium II appears (lines at 3897 and 3530 A). In the last photographs, lines of potassium III, with two electrons removed, may be seen (as at 3364 A). Photographed by William F. Meggers, United States Bureau of Standards.

mum energy, though less than the amount required to detach them from the atom. The atom is then said to be in an *excited state*. The amount of energy that will bring an electron from the ground state to a given excited state is called the *excitation potential* for that state. Excitation potentials, like ionization potentials, are expressed in electron volts.

The excitation potential of a state is almost always less than the ionization potential of the corresponding atom.* The Balmer Series of hydrogen, for instance, has an excitation potential (for absorption) of 10.15 electron volts, whereas the ionization potential of hydrogen is 13.54 electron volts.

Atoms may be excited or ionized by energy that falls upon them. In many cases the energy comes in the form of radiation, and the higher the temperature of the radiating source, the more energy will be received (the Stefan-Boltzmann Law), and the greater will be the excitation or ionization. There are other methods of ionization and excitation, however, such as the reception of energy from collisions with other particles (atoms, ions, or electrons).

The great majority of atoms that are encountered in astronomy are single, not in chemical combination, because of the high temperatures of most stars. Chemical compounds, however, are observed in the envelopes of the cooler stars, on the surfaces of planets, and in interstellar space. Here we observe atoms that are combined to form *molecules;* a molecule may be put together from two atoms (as in common salt, sodium chloride), or from many, as in complex organic substances.

In the stars, only the simplest molecules are observed, such as cyanogen (CN) or metallic oxides (TiO, ScO).† A few polyatomic molecules, such as the simple hydrocarbon, CH_2, are observed in cool stars and in interstellar space. Planetary atmospheres contain more complex molecules, such as those of methane (CH_4) and ammonia (NH_3); a large variety of solid compounds make up the planetary surfaces and probably the grains of interstellar space.

The spectra of molecules are even more complicated than those of atoms. Besides the energies associated with their individual electrons, molecules possess arrays of energy levels both of vibration and rotation; and both kinds are subject to restrictions. Molecules, therefore, display arrays of closely crowded spectrum lines (banded spectra) rather than the individual lines found in atomic spectra. The molecular spectra are difficult to analyze, and much work still remains to be done before they can be fully exploited by astrophysicsts.

Temperature and Its Significance. In our survey of the properties of planets, stars, and nebulae, we shall continually express them in terms of mass, size, density, brightness, and temperature. Most of these quantities involve well-known concepts, but although temperature is a term in everyday use, it is hard to define without a theoretical groundwork that is beyond the scope of an elementary text. A good simple working definition is that

---

* Occasionally, two electrons are excited by amounts that add up to more energy than that of ionization from the ground state. In such cases the atom has more than one ionization potential, according to the distribution of energy among its electrons.

† It should be noticed that these are the simpler radicals that correspond to the oxides of titanium (TiO_2) and scandium (ScO_2) on the earth.

temperature expresses the thermal state (or state of heat content) of a body, considered with reference to its power of communicating heat to other bodies. A thorough understanding of temperature reposes on the study of *thermodynamics,* the study of the flow or transfer of heat.

In everyday life we measure temperature with thermometers (either of bulb form, like the mercury-in-glass thermometer, or of dial form, like the common oven thermometer). These instruments measure essentially the *energy of agitation* of the atoms; for the bulb thermometer, an increase in the energy of agitation causes the atoms to move more rapidly and therefore occasions an expansion of the mercury. The oven thermometer works by the bending of a strip made of pieces of two different metals which expand at different rates with rising temperature. Both kinds of thermometer are standardized empirically.

Almost all our estimates of temperature on the earth depend on the energy of agitation of the molecules or atoms, though for very high temperature the properties of the radiation may be employed.

On the earth we usually receive heat by direct transmission of the energy of agitation of the particles of a hot body (i.e., from a hot water bottle, in cooking, ironing, etc.). Sometimes we receive it by circulation of heated material (motion of water in a domestic heating system; motion of air masses, which bring a change of weather). Sometimes, however, we receive it by direct radiation (e.g., the heat of a fire—in part at least this is radiation energy—and the heat of the sun, which, as it is transmitted through a vacuum to us, is entirely radiation energy). We have already considered the relation of continuous radiation to the temperature of the radiating body.

The energy content of a radiating body determines the amount and quality of the radiation that it gives out. But energy content expresses itself in other ways. The energy content of the atoms and molecules of a substance governs its energy of agitation, the property by which we measure temperature with a mercury thermometer. The energy content of a mass of gas governs the degree to which it is *ionized,* i.e., the atoms are robbed of one or more electrons; or *excited,* i.e., the atoms or molecules are raised to energy states higher than the lowest, or ground, level. The energy content of an assembly of molecules governs the degree to which they are dissociated—resolved into their constituent atoms. Insofar as these conditions—degree of agitation, ionization, excitation, or dissociation—can be determined by observation, they can be used as measures of, and therefore expressed in terms of, temperature.

In using these effects of temperature as measures of temperature, however, we must not lose sight of the fact that the "temperature" so determined has a much more general sense than we ordinarily give to it. Other effects than heating may contribute to ionization, dissociation, or the agitation of atoms—collisions with other particles, for example. There are times when temperatures determined by several of the means just mentioned would differ greatly. For example, a small particle at rest in interstellar space would be very cold—near to absolute zero—but yet the temperature of the radiation in the same district (as measured by its energy distribution) might well be 10,000°K. A gaseous nebula probably consists of a mixture of solid particles (therefore comparatively cold) and ionized atoms, such as those of Oxygen III, which, if they had been ionized by radiation,* would be very hot indeed.

Temperature and Energy of Agitation. Stars consist of atoms and occasional molecules. Only their outer envelopes can be observed, down to the level at which the gases become opaque, but even so, they present for observation enormous assemblages of particles. Because of their great scale, they permit us to observe the average behavior of a vast number of particles, all roughly under the same general conditions.

The statistical view of an assemblage of atoms may be understood by the analogy of a swimming pool. Suppose we survey a swimming pool throughout the course of a day. At first it is empty; soon a few people enter, then more and more. Finally the "population" of the pool attains a "peak." As the day wears on, the people thin out, and by the end, the pool is empty again. When we look at the situation closely, however, we see it as more complicated. At all times there will be *some* people entering the pool, some leaving it, and near the "peak," when about the same number are coming in and out, the "population" will be steady, even though different individuals are in and out at different times. We may think of the pool as being in "statistical equilibrium" at such a time.

Furthermore, at all times there will be a variety of speeds represented by the swimmers; some go fast, some slowly. Probably the average speed of all the swimmers in the pool at one time will remain about the same. If we think of an assemblage of particles as swimmers in a pool, *the average speed of the swimmers corresponds roughly to the temperature* of the assemblage. If faster swimmers are entering, the "temperature" will rise; it will also do so if slower swimmers are leaving, or if for some reason everybody speeds up.

Furthermore, if there are few swimmers in the pool, their chances of colliding one with another are smaller than when the pool is fullest. Chances of collision will also depend to some extent on the speeds of the swimmers, and also (if some are unusually large or small) on their dimensions. All these conditions provide useful analogies for the motion in an assemblage of atoms.

As mentioned earlier, the temperatures measured on earth usually de-

---

* The physical state of the gaseous nebulae will be discussed in Chapter XIII.

pend on the energy of agitation of particles. In theory one could determine
temperatures of stars from the agitation of the atoms in their atmospheres.
This energy can in fact be measured (by methods to be described later),
and we find in many stars (e.g., the sun) that the atoms at their surfaces
are indeed in thermal motion. But the effects of this motion are so small, in
comparison with other effects that produce very similar appearances, that
we do not usually measure temperature in this way.

The atmospheres of the sun and stars contain, in addition to atoms
(and some molecules), a large number of electrons, and the energy of agitation of these electrons also produces observable effects. Temperatures of
agitation of electrons are very useful in studying the physical state of the
nebulae, which consist of diffuse clouds of gas (mixed with dust), and in
some such objects the "electron temperature" may be very high indeed,
though at the same time the dust is near absolute zero.

Temperature from Statistical Equilibrium. In many of the cooler stars
we observe atmospheric molecules, which consist usually of two atoms in
chemical combination, such as CN (cyanogen). Owing to the high temperature, these molecules are continually flying apart and becoming their
component atoms, and whenever two such atoms meet (with suitable
speeds), they will re-form into a molecule. To go back to the swimming
pool analogy, many are leaving the pool all the time, many re-entering.
If we assume an "equilibrium state" (i.e., equal numbers combining and
separating in a given time), we can calculate the temperature (which is
related to the tendency to fly apart) if we know the density (which is related
to the likelihood that there be an encounter). As in the swimming
pool, when the pool is fullest, collisions will be most frequent.

Exactly similar conditions apply to the separation of their external electrons from the atoms (ionization) in the stellar atmosphere. The temperature will govern the tendency to fly apart, and the density will govern the
tendency to recombine. Thus if we know the density and the properties of
the particular atom, its presence and its commonness will enable us to tell
the temperature. This is the most important of all methods of determining
the temperatures of stars.

An atom is a complex assemblage of electrons around a nucleus, and
the temperature of the atom, by governing the excitation, will determine
how these electrons are arranged. Thus if we can determine how they are
arranged (which is deducible from the spectrum) we can infer the temperature. This method depends on temperature alone, and does not involve the
density (at least not to an important extent).

Precisely similar considerations apply to the molecule, which also gives
spectroscopic clues to the arrangement of its electrons. This method of
determining temperature is important for low-temperature stars and interstellar material.

4. ASTRONOMICAL INSTRUMENTS

The instruments used in astronomy are adapted to the kind of information that is sought. The subject is large and technical, and will be considered here only in the broadest outline.

Astronomical observations can be roughly classified as: measures of position; measures of brightness; and analysis of light, i.e., the qualitative and quantitative study of the distribution of energy among different wave lengths. The corresponding branches of astronomy are: positional astronomy, photometry, spectroscopy, and spectrophotometry. The detailed study of the sun is a specialized subject with its own instrumentation.

Positional astronomy and photometry were studied in early times without optical aids, but cannot go far without them, and spectroscopy requires them. The chief optical devices used are the mirror, the lens, the prism, the diffraction grating, and the interferometer (p. 283). From these components, or modifications of them, the most elaborate astronomical instruments are built.

The *lens* focuses light from a distant source. If the source is at infinity, the distance from lens to focus is the *focal length*. If the lens is thicker at the middle than the edges, the image is magnified (double-convex, plano-convex, or positive meniscus); if the center is thinner than the edge, the image is virtual and diminished (double-concave, plano-concave, negative meniscus), see Figures 3.16, 3.17. Lenses were known in early

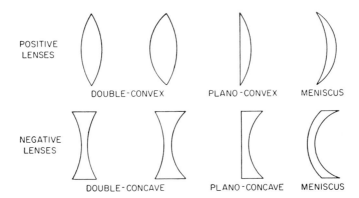

Fig. 3.16. Positive lenses (thicker in the middle) and negative lenses (thicker at the edges).

times. Nero is said to have worn a sapphire monocle. Not until Renaissance times, however, is there a definite record of the use of two lenses (long-focus objective and short-focus eyepiece) to form a *telescope*. Galileo's tiny telescopes (one of which is to be seen in the Museum of Science in Florence) opened the gates of modern astronomy.

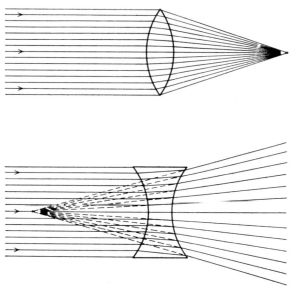

Fig. 3.17. A positive lens produces a *real image;* a negative lens gives a *virtual image* (convergence of the broken lines). When parallel light falls on a lens, the distance from lens to image measures the *focal length.*

If the purpose of a telescope is to collect as much light as possible, the aperture of the objective is made as large as possible. If large-scale photographs are to be made with an instrument, the focal length must be

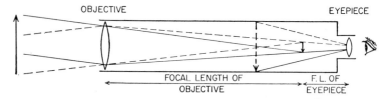

Fig. 3.18. A visual telescope. An inverted image of the object is formed at the focus of the objective; the *eyepiece* produces an enlarged version of the image. The figure shows a positive lens at the eyepiece; a negative lens would invert the image again. In a photographic telescope a plate is placed at the focal distance of the objective.

large, for scale is proportional to focal length. Telescopes of long focal length are needed for surveying planetary surfaces, for measuring relative stellar positions and motions, and for studying regions dense with stars, such as the neighborhood of the galactic center. But long focal length is not a sufficient condition for resolving dense regions and revealing planetary detail, for even the image of a point is spread over an area by diffraction. The angular size of this area is $4''.5/a$, where a is the aperture in inches (Dawes' Rule). For high resolution this fraction must be made as small, and the aperture as large, as possible. Similar requirements apply to telescopes that are to be used visually. They led to the construction of such large-aperture, long-focus refractors as those at Lick and Yerkes observatories, with apertures of 36 and 40 inches, and focal lengths of 57.8 and 63.5 feet, respectively.

If, on the other hand, the greatest *speed* in photography is desired for the study of faint surfaces such as nebulae, the *focal ratio* (focal length/aperture) must be made as small as possible. The focal ratios of the large refractors are between $f/15$ and $f/20$. A refractor of small focal ratio is difficult to make because the image defects mount up, with a given focal length, for larger apertures. If the focal length is kept small, the scale will be small and the field large. The large (4-component) Ross lens of the Lick Observatory has an aperture of 20 inches, a focal length of 144 inches, and accordingly, a focal ratio about $f/7$; it has a field $20°$ in diameter, whereas the Yerkes 40-inch refractor, used visually with a 1-inch eyepiece, covers a field of less than $3'$.

Reflecting telescopes can be made with smaller focal ratio than refractors; the inevitable image defects (*aberrations*) affect them also. The focal length of a mirror, unlike that of a lens, is independent of the color of the light, and mirrors accordingly bring light of all colors to the same focus. The 100-inch reflector of Mount Wilson Observatory has a focal length of 508 inches, and a focal ratio $f/5.1$. The corresponding quantities for the 200-inch on Palomar Mountain are 660 inches and $f/3.3$.

The smallest focal ratios and the greatest speeds are attained by the Schmidt telescope and its modifications. The 48-inch Schmidt on Palomar Mountain operates at $f/2.5$, and the 18-inch Schmidt, at $f/2.0$. The Harvard meteor cameras have focal ratio $f/0.65$ (optical) or 0.82 (effective), and fields of $55°$ and $52°$, without, and with, the rotating shutter (p. 246). Focal ratios as small as $f/.33$ are used in the "solid Schmidts" employed in some nebular spectrographs (the use of a glass, rather than an air path within the camera, theoretically increases the speed).

Except for the study of planetary surfaces and of comets, and most meridian observations, telescopic work is now done mainly by photography or by some such device as the photoelectric cell, the lead sulphide cell, or the thermocouple.

The first telescopes had lenses (*refractors*). Newton realized that a concave parabolic mirror, which brings parallel light to one focus (which a spherical mirror does not do) could be used as a telescope. He made the first *reflector,* four inches long with a 1-inch mirror, the remote ancestor of the 200-inch.

The mirror of a reflector throws back the light toward the object.

Newton brought it to the side by a small tilted mirror within the tube. Other arrangements for making the image accessible were used by Cassegrain, Gregory, and Herschel. With any one mirror, the Newtonian, Cassegrain, and Gregorian arrangements give different focal lengths. Many large reflectors are constructed to be used alternatively at the prime focus, the Newtonian and the Cassegrain focus.

Reflectors and refractors have different advantages. A simple lens (made of the same substance throughout) refracts light of different colors through different angles. Therefore its focal length is not the same for various wave lengths, and if one color is in focus, the others are not. Since the light of most objects has a large range of colors, the result is a blurred image (*chromatic aberration*).

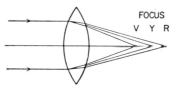

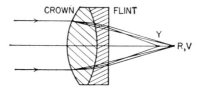

Fig. 3.19. Chromatic aberration. The upper lens is a *simple lens,* made of one kind of glass. The *refractive index* (which determines the amount by which light of different colors is bent on passing through the lens) is different for light of different colors. Red light comes to a focus further from the lens than violet light.

The lower figure shows a *compound lens,* two lenses of different refractive indexes cemented together. The glass has been so chosen that the differences of refractive index (and therefore of focal length) compensate one another for red and violet light. Thus red and violet light are brought to a focus, but yellow light is not. The use of several kinds of glass of different refractive indexes could bring more colors to a focus. All astronomical objectives are compound lenses; some have as many as six components, not necessarily in contact.

By building up a compound lens from two or more kinds of glass with different refracting properties, it is possible to correct the chromatic aberration for a limited series of wave lengths. Most astronomical objectives have compound lenses. The calculation of the surfaces of a compound lens system is a highly specialized science, almost an art. The number of surfaces that have to be fashioned increases the cost of such a lens and also cuts down the light that it transmits, because some is lost by reflection at each face. The largest compound lens in a telescope is that of the Yerkes 40-inch. A larger one would be expensive to figure, but probably the most effective restraint on making one is not the cost of construction, but the difficulty of getting a large enough block of optical glass sufficiently free of flaws.

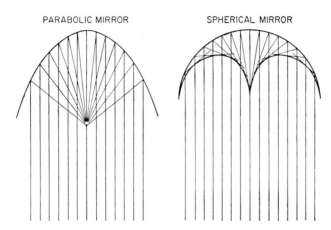

Fig. 3.20. A parabolic mirror brings parallel light to a focus; a spherical mirror does not. The rays reflected from a spherical surface envelop a special curve, the *cardioid*. Light reflected from the edge of a coffee cup on the surface of the coffee can be seen to concentrate along the edge of such a curve: the rim of the cup acts as a spherical mirror.

Reflecting telescopes bring light of all colors to the same focus, and in this they have the advantage over refractors. The images that they produce, however, suffer from greater loss of light and greater image distortion, away from the axis, than refractors. Although they have only one (parabolic) surface to be figured, the figuring must be more accurate than that for the surfaces of a lens, and this is the chief difficulty in their construction.

Newton's tiny reflector, and the much larger ones (up to six feet in diameter) made by William Herschel, were of highly polished speculum metal (an alloy of copper and tin). Modern reflectors have glass surfaces coated with a metallic film (formerly silver, now usually aluminum). Since only the surface of the mirror is used, the glass need not be so flawless as that used for a lens. The mirror must retain its shape under changes of temperature; fused quartz mirrors would be better than glass for this reason, but a very large one has not been successfully made. The 200-inch mirror is of Pyrex, whose heat-resistant properties are well known in the kitchen.

The great weight of the mirror (which must be thick enough to bend as little as possible under its own weight) makes the operation of a modern reflector as much a problem in engineering as in optics. The mirror must

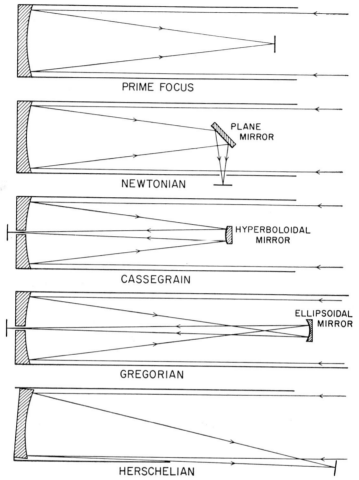

PRIME FOCUS

PLANE
MIRROR

NEWTONIAN

HYPERBOLOIDAL
MIRROR

CASSEGRAIN

ELLIPSOIDAL
MIRROR

GREGORIAN

HERSCHELIAN

Fig. 3.21. Several arrangements of a reflecting telescope. In the first, light is reflected directly to the focus of the parabolic mirror (*prime focus* arrangement). The
Newtonian arrangement turns the beam of light through a hole in the tube by means
of a flat mirror placed within the tube *between* the mirror and the prime focus. *The
Cassegrain* form reflects the light back, through a hole in the main mirror, from a
convex hyperboloidal mirror placed *between* the large mirror and the prime focus.
The *Gregorian* form reflects the light back through a hole in the main mirror, from
a concave ellipsoidal mirror placed *beyond* the prime focus of the main mirror. In the
Herschelian form, the main mirror is slightly tilted so that the image at the prime
focus falls just outside the tube. The most usual arrangements are the prime focus,
the Newtonian, and the Cassegrain. Many large telescopes are so arranged that all
these forms can be produced by the removal or interchange of the secondary mirror.

be supported so that it keeps its shape accurately in all positions. The
200-inch mirror weighs about 17 tons, the mounting, 500 tons.

The most powerful type of modern telescope (the *Schmidt telescope*)

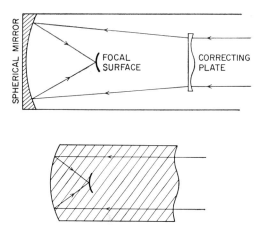

Fig. 3.22. Principle of the Schmidt telescope. The spherical mirror is "parabolized" by the differences of light-path through different parts of the transparent *correcting plate*. The light is brought to a focus on a curved surface, and photographs made with Schmidt telescopes require curved plates or films that conform to the focal surface. The lower part of the diagram shows the principle of the *Solid Schmidt* telescope, which is a single block of glass. The front surface forms the correcting plate, the rear surface, the mirror (silvered or aluminized). The photographic film must be inserted in a small slit in the block of glass. The speed of a solid Schmidt is increased in proportion to the square of the refractive index of the glass.

Fig. 3.23. The 48-inch Schmidt telescope on Palomar Mountain.

Fig. 3.24. A reflecting telescope (the 100-inch telescope of Mount Wilson Observatory). The mirror is 100 inches in diameter.

Fig. 3.25. The 200-inch reflector, largest telescope in the world, on Palomar Mountain, California.

Fig. 3.26. A refracting telescope (the 15-inch telescope of the Harvard Observatory), so called because the diameter of the objective is 15 inches. The telescope is *equatorially* mounted; it swings around an axis that points toward the celestial pole (the *polar axis*) and is pivoted to move perpendicular to this axis. Rotation around the polar axis moves the telescope in *Right Ascension;* motion perpendicular to the axis, in *Declination.* Telescopes are driven around the polar axis by a clock that compensates for the apparent diurnal motion of the heavens.

uses a spherical mirror, whose behavior is "parabolized" by a glass plate through which light passes in such a way that the graduations in thickness compensate for the difference between the spherical mirror and the desired paraboloid. Schmidt telescopes can be made large, of wide field, and great speed; the Harvard meteor cameras mentioned earlier are of Schmidt type; so is the 48-inch telescope on Palomar Mountain. For problems that require scale, field, and speed, the Schmidt, or some modification of it, has a great future.

All the telescopes for the general study of the sky are mounted so that they can be set in two co-ordinates: declination and hour angle. The arrangement is known as the *equatorial mounting*. The instrument swings around an axis (the *polar axis*) directed toward the pole of the sky (and graduated in hour angles) and can pivot up and down perpendicular to the equator (in declination). It is driven in right ascension by a sidereal clock that keeps them pointed toward the desired right ascension (and declination). For all but the smallest instruments and the shortest exposures, the instrument must also be guided manually by the observer in order to remove effects of variable driving rate or shifts of apparent star position by refraction. Various types of equatorial mounting may be seen in Figures 3.23 to 3.26.

5. ACCESSORY EQUIPMENT

a. The Photographic Plate. The modern panorama of the stellar universe was opened up when astronomy adopted photographic methods. The charting of stars, measurement of their positions, motions, and spectra, and much of the study of their brightness is now done by using the photochemical process known as photography. It has many advantages: the record is permanent; increasingly faint stars can be reached by long exposures (within the limits imposed by the brightness of sky background); and the modern sensitized plate throws open spectral regions beyond the range of the eye.

Astronomical photography, however, is beset with difficulties. Photographic images are blurred by scattering of light in the emulsion, and distortions of the emulsion itself may affect accurate measures of position. In recording brightness, the photographic plate has the disadvantage of non-linear response to light intensity, variable with conditions and difficult to calibrate. Even on different parts of one plate the emulsion may not respond uniformly. Nonetheless, photography is now, and will long remain, our major tool for securing astronomical records. If a way could be found of doubling the sensitivity of photographic emulsions, the effect would not be far different from that of building a telescope that would reach stars twice as faint as reached heretofore.

b. The Photoelectric Cell. Accurate measures of stellar brightness and color are made by using the photoelectric effect. When light falls on certain surfaces, its energy is absorbed and electrons are ejected in numbers proportional to the intensity of the light. These electrons are counted, and the amount of incident light can thus be determined. The photoelectric cell is more accurate than the eye or the photographic plate as a photometric instrument, and very precise work in stellar brightness and variation is beginning to rely on it almost exclusively.

c. The Spectroscope. Light passed through a prism is deviated, and dispersed into the colors of the spectrum, as was first shown by Newton. The *dispersion* is accounted for by the fact that the *deviation* varies with wave length (though not linearly), and violet light is deviated more than red. The astronomical spectroscope is adapted to photography and is used in conjunction with a telescope.

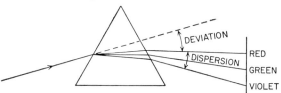

Fig. 3.27. Refraction of light through a prism. The *deviation* is the angle through which the ray of light is bent. Deviation varies with wave length, and the variation produces the *dispersion,* or separation of light of different colors.

The spectrum of a star usually consists of a colored band of light from which certain narrow ranges of color (*spectrum lines*) are partially removed by the action of the atoms in the star's atmosphere (p. 72). From the wave lengths of these lines can be deduced the chemical composition of the star's atmosphere, the physical condition of its surface, and the speed of the star away from, or toward us (p. 289). The latter is derived by the use of the spectroscopic Doppler effect (p. 290), the change in the wave lengths of the spectral lines of the star caused by the speed in the line of sight. The Doppler effect is measured by means of a comparison of the wave lengths of these lines with those of the same substance, photographed simultaneously with the stellar spectrum, from a terrestrial source.

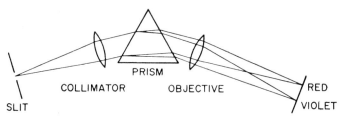

Fig. 3.28. Principle of the prism spectroscope. Light is passed through a narrow slit and rendered parallel by a lens (the *collimator*). The light then falls on the *prism,* which disperses it. The *objective* on the far side of the prism brings the light to a focus, and thus produces an image of the slit in light of each individual color. A photographic plate may be placed at the focus. Many spectroscopes employ several prisms, through which the light passes successively. Both collimator and objective are compound lenses.

The dense glass used for prisms absorbs ultraviolet light; some ultraviolet light is transmitted by light crown glass. Ultraviolet stellar spectra, as far as the ozone limit, are obtained with a quartz spectrograph. Prisms for use in the infrared are of rock salt.

A very simple device for photographing stellar spectra is the *objective prism:* a prism is placed in front of the objective, and each image in the field is spread into a spectrum. Only by special devices can objective prism spectra yield the speed of the stars; their chief value has hitherto lain in recording simultaneously the spectra of a large number of faint stars. They are less wasteful of light than the slit spectrograph, but as a record of detail they are far inferior.

The *diffraction grating,* made by ruling a series of fine parallel lines on glass or metal, produces a series of spectra that have equal dispersion at all wave lengths, whereas the dispersion of the prismatic spectrum falls off toward the red. The earliest gratings were very wasteful of light, and their use was confined to the study of very bright objects such as the sun. But the grooves of modern gratings are so shaped that the light is more efficiently used, and stellar spectroscopes of the future will use them increasingly.

The Transit Instrument. The fundamental work of positional astronomy is done with the transit instrument; positions and motions measured photographically are only relative to nearby stars in the field. The transit instrument fixes the absolute foundation to which the relative measures can be referred.

The transit is a telescope fixed to move only along the meridian, and it is used to determine the exact sidereal time of the star's *transit,* or crossing of the meridian; this sidereal time is the right ascension of the star (p. 50). The *meridian circle* is a transit constructed for the measurement of the declination of a star as it crosses the meridian. The right ascension and declination so measured are the fundamental values to which photographically measured positions and motions must ultimately be referred.

Observations with transit and meridian circle are among the few in astronomy that are still made visually. They require the utmost refinement in construction and in observing methods, in order to eliminate errors of instrument and observer.

Instruments for Solar Research. The visible disc of the sun and the intensity of its light make possible studies of a kind that can be made for no other star. A combination of telescope and spectroscope, in which the latter is used to sort out light of the desired wave length (the *spectroheliograph* and the *enregistreur des vitesses*), permits the study of details of the solar surface in light of one color (see Figures 4.1 to 4.4).

The Solar Towers at Mount Wilson Observatory use a *coelostat,* or moving mirror, at the top of the tower; this throws the sunlight down the tower through an objective that focuses it on the slit of a grating spectroscope, which is at the bottom of a pit below the tower.

The most versatile instrument for solar research is the *coronagraph,* which produces artificial eclipses of the sun by interposing an artificial moon of the right size in front of the solar disc. The first coronagraph was made by the French astronomer Lyot, who overcame the difficulties that had

Fig. 3.29. The 150-foot Tower Telescope of Mount Wilson Observatory, designed for solar research.

stopped previous workers: scattering of the sunlight both inside and outside the equipment. The light scattered inside was eliminated by treatment of the optical surfaces; atmospheric troubles were minimized by the construction of an observatory on a high mountain, the Pic du Midi, in the Pyrenees.

The coronagraph makes possible daily studies of phenomena that were previously available in the few precarious moments of a total eclipse. The structure of corona and prominences can be continuously recorded on movie film, and the spectra of the sun's outer layers followed from day to day. The coronagraph has permitted the study of the short-lived solar spicules, which rush up from the sun's surface and subside again within a few minutes. Several coronagraphs are now in operation in various parts of the world, and more are in course of being built. Our detailed knowledge of the nearest of the stars is advancing rapidly as a result of Lyot's pioneering.

THE SUN

Study is like the heaven's glorious sun,
That will not be deep-searched with saucy looks;
Small have continual plodders ever won,
Save base authority from others' books.
SHAKESPEARE,
Love's Labour's Lost, I, i

Shakespeare is still right: the sun is inscrutable. Both as the central body of the solar system and as the nearest of the stars, it still presents many unsolved, if not insoluble, problems. We shall often find in astronomy that the most thoroughly observed phenomena present the most puzzles: the more we know, the less do we find that we understand. The sun is no exception.

Our sun has a double importance in astronomy. It is the solar system's central body and, as such, contains 999/1000 of the mass and therefore effectively controls the motion of all the other bodies in the system. It is also the nearest of the stars, the only one that shows a disc, and hence the only one of which we directly see the surface detail. The sun shows of what a star is capable; however, we must keep in mind that it is a relatively small, faint, and cool star, and that (as stars go) its behavior may be comparatively mild.

1. DISTANCE, SIZE, AND MASS
OF THE SUN

The average distance of the earth from the sun is about 93 million miles; because of the ellipticity of the earth's orbit it is about three million miles less in January than in July. The sun's distance is measured by triangulation, the fundamental step in the survey not only of the solar system, but of the rest of the universe as well. The process will be described in Chapter VII.

The disc of the sun has, on the average, an apparent diameter of a little over half a degree ($31'59''.30 \pm 0''.01$) and differs slightly with our position in our orbit. Its actual size is 1,393,000 kilometers, or 109.30 times the mean diameter of the earth. Thus the sun has more than a million times the earth's volume. For comparison, we may note that the diameter of the orbit of the moon is less than the sun's diameter; the entire orbit of the moon could fit inside the sun, with room to spare.

The mass of the sun is calculated by its accelerating effect on the earth. If the earth's orbital motion were suddenly stopped it would fall less than one-ninth of an inch toward the sun in a second. The curvature of the earth's orbital path is, in fact, caused by a deviation from straight-line motion by exactly this amount every second. From the laws of dynamics (which state that a falling body covers in the first second a distance numerically equal to half the acceleration, and that the acceleration of a body is equal to the ratio of the force acting on it and its mass), the mass of the sun is found to be 333,420 times the mass of the earth, or 1.992×10^{33} grams.

The volume of the sun is more than a million times that of the earth, but its mass is little more than 300 thousand times the earth's mass; therefore, the sun must be far less dense than the earth. The sun's density is, in fact, about a quarter of the earth's, and its specific gravity, 1.41 (in terms of the density of water under normal conditions).

Even if the sun were of the same density throughout, the pressure at its center would be a billion atmospheres because of the weight of the superincumbent layers; actually, it is believed to be far greater, for we know that the density of a star increases rapidly from outside to center. The sun is very hot; even at the surface the temperature is about 6000°K, and it increases rapidly inward to about 20,000,000°K. At such temperatures the sun must be completely gaseous all the way through, despite its high central pressure and density.

The outer temperature of other stars may be as great as half a million degrees or as low as two thousand; however, the internal temperatures of most stars are very likely nearly the same as the sun's. On the other hand, the mean densities of many other stars differ greatly from that of the sun: some (the supergiants) have only a millionth of the sun's density, others (the white dwarfs) are several hundred thousand times as dense as the sun.

The value of gravity at the surface of a body is proportional to its mass divided by the square of its radius. For the surface of the sun, gravity is about 28 times as great as on the earth, and an object that weighed 10 pounds here would weigh 279 pounds on the sun. The *surface gravity* has not merely the science-fiction interest of determining the nature of the athletic feats that would be possible on other bodies. It has great scientific interest in defining the speed an object must have to escape from gravitational attraction at the surface and fly off into space. The *velocity of escape* (proportional to the square root of the ratio of mass to radius) is 11 km/sec at the earth's surface, 618 km/sec at the sun's. Surface gravity plays an important part in governing the ability of a planet to retain an atmosphere, as we shall see later.

2. THE FACE OF THE SUN

The uniformity of the sun's disc is illusory. No part of the surface is quiescent. Its swirling, spurting, boiling motions are the more impressive when one realizes that the surfaces of all stars must be the scene of similar motions, many of them even more violent.

Sunspots were the first markings to be observed on the face of the sun. When Galileo saw them with his baby telescope, many people were scandalized at the suggestion that the pure disc of the sun could be flawed. Less obvious than sunspots are the large mottlings or *faculae,* which are often found near sunspots, the smaller *flocculi,* and the very small *granulations,* which probably include markings too small to be individually seen. Sunspots are slightly less brilliant than the solar surface and look like dark markings; the other features mentioned are slightly brighter; and all of them last for hours or days. Sometimes very short-lived, very bright markings are seen on the sun's face: *flares* are brilliant localized areas, usually associated with sunspots, and *spicules* are tongues of glowing gas that shoot up near the sun's poles and disappear within minutes.

Besides the markings on the solar surface, clouds of glowing gas are often seen poised far above it—some nearly stationary, some spurting upward, some falling downward. These are the *prominences,* full of clues as to conditions near the sun's surface, and also rich in unsolved problems. Certain types of prominence are often connected with sunspots.

The visible disc of the sun, on which the surface markings appear, is known as the *photosphere* (sphere of light). The part of the sun beneath the photosphere is not visible. The photosphere is opaque, and its opacity is caused by the presence of large quantities of hydrogen atoms that have temporarily acquired an extra electron.

The photosphere is not equally bright all over. The edge, or *limb* [Latin: *limbus,* a border], looks darker than the center. The effect can easily be seen with the naked eye near sunset. The "darkening at the limb" of the sun is a consequence of our seeing through deeper layers of its atmosphere at the edge than at the center.

Above the photosphere lies the sun's "atmosphere," which is fairly transparent. The lower part of the atmosphere, just above the photosphere, is called the *reversing layer,* and here the Fraunhofer spectrum of the sun (consisting of the absorption lines of the spectra of atoms and molecules) is formed. This absorption spectrum (often known as "*the* solar spectrum") is of enormous intricacy and interest; on analysis it reveals both the chemical composition of the sun, and the physical condition at the surface. The reversing layer is a few hundred kilometers deep.

Just above the reversing layer, and merging into it, is the *chromosphere* (sphere of color), so called because it has a pinkish tinge due to hydrogen, when seen above the obscuring rim of the moon during a total solar eclipse. It extends for several thousands of kilometers.

The outermost envelope of the sun is the *corona* [Latin: a crown], which appears as a pearly halo during a total eclipse; it may extend more than a million kilometers from the sun.

3. SPECTROHELIOGRAMS

The detail on the sun's surface, already visible with the telescope in light of all colors, is greatly accentuated when single colors are isolated. The spectroheliograph isolates light of one wave length, which therefore comes from a single kind of atom, and permits us to obtain a *monochromatic* (one-color) photograph of the sun.

Surveys of the solar surface in the light of different atoms exhibit the surface features in great detail, which differs from atom to atom. There are several reasons for this. As we have already learned, different atoms, and different states of the same atom, require various degrees of *excitation*. If, therefore, conditions (such as temperature) vary from point to point on the surface of the sun, the response of the atoms will vary also, and they may radiate (or absorb) with greater intensity at some places than others. The atom of ionized calcium, for example, requires no energy to put it in the condition to absorb the H and K lines at 3967, 3933 Angstroms, provided there are ionized calcium atoms present (which, at the solar temper-

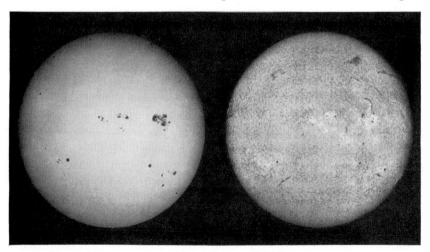

Fig. 4.1. Direct photograph of the sun (left), and photograph taken in the light of hydrogen (Hα). Notice the variegation of the hydrogen picture, some of which is faintly visible on the direct photograph. Notice also the dark filaments of hydrogen, and compare their distribution with that of the sunspots. (Photographed August 12, 1917, at Mount Wilson Observatory.)

ature, there will always be). The line of ionized calcium at 8542 Angstroms is less readily excited. On the other hand, atoms of hydrogen require considerable excitation before they can absorb the Balmer lines. Variations in conditions from place to place may therefore be expected to result in variations in the brightness of the surface in the light of ionized calcium and hydrogen atoms, and to show up more conspicuously in the former than in the latter.

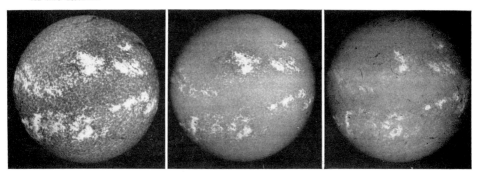

Fig. 4.2. The sun, photographed on August 6, 1937, in the light of three atoms. The first picture is taken in the light of the K line, the ultimate line of ionized calcium; the second is in the light of an infrared line of ionized calcium, somewhat harder to excite; the third is in the red light of hydrogen (Hα), still harder to excite. Notice that the photograph in the light of the most easily excited line shows more conspicuous variegations. Pick out the sunspots, and note their relation to the locations of the variegated areas. Note also the "dark" filaments, silhouetted against the brighter surface, and visible on all three photographs (Photograph by d'Azambuja, Meudon Observatory.)

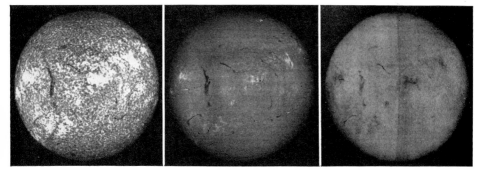

Fig. 4.3. The sun, photographed in the light of three atoms on September 17, 1938. The first picture was taken in the light of the K line, one of the ultimate lines of ionized calcium; the second, in the red line of hydrogen; the third in the light of an infrared line of neutral helium. The calcium atom is the easiest of the three to excite; the helium, the most difficult. Compare the detail shown by the three atoms. The variegations that are conspicuous in calcium light are far less so in the light of hydrogen, and appear as dark areas in helium light. This is the consequence of the difference in excitation energy of the three lines, which differentiate to different degrees the disturbed areas on the sun's surface. (Photograph by d'Azambuja, Meudon Observatory.)

The variations of conditions and differences of response are illustrated in Figure 4.2, which shows spectroheliograms made almost simultaneously by the light (violet) of ionized calcium, by the (infrared) light of the line at 8542 Angstroms, and by the (red) light of the first line of the Balmer series. All three show very similar surface patterns, but the bright areas are most strongly accentuated in the light of the easily excited ionized calcium (K) line, least so in hydrogen light. Some of the bright features in the "K" picture show as dark markings in the light of hydrogen.

Figure 4.3 shows the same effect even more strongly. Here we see simultaneous pictures of the solar surface taken in the light of easily excited ionized calcium (violet), more difficult hydrogen (red), and neutral helium (infrared), which is still harder to excite. Notice that the calcium picture is covered with bright mottlings and has but few dark markings; the mottlings in the hydrogen picture are less bright and the dark markings, if anything, are stronger; finally, the helium picture shows practically no bright mottlings, and several of the brightest areas in the two other photographs appear in this one as dark markings.

4. PROMINENCES

The dark filaments that appear on spectroheliograms take on new meaning when they are seen to cross the edge of the sun's disc. They are often observed as *bright* prominences when seen beyond the sun's edge; their dark appearance against the disc is an effect of contrast. Figure 4.4 shows changes in the dark markings on the solar surface over a three-day interval. The great plume at the top (south pole) of the disc erupted the following day into the largest prominence that has ever been photographed, and spiraled off into space within about an hour and a half.

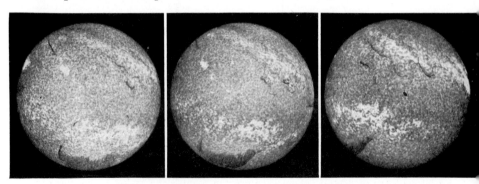

Fig. 4.4. A great prominence, silhouetted against the surface of the sun. The photographs were taken in calcium light, on May 30, June 2, and June 3, 1946. Notice that the features (including the prominence) are carried around by the rotation of the sun. On June 4 the prominence erupted, and was seen to spiral off above the solar surface (Figure 4.5). (Photograph by d'Azambuja, Meudon Observatory.)

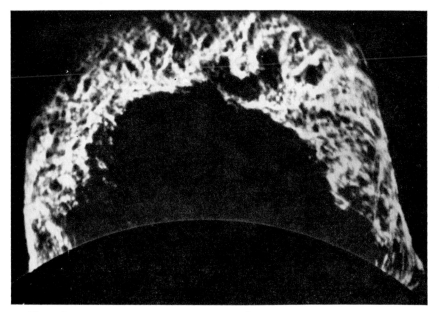

Fig. 4.5. The great prominence that spiraled off from the surface of the sun on June 4, 1946. Compare Figure 4.4. (Photograph, Climax Station, Harvard Observatory.)

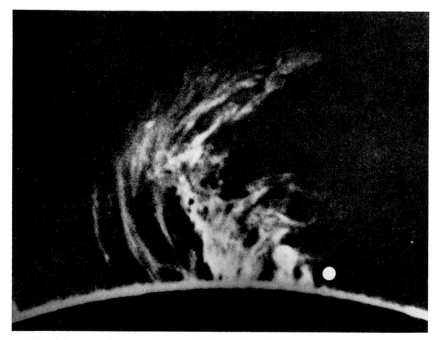

Fig. 4.6. An active prominence, 140,000 miles high, photographed July 9, 1917, in calcium light. The white disc shows the size of the earth. (Mount Wilson Observatory)

All prominences are not like this gigantic plume; rather, they display a great variety, both in form and motion. Another spectacular eruptive prominence is shown in Figure 4.6, and a developing loop, in Figure 4.7.

The motions of prominences are not completely understood. Local magnetic fields probably play some part, and, as we shall see, the surface of the sun can be the seat of enormous magnetic fields.

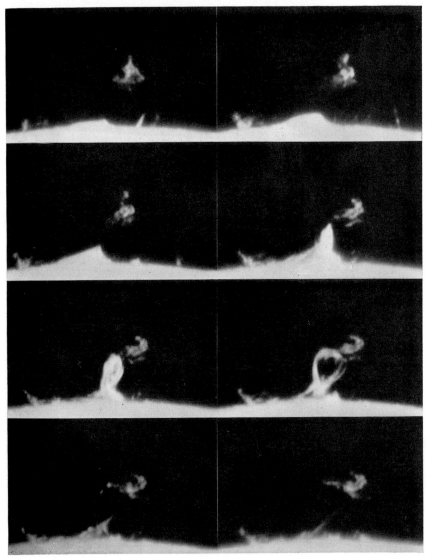

Fig. 4.7. A surge prominence that developed into a loop. The disconnected bright patch is probably independent of the explosion. (McMath Hulbert Observatory)

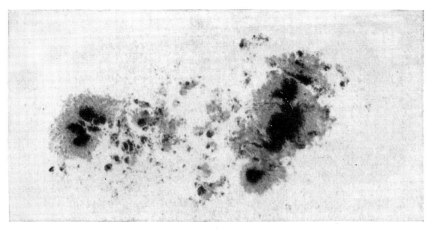

Fig. 4.8. Photograph of a large group of sunspots, July 4, 1947. Compare it with Figure 4.9. Note the structure of the penumbra and the surface granulations of the sun. (Mount Wilson Observatory)

Fig. 4.9. Drawing of a large sunspot group. Notice the penumbral filaments that extend into the umbra, and the surface granulations. (Langley, 1873)

5. SUNSPOTS

The sunspots are gigantic areas of the solar surface, 800 to 80 thousand kilometers across, that are less brilliant than the surrounding regions. They often occur in groups. A sunspot has a complicated structure: a central dark region [the *umbra,* Latin: a shadow] surrounded by a striated, less dark zone (the *penumbra*), which may be as much as 240 thousand kilometers across. Neither umbra nor penumbra is really dark; they only seem so by contrast with the brighter photosphere. Actually, the surface of the umbra is brighter than most terrestrial sources of light.

Most sunspots last a few days, about half exist for less than four days, but a few (about 2/10 of 1%) persist 100 days or more. Bright patches, or faculae, appear as the forerunners of sunspots and remain for a time after the spots have died away. The sunspots are probably disturbances that break through the photosphere; some faculae are unaccompanied by sunspots and furnish evidence of disturbances that never get to the surface as sunspots.

That sunspots resemble tornadoes is suggested by the lower temperatures within them, deduced from spectroscopic studies. Intense magnetic fields at their center are detected and measured by the spectroscopic "Zeeman effect." The measured magnetic fields are enormous and must be caused by electric currents that flow in circles about the spot, and produce strong magnetic fields. Sunspots, however, present many problems; for example, no striking motions of the surface are seen near them, though if the analogy with a tornado were strictly correct we might expect to observe them.

Sunspots are not always present. At times there may be a hundred at once; at other times there are none for several weeks. Nor is the distribution haphazard. Great and small numbers alternate at intervals of about 11.13 years (the *sunspot cycle*). The degree of spottedness differs from one spot maximum to another, but the regularity of the cycle has been demonstrated by careful observations since about the year 1700.

Sunspots are almost confined to the intermediate solar latitudes. They are rare at the solar equator and unknown at the poles (as defined by the sun's rotation). Moreover, they appear at different latitudes in different parts of the spot cycle. At the beginning of a cycle, when the number of spots is small, they occur at high latitudes; as the number increases to maximum and beyond, the "spot zone" drifts steadily toward lower latitudes. At spot minimum, the spots begin to break out again in high latitudes, and the cycle is repeated. Neither the sunspot cycle nor the drifting of the spot zones during the cycle is satisfactorily understood.

There is no doubt that sunspots possess powerful magnetic fields, but the intensity of the magnetic field *as a whole* is very uncertain. The most recent measurements have failed to reveal a steady one; probably the sun has a small but variable magnetic field. Variable magnetic fields as large as, or larger than, those found in sunspots have been observed for a number of other stars.

The form of the sun's outer corona is found to be closely related to the sunspot cycle. The corona tends to be circular at spot maximum, and seems to have axial symmetry near minimum, as though it responded differently

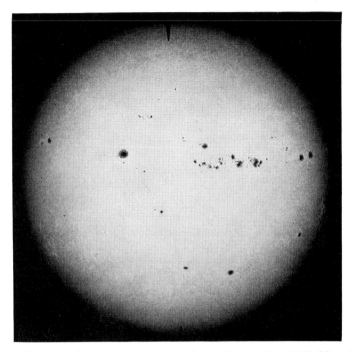

Fig. 4.10. The spotted sun near sunspot maximum (August 14, 1947). Note that the heavily spotted zone is near the solar equator. (Mount Wilson Observatory)

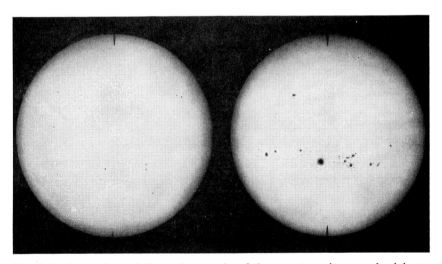

Fig. 4.11. Comparison of direct photographs of the sun at maximum and minimum spot activity (November 30, 1929 and June 22, 1931). (Mount Wilson Observatory)

to conditions over the sun's poles and equator. Probably magnetic forces play a large part in determining the shape of the coronal streamers.

Sunspots, or features associated with sunspots, have definite effects on the earth, the results of ultraviolet radiation and of high-speed particles that are ejected from the sun. The particles are electrically charged and comprise electrons, protons (hydrogen nuclei), and some heavier nuclei. The particles travel with great speeds (800 to 3000 km/sec) and take a day or two to reach us. When they approach the earth, they are affected by the terrestrial magnetic field, and spiral inward about the lines of force that radiate from the magnetic poles. For this reason, the effects are greatest near the poles and small in equatorial regions.

The charged particles produce large magnetic and electrical disturbances on the earth and sometimes disrupt telegraphic service and interfere with radio and television reception by their influence on the layers of ionization in our atmosphere (Figure 2.4). At the same time, they cause excitation of the atoms in the high atmosphere and produce auroral displays. All these effects are actually greatest at times of sunspot maximum and are therefore to be associated either with the spots themselves, or with some activity connected with them.

Many attempts have been made to connect weather cycles, business activity, the occurrence of wars, the fertility of wild animals, and other matters of human concern with the sunspot cycle.

These problems depend on the discrimination of so many intangible factors that most of the results must be classed as highly speculative.

6. THE SUN'S ROTATION

The sun rotates, in the same direction as the planets revolve, about an axis inclined at an angle of 82°49'.5 to the ecliptic (the plane of the earth's orbit) and with an average period of a little less than a month. Unlike the earth, however, it does not rotate like a solid body: the surface near the equator turns faster than regions nearer the poles.

The period of rotation is easily measured in the latitudes where sunspots occur (from about 20° to 35°). Up to about 40°, the rotation can be measured by means of the faculae, and for latitudes less than 20° and greater than 40°, only by the spectroscopic Doppler effect.

The periods of rotation at various latitudes, pieced together from the sources mentioned, are as follows:

Latitude	Period days	Latitude	Period days
0°	24.65	40°	27.48
20	25.19	60	30.93
30	25.85	75	33.15
35	26.63	90	34:

SUN-SPOT GROUP
July 24-30, 1946

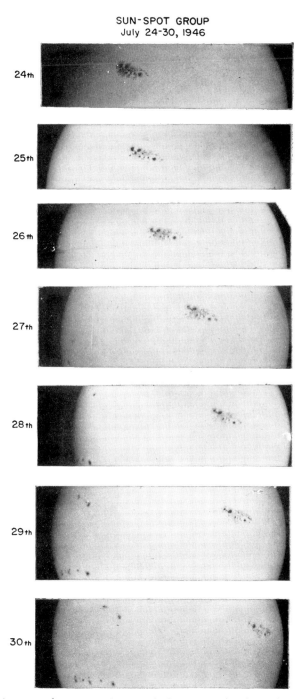

Fig. 4.12. A group of sunspots, photographed on successive days, shows the rotation of the sun between July 24 and July 30, 1946. (Kearons)

105

Thus the time taken by the polar regions of the sun to make one turn is about 40% greater than the time at the equator! The equatorial regions are continually pulling ahead. These figures refer only to the outer skin of the sun; we do not know how fast the interior is rotating, and it may be going much faster. Nor do we understand the surface differences. They may be caused in some way by the braking effect of outflowing radiation, or they may represent the remains of a slowly dying current that was set going long ago.

The nonuniformity of the sun's rotation raises interesting qestions about the rotation of other stars, some of which are spinning very much faster. Perhaps their rotation, also, is nonuniform.

If the rate of rotation of the sun's surface gives even an approximate idea of the rotational period of the whole sun, then a remarkable fact emerges. The energy of rotation of the sun is extremely small in comparison to the total rotational energy of the solar system. This fact, as we shall see, has an important bearing on the problem of how the system originated.

7.   THE SPECTRA OF THE SUN

> . . . a spot like which perhaps
> Astronomer in the sun's lucent orb
> Through his glazed optic tube * yet never saw.
> The place he found beyond expression bright,
> Compared with aught on earth, metal or stone;
> Not all parts like, but all alike informed
> With radiant light, as glowing iron with fire;
>
>   *   *   *
>
> What wonder then if fields and regions here
> Breathe forth elixir pure, and rivers run
> Potable gold, when with one virtuous touch
> Th'arch-chemic sun so far from us remote
> Produces with terrestrial humor mixed
> Here in the dark so many precious things
> Of color glorious and effect so rare?
>
>   MILTON
>   *Paradise Lost*, III, 588–594, 606–612

The spectra of individual regions of the sun can be studied directly, and they display a rich, fascinating variety. The spectrum of the whole sun is crossed by many thousand absorption lines and bands (the Fraunhofer spectrum), of which all the strongest have been identified with lines of atoms or molecules known on earth. A number of the faintest lines and bands still elude analysis, but we are confident that they all are unidentified lines of atoms known terrestrially. The solar spectrum actually shows some

* Milton saw the first telescope, the "glazed optic tube," in the hands of Galileo.

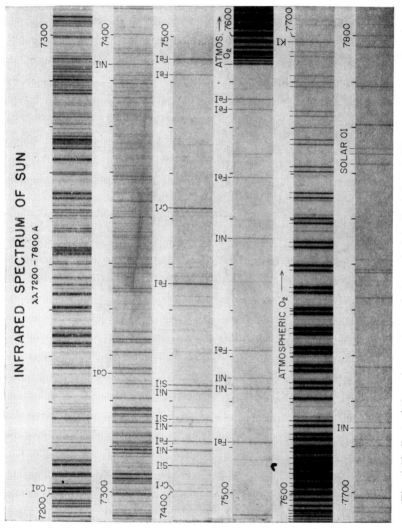

Fig. 4.13. Part of the solar spectrum photographed in the infrared. The great system of molecular bands is produced by the O_2 molecule in the earth's atmosphere. The three lines of neutral oxygen (O I) come from atoms of oxygen in the atmosphere of the sun itself. (Mount Wilson Observatory)

107

atomic spectra in more detail than can be observed in the laboratory. For example, the wave lengths of many hundreds of the lines of iron, too faint to be seen in the laboratory, may be calculated from the known rules of spectral structure and the observed energy levels for the iron atom. A large number of these lines have been photographed in the spectrum of the sun, although they have never been observed in the terrestrial laboratory.

The atoms present in the sun's atmosphere can be ascertained by the presence of their lines in its spectrum. All the atoms known on earth are either found at the sun's surface, or else their lines are absent from its spectrum for good and sufficient reasons (such as that they lie in inaccessible regions of the spectrum, or that the element is so rare on the earth that we should not expect its lines to be detectable on the sun, which has much the same composition as the earth).

If the correct allowance is made for the effects of excitation (p. 76), the strengths of the lines observed in the spectrum of the sun can be used to make a quantitative analysis. With some notable exceptions, the relative numbers of atoms are remarkably close to those observed on the earth.

Not all the absorption lines that we observe in the spectrum of the sun are produced in the solar atmosphere. Sunlight traverses our own atmosphere before it reaches us, and atoms in the earth's own envelope cut their characteristic patterns in the sunlight. Figure 4.13 shows part of the infrared solar spectrum that is crossed by the banded spectrum of the oxygen molecule in our atmosphere (O_2). Three lines in the lowest strip, however, are lines of the oxygen *atoms* that are actually in the sun's reversing layer. Water vapor, ozone (O_3), and carbon dioxide in our atmosphere are responsible for strong absorptions in the infrared and the ultraviolet. Ozone, in fact, prevents light with wave length less than 2900 Angstroms from reaching the earth's surface. Shorter wave lengths can be studied only by means of rocket-borne spectrographs.

Among the strongest lines observed in the solar spectrum are the H and K lines (3967 Angstroms and 3933 Angstroms) of ionized calcium (Figure 4.15), the Balmer lines of hydrogen (Figure 4.14), and the ultimate lines of neutral and ionized magnesium (observed in rocket spectra, Figure 4.16).

The H and K lines of ionized calcium, from the ground state of that atom, are strong and broad because almost all the calcium atoms in the solar atmosphere are able to absorb them. The Balmer lines, on the other hand, are hard to excite, and only about one hydrogen atom in a hundred thousand is in a state to absorb them at a given time. Hence, although hydrogen is in fact more abundant in the sun's atmosphere than calcium, its lines are much weaker.

The Balmer lines may be noticed to be wide and fuzzy in the spectrum of the sun. This is a consequence of the high speed of agitation of the hydrogen atoms. If (as is roughly the case) the energies of agitation of all

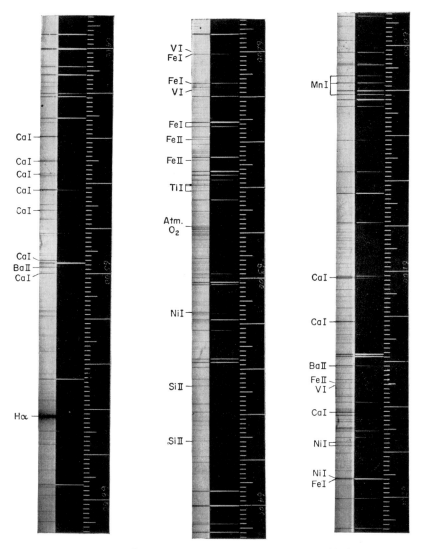

Fig. 4.14. Spectrum of the sun from wave length 6000 to 6600 Angstroms. The comparison spectrum is that of neutral iron. Notice that the iron lines can all be traced in the solar spectrum. Lines of several atoms are marked. Notice the strength and width of the line of hydrogen (Hα). (Mount Wilson Observatory)

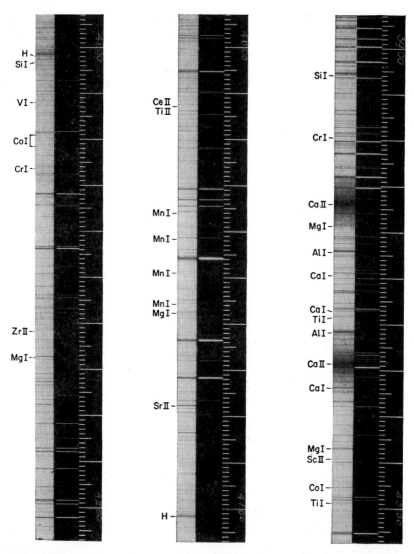

Fig. 4.15. Spectrum of the sun from wave length 3900 to 4200 Angstroms. Notice the great strength of the H and K lines of ionized calcium, and the relative weakness of the hydrogen line at 4101 A (Hγ)

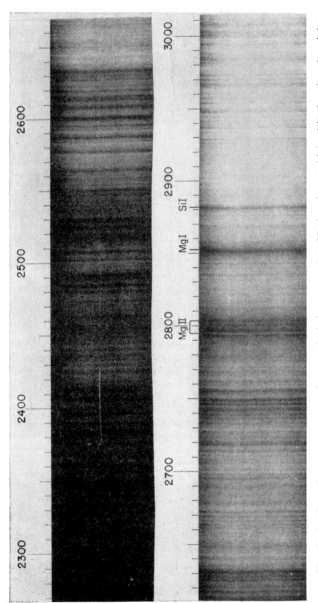

Fig. 4.16. The solar spectrum as photographed from a rocket. Notice that at high altitudes the ultraviolet part of the spectrum appears, because the rocket rose above most of the layer of ozone. Notice the very strong lines (ultimate lines) of neutral and ionized magnesium (Mg I, Mg II).

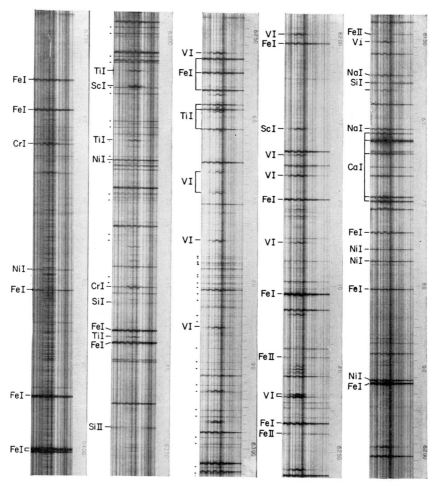

Fig. 4.17. Spectrum photographed across the core of a sunspot. The zigzag appearance of the lines is caused by a polarizing device, which transmits alternately the red and violet components of the broadened and split-up lines. Notice the great intensity of some lines in the core of the spot that are inconspicuous at the edges (which show the solar disc). These lines are of very low excitation potential (especially conspicuous are those of neutral vanadium, V I) and are weaker in the hotter disc than in the cooler spot. Notice that some lines of ionized iron (Fe II) are weaker in the spot spectrum than in that of the disc, again because of the difference in temperature. Compare Figure 4.14, part of the same region. (Mount Wilson Observatory)

atoms in the sun's atmosphere are equal, the least massive ones are moving most rapidly.* Thus the "thermal speed" is largest for the lightest of all atoms, hydrogen.

The spectrum of a sunspot is very different from that of the solar surface. Lines that require little excitation are stronger in the sunspot spectrum, lines of high excitation, weaker. The difference may be seen in striking fashion in Figure 4.17. Some lines that require very little excitation, such as those of neutral vanadium, are scarcely to be seen in the spectrum of the solar disc, but are very strong in the sunspot. Other lines, such as those of ionized iron and hydrogen, are much weaker in the spot than in the disc spectrum. The difference is undoubtedly due to considerably lower temperature (about 4000°) within the sunspot.

The most remarkable effect shown by the sunspot spectrum, however, is the evidence it furnishes of large magnetic field within the spot. When a

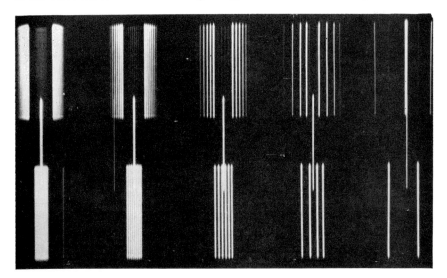

Fig. 4.18. Effect of a magnetic field on some lines of neutral vanadium. The central strip shows the lines as they appear without a magnetic field. Each line is split into a number of components by the large magnetic field (about 13,500 Gauss); they are polarized either perpendicular to (above) or parallel to (below) the field.

All the lines shown come from similar transitions, which differ in the quantum numbers that are involved; the variety in the numbers and spacing of the components is in full accordance with quantum theory. For more moderate fields the patterns (called Zeeman patterns after their discoverer) are symmetrical; the field used here was so large that some of the patterns are slightly asymmetrical, again in accordance with quantum theory. Zeeman patterns of this kind are responsible for the changing appearance of spectrum lines in the core of a sunspot (compare Figure 4.17). (Massachusetts Institute of Technology)

* The energy of a moving particle is proportional to $\frac{1}{2}mv^2$, where m is its mass, v its velocity. Thus, for equal energy, v is large if m is small.

radiating or absorbing atom is in a magnetic field, its line spectrum is greatly modified; individual lines split up into a number of separate components. This phenomenon, called the *Zeeman effect* after the Dutch physicist who first observed it, is very complex and varies from atom to atom and even from line to line of a given atom. The interpretation of the Zeeman effect is a beautiful application of the quantum theory of the spectrum. The splitting increases with the magnetic field, which it can therefore be used to measure. The magnetic fields in sunspots are often very large—many thousands of gauss (p. 479); for comparison we note that the magnetic field of the earth, which affects the compass, is but a fraction of a gauss.

The magnetic field is directed along the axis of the sunspot, radially away from the sun. Some sunspots are "north polar," some, "south polar" and, when (as often happens) sunspots occur in pairs, the polarities tend to be opposite. The magnetic fields within sunspots are thought to be the results of circulating currents of electrons; like many other solar phenomena, however, they are incompletely understood. If the whole sunspot were a whirling vortex, we should expect to observe rapid surface motions near to spots, but such motions have not been recorded.

At the time of a total solar eclipse, the spectrum of the upper reversing layer of the sun (the *chromosphere*) may be observed around the edge of the obscuring moon. It differs strongly from the spectrum of the sun or of sunspots, which consist of absorption lines. The chromospheric spectrum is a *bright-line spectrum*: it comes from atoms that are radiating the characteristic lines of the constituent atoms. It shows the presence of the same atoms that the spectrum of the solar disc reveals, but they are in a different condition, which points to lower density (as might have been expected) and also higher temperature (which is somewhat surprising). The temperature of the chromosphere, deduced from ionization and excitation, is about 20,000–30,000°K. Bright lines of helium and even of ionized helium are present, and one would hardly have expected to observe ionized helium, even at these high temperatures.

The outer envelope of the sun, the *corona,* shows a very different spectrum, which for many decades defied analysis, and was supposed to come from an otherwise unknown substance, "coronium." However, the Swedish physicist Edlen recently showed that it comes from known substances in a very highly ionized state, notably, Fe X, XI, XIII, and XIV; Ni XIII and XV; and Ca XV. Some of these atoms have ionization potentials of about 500 electron volts, and the temperature required for this degree of ionization is about a million degrees.

The picture of the sun, deduced from its spectra, is unexpectedly complex. The temperatures of the successive outward layers increase to a very large value for the corona. Very likely the observed spectra of the stars come from the integration of an equally complex series of layers, and for some of them, too, the temperature probably increases outward.

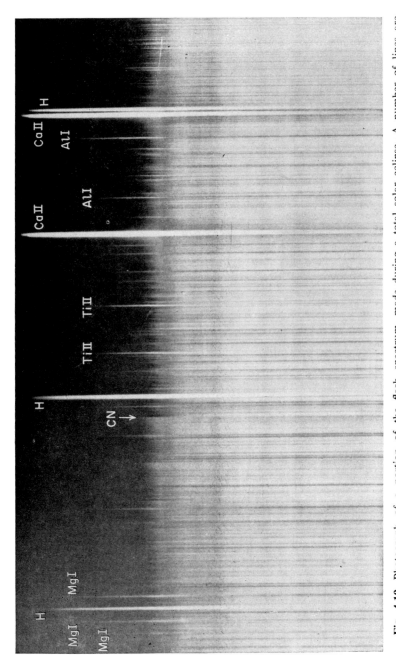

Fig. 4.19. Photograph of a portion of the flash spectrum, made during a total solar eclipse. A number of lines are marked. Notice the great length of the lines of hydrogen and ionized metals. The plate was moved uniformly during the exposure, so that the height of the lines shows the actual height above the photosphere at which they were produced. At the bottom of the photograph, the flash spectrum merges into the absorption spectrum of the edge of the sun's disc.

Fig. 4.20. The corona of the sun, photographed during a total solar eclipse. Notice the radial distribution of the coronal streamers.

8. SUN AND EARTH

Not only does the sun, by reason of its great mass, control all the motions of bodies in the solar system. It is virtually the whole source of heat (and light) within the system. We receive a little light from the stars, and radioactive processes contribute somewhat to the earth's internal heat. But our main source of heat and light is the sun itself, the only self-luminous body in the solar system.

The sun radiates energy at the enormous rate of 3.79×10^{33} ergs a second. The earth, being far away and small, intercepts only a small fraction (about one two-billionth) of this energy flux, but even so, the amount is enormous: 4,690,000 horsepower per square mile. About 30% of the energy is absorbed, or scattered backward, in passing through our atmosphere, including (luckily for us) the ultraviolet light. But the amount that strikes the surface is still a staggering one, and it is a pity that man has made so little progress in turning it to direct use. In most of our ways of using solar energy (the burning of wood, coal, and oil; the drawing of power from rivers; or the eating of organic substances) we depend on letting plants or animals do the work of "trapping sunlight" for us, or on making use of the water cycle.

The sun's role in warming the earth is of the first importance in making life possible on the surface; life, of the kind we know (the most likely, if not the only possible kind), depends critically on the presence of water. Moreover, the water must not be above boiling point (100°C, or 373°K) or permanently below freezing point (0°C, or 273°K). The temperature at the surface of the earth, or any other planet, depends on its distance from the sun (which determines how much radiation it intercepts), its absorbing and reflecting ability, and the blanketing effect of its atmosphere. The first of these factors is by far the most important.

The sun's heating effect is, of course, greatest, and the temperature therefore highest, for the planets with smallest orbits, and will fall off farther out in the system. The temperatures calculated for the planets (with proper allowance for distance, reflecting ability, and atmosphere) are very similar to those that have been measured from the quality of the reflected light. The measured temperatures (absolute scale) are as follows:

Mercury (sunlit side)	690°K
Venus (sunlit side)	330
Earth (average)	287
Mars (warmest parts)	285
Jupiter	135
Saturn	120
Uranus	< 90

With such temperatures, only Venus, Earth, and Mars provide surface conditions between the boiling and freezing points of water. Their temperatures will be discussed in more detail in Chapter VI. These general ideas suggest that life can occur only within a restricted zone in the solar system.

The surface temperatures of the planets play an important part in determining the possibility of the retention of atmospheres. Gravity at the surface specifies the velocity of escape from a planet, and surface temperature specifies the actual velocities of the atmospheric particles. A combination of temperature and surface gravity therefore determines whether a given planet can retain an atmosphere or not. We shall see that the moon and Mercury cannot hold atmospheres, whereas the other planets are able to do so.

THE MOON

That orbéd maiden
With white fire laden
Whom mortals call the Moon.
SHELLEY

This exquisite poetic description of the moon is very wide of the mark. As we shall see, the moon is no maiden, but a scarred and wrinkled crone; she is not white, and she bears no fire.

The moon, our one satellite, is large as moons go, the fifth in diameter among planetary satellites, more than two-thirds as large as Mercury, and more than three times the diameter of any asteroid. In fact it is over one-fourth of the size of the earth, with diameter 3476 kilometers or 2160 miles.

1. THE MOON'S DIMENSIONS

The mass of the moon is $1/81$ of that of the earth. It is measured with the aid of the corollary of the law of gravitation (p. 170), which states that two bodies move under their mutual attraction in similar orbits around each other and their common center of gravity, and that the sizes of these two orbits are in inverse proportion to the masses of the bodies. Earth and moon move in nearly circular orbits around their common center of gravity; and the size of the moon's orbit bears the same ratio to that of the earth's orbit as the mass of the earth bears to the mass of the moon.

The size of the orbit described by the earth under the moon's gravitation must be measured by reference to something *outside* the earth-moon system. Although the motion is small, it can be detected by means of small monthly displacements of the nearest bodies of the solar system, such as Mars, Venus, and close-approaching asteroids. The radius of the earth's orbit around the center of gravity of earth and moon is 4635 kilometers, $1/82.9$ of the distance from earth to moon. Thus the moon's orbit is 81.9 times the earth's (Figure 5.1), and the moon's mass is $1/81.9$ of the earth's. Dynamical methods of measuring the moon's mass, based on the theory of the moon's motion, give similar results.

The moon is 0.273 of the size, and 1/81.9 of the mass, of the earth. Thus the moon's density is about 0.61 of the earth's. The surface gravity is 0.165 of that at the earth's surface, and the velocity of escape is only 2.4 km/sec.

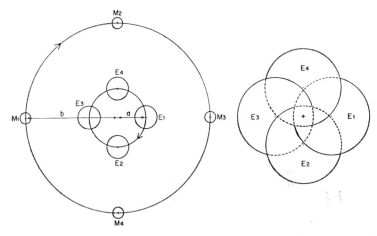

Fig. 5.1. How the mass of the moon is determined. The figure represents the orbits of the earth (E) and of the moon (M) around their common center of gravity. Neither the bodies nor their orbits are drawn to the correct scale. Positions of earth and moon at four times are numbered 1, 2, 3, 4. Notice that at any one time they are on opposite sides of the common center of gravity.

At time 1, the distance earth-center is a, the distance moon-center is b, and the distance earth-moon is $a + b$. The ratio of the distances (moon-center/earth-center) is b/a. By measurement of the orbit of the earth around the center, $a = (a + b)/82.5$, so that $b = 81.5a$.

Actually the earth's orbit around the center is so small that the center of gravity of the earth-moon system is always inside the earth. The motion of the earth about the center of gravity of the system (cross) is shown to scale on the right.

Since the moon is our nearest neighbor, the distance can easily be measured by geometrical methods. The average is 384,403 kilometers, or 238, 857 miles, and is known with a precision of 1 in 300,000. The distance varies somewhat because the orbit is elliptical.

Next to the sun, the full moon is the brightest object in the heavens. Expressed in stellar magnitudes (p. 266), its brightness is -12.55, as compared to -26.72 for the sun; in other words, the sun is about 400,000 times brighter.

The surface of the moon is rough and brownish and reflects light very poorly. In fact, the moon is about the poorest reflector in the solar system. The reflecting coefficient is expressed by the *albedo* [Latin: *albus*, white], which is the ratio of light reflected to light received. The moon reflects only 7% of the sunlight that falls upon it, so the albedo is 0.07.

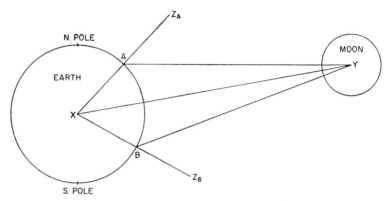

Fig. 5.2. How the moon's distance is measured. Simultaneous observations are made from two stations (A, B) on the same parallel of longitude. The distance of the moon from the zenith is measured at both stations, which gives the angles $Z_A AY$ and $Z_B BY$. The angles XAY and XBY are the supplements (180° minus) of these angles. The angle AXB is the numerical difference of latitude between the stations A and B. Thus the *shape* of the quadrilateral $XAYB$ is completely determined. The *size* of the quadrilateral is determined by the fact that $AX = BX =$ the earth's radius. Therefore the diagonal XY, the distance between the centers of earth and moon, can be determined.

2. THE MOON'S PHASES

The phases of the moon are caused by the changing relative positions of earth, moon, and sun.

The moon goes around the earth in $27^d7^h43^m11^s.47$ (27.32166 days); this is the average time taken by the moon to make an apparent circuit of the sky, from one star back to the same star. Therefore this month is known as the *sidereal month*. Because the moon's motion undergoes many disturbances (*perturbations*), the sidereal month may vary by as much as seven hours.

The month that is more in tune with human affairs is the interval between similar phases of the moon (new moon to new moon, for instance). It is known as the *synodic month,* and is more than two days longer than the sidereal month, with an average length of $29^d12^h44^m2^s.78$ (29.53059 days). The length of the synodic month may vary by as much as thirteen hours, chiefly because of the eccentricity of the orbit and the consequent nonuniformity of motion (see p. 165).

The sidereal and synodic periods (months) are related in a simple way. If M is the sidereal period, the moon goes through $1/M$ of the way around the earth in a day. Similarly, if E is the earth's sidereal period around the sun, the sun appears to move $1/E$ of the circuit of the sky in a day. The difference is the amount by which the moon gains daily on the sun. The moon *gains* because M is smaller than $E,$ and $1/M$ therefore larger than

$1/E$. The moon gains a whole revolution in one synodic month (S), for example between one new moon and the next, when the moon is in line between sun and earth. Thus

$$1/M - 1/E = 1/S, \qquad \text{or} \qquad S = EM/(E\text{-}M)$$

The moon pulls 12.2° eastward of the sun, an average of 360°/S, each day.

Half the moon is always in sunlight, half in shadow. The edge of the shadow (the *terminator*) is therefore a semicircle seen obliquely and is hence an ellipse (the projection of a circle). The edge of the terminator and the edge of the bright moon (the *limb*) are slightly serrated by the lunar mountains. The "horns" of the moon (the *cusps*) are at the ends of

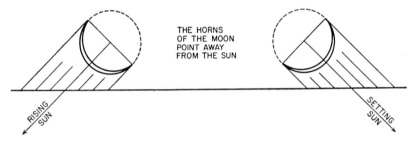

Fig. 5.3. The "old moon," just before sunrise, and the "new moon," just after sunset. Notice the direction of the cusps. The word "crescent" means "growing" [Latin: *cresco*, I grow] and should be strictly applied only to the new moon, but common usage applies it to the *shape* of the illuminated portion, irrespective of the moon's age. More correct terms for the new and old moon are "waxing" [cf. German: *wachsen*, to grow] and "waning."

the terminator; a straight line joining them is always perpendicular to the line joining sun and moon, and the horns always point away from the sun. At the equator the new moon may seem to be "lying on its back"; from the polar regions it will appear nearly vertical. It is an amusing pastime to note the "impossible moons" portrayed by some artists: a new moon high in the northern sky, for instance; a full moon near sunset in the west; or a crescent with horns pointed downward.

The "old moon in the new moon's arms" shines with a bluish light, reflected from the earth's atmosphere. From the brightness of "earthlight" the albedo of the earth is found to be 0.29—the earth being a fairly good reflector. An observer on the moon would see "full earth" 40 times as bright as we see full moon—the earth reflects better than the moon and is also larger. A lunar observer would see "full earth" when we see new moon, and the opposite.

The moon has no observable atmosphere. All the surface features are sharp; there are no haze, no clouds, no evidences of storms (such as are seen on Mars). There is no twilight arc at the cusps (like that seen on

Venus). When the moon passes across a star, the star vanishes abruptly; there is no progressive dimming and no refraction. Furthermore, no atmosphere would be expected. The velocity of escape (2.4 km/sec) is so small

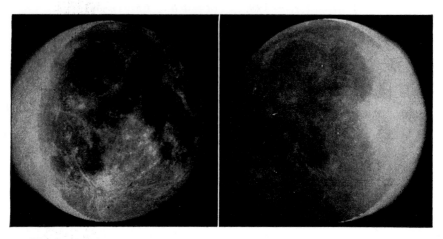

Fig. 5.4. "The old moon in the new moon's arms": the dark portion of the moon is faintly illuminated by the sunlight reflected from the earth's surface. The brightness of this "Earthlight" measures the reflecting power (albedo) of the earth. The two pictures show the phenomenon just before, and just after, new moon. (Harvard Observatory)

that the natural (thermal) speeds of any atmospheric particles it may once have had (about a kilometer a second) would have caused them to evaporate from the surface within a few thousand years. For the same reason the moon could not retain surface water.

The temperature on the moon's surface is about 373° absolute at lunar midday, 263° at sunset, and 193° at midnight. Unlike the earth, it has no blanket of atmosphere, so the surface adjusts itself very fast to the sunlight it receives. The moon, however, does have a blanket of sorts—probably a layer of volcanic dust several inches deep. Under this blanket the temperature shows a lag resembling that shown by the earth's surface, as can be seen by studying the moon's face with very long microwaves (radar) which penetrate the dust layer, as visible light cannot. It should be noted that although the moon always presents the same face to the *earth,* the sun shines on all sides of it in turn.

3. THE SURFACE OF THE MOON

The moon's surface, when seen through even a small telescope, shows great variety. Elaborate maps have been made of the moon, and several thousand selenographic features have been named.

The *maria* (seas) were so called when they were thought to be bodies of water; we know now that they are smoothish brown areas, perhaps hardened lava pools. They cover about half the surface that we see.

The *mountains* are evidently not produced by folding, like many of the earth's mountain ranges. Their heights are easily measured by the lengths of their shadows. Some are as high as Mount Everest. The great mountain wall that bounds the *Mare Imbrium* is 20,000 feet above the plain.

The *craters* are scattered over the surface, and more than 30,000 have been mapped. Some are as much as 150 miles across, others a few miles or less. Many, though not all, have central bosses that look like those shown by some volcanoes. Not only are there craters within craters, but craters are seen overlapping, and small craters are found on the walls of big ones.

Rays radiate from craters and are brighter than the surface that they overlie. Some rays are 1500 miles long and ten miles wide. They strongly

Fig. 5.5. The full moon. Compare with the views of Figure 5.6 and Figure 5.8, and notice the slightly different aspect of the moon's surface, the consequence of libration. (Mount Wilson Observatory)

Fig. 5.6. The moon at ages of five and three days. The dark oval on the face of the five-day moon is the Mare Crisium; below it is part of the Mare Fecunditatis. Since the moon always turns nearly the same face to us, the illuminated edge shows the same features in both pictures. Notice the great difference produced by illumination at different angles. (Mount Wilson Observatory)

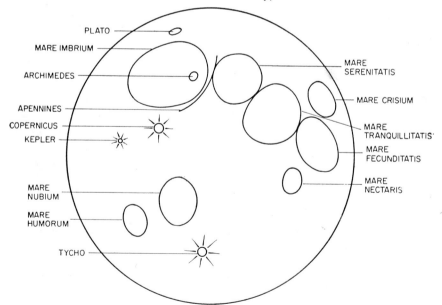

Fig. 5.7. Names of some prominent lunar features. Compare Figure 5.5.

Fig. 5.8. The old moon, age 26 and 23 days. Compare with Figure 5.5 and notice the effects of varying illumination and of libration. (Mount Wilson Observatory)

suggest splashes of fine dust that radiate from a center, such as a large crater.

Rills are narrow, sharply defined features that extend across the surfaces of *maria*. They may be wrinkles or cracks in dried lava beds.

The history of the moon is written on her face, and many attempts have been made to decipher it. An enormous variety is evident in the lunar landscape, from the high, craggy mountain ranges to the relatively flat surfaces of the *maria*. The many thousands of craters present infinite variety: no two are alike. Some have central peaks, or groups of peaks; others have flat interiors; and the flat floors of some fill them almost to the brim. The interiors of many craters are pitted; others show smaller craters that break through their walls. Many of the largest craters are the centers of great, radiating systems of rays.

The craters are a starting point for the understanding of the moon's history. The central humps seem to suggest an analogy with terrestrial volcanoes, but many lunar craters show no such humps, and in scale and structure they are all totally unlike that of any of our volcanoes. An extremely important fact concerning the lunar craters is that the volume of their interior is about equal to the volume of the material in the walls—

they are not as deep as they appear to the eye. Such a general relationship can readily be explained if all the craters were produced by the impact of missiles on the surface of the moon. The great ray systems that radiate

Fig. 5.9. Northern portion of the moon, age eight days. Compare with Figure 5.5 and notice the effect of illumination. Notice the tiny craters on the floor of the Mare Serenitatis and within some of the larger craters. (Mount Wilson Observatory)

from some craters might very well be dust splashes produced by the same events. Moreover, if the craters were produced by impacts, there is nothing strange in finding craters that intersect one another.

An explanation that accounts for the craters and rays of the moon must be expected also to explain the other features of the lunar surface, notably the *maria* that cover so large a part of it, and the rills (cracks or wrinkles) that cross them, as well as the mountain ranges. The color and texture of the *maria* suggest that they are covered by lava flows, and they may very likely be the sites of far larger explosive impacts than those that were responsible for the smaller craters. As a result of these impacts,

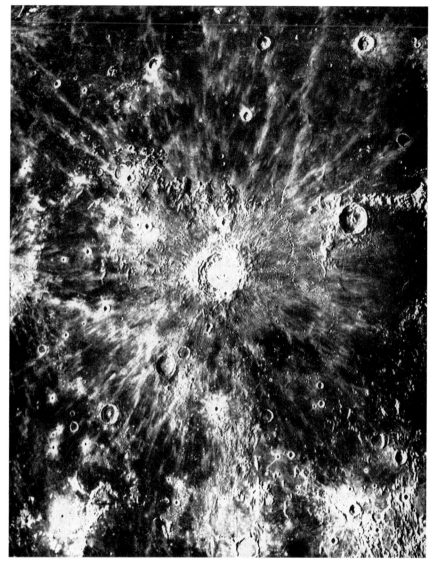

Fig. 5.10. The great lunar crater Copernicus. Notice its inner structure and the rays that spread out in all directions from it, and the ripples on the floor of the Mare Imbrium above it. Compare this aspect of Copernicus with that shown at full moon, Figure 5.5. (Mount Wilson Observatory)

lava is thought to have welled up from the moon's interior and to have covered the large, often nearly circular, areas that constitute the *maria*. The rills may be cracks or faults in the solidified surface of the lava flows. The mountains that rim the *maria* are seen as results of the same explosive impacts as the *maria* themselves.

If explosive forces have played the major part in fashioning the surface of the moon as we see it today, evidence of similar processes would be expected in other parts of the solar system. The agents must be sought in

Fig. 5.11. Southern portion of the moon at last quarter. Notice the large number of intersecting craters and the floor of the *Mare Nubium*. (Mount Wilson Observatory)

large and small meteorites, even perhaps objects of the size of small asteroids, that struck the surface of the moon with great enough violence to produce the craters and even the *maria*. On the earth's surface we know of a number of undoubted meteor craters that resemble those of the moon in general structure and many more craters that may well be of a similar nature. But only the most recently formed craters would be observable on the earth, where the surface is continually being smoothed off by erosion and obliterated by sedimentation. The earth differs from the moon in that it has an atmosphere, which is the seat of continual weather changes. On the lunar surface there are no such forces, able to obliterate the effects of the impacts of meteorites.

The planet Mercury, which (as far as we can see it) possesses a surface that seems to resemble that of the moon in structure, may be regarded as an oversized moon. Mercury has no atmosphere. Probably the similarity of surface is an indication that the smallest of the planets, like our satellite, bears on its face the evidences of meteoritic bombardment.

4. THE MOON'S MOTION

The orbital motion of the moon, and the motion of the orbit itself, are an excellent, nearby example of the kind of motion found throughout the solar system. They may seem, and are, extremely complex—indeed this subject, known as *lunar theory,* is one of the most specialized and difficult branches of dynamical astronomy. An elementary book cannot even indicate its scope. But we can give a descriptive idea of some of its main features, which are direct consequences of the theory of gravitation.

The moon appears to move around the celestial sphere very nearly in a *great circle* which does not lie in the ecliptic, but is inclined to it on the average by 5°8', because the orbit of the moon around the earth is not in the same plane as the orbit of the earth around the sun. Just as the celestial equator cuts the ecliptic at the equinoxes (p. 51), the moon's path on the celestial sphere cuts the ecliptic at the *nodes*. The node where the moon crosses the ecliptic from south to north is called the *ascending node;* the other, where the moon crosses from north to south, is the *descending node*. Only when the moon is at or very near the nodes can eclipses occur, for only then can sun, moon, and earth be suitably lined.

The moon's orbit is an ellipse of average eccentricity 0.055, hence it is three and a half times as elliptical as the earth's. The apparent changes of the moon's size are therefore larger than those of the sun's size; the angular diameter of the moon varies between 33'30" when it is nearest, to 29'21" when it is farthest, from us. The positions in the orbit when the moon is nearest to us and farthest from us are called *perigee* and *apogee* respectively [Greek: *gē,* the earth; compare *perihelion, aphelion*]. The line connecting apogee and perigee is called the *line of apsides*.

If the solar system consisted of only two perfectly spherical, homogeneous gravitating bodies, the motions would be simple and would repeat themselves exactly. But if there are even three gravitating bodies, the problem is made very difficult by the interplay of mutual attractions. In fact it is too difficult to be solved completely except in simplified cases (for example, if one of the bodies has negligible mass). The difficulty lies in the continually changing variety of possible configurations. The solar system contains a very large number of gravitating bodies, and the fact that most of them are neither homogeneous nor spherical introduces further complications. The changes in the simple motion of two bodies that are produced by all these effects are called *perturbations*.

We shall mention some of the principal perturbations of the moon's motion as illustrations of its complexity. They are mainly caused by the gravitational action of the sun on the earth-moon system.

a. The *evection* is a periodic change in the orbital eccentricity of the moon, which repeats at intervals of 31.8 days. It was discovered more than two thousand years ago by Hipparchus.

b. The *variation* is an effect that makes new and full moon occur too early, and half moon too late, in the cycle.

c. The *annual equation* is a result of changes in the sun's disturbing force as the earth travels its elliptical orbit, and our distance from the sun changes. The variation and the annual equation were first observed by Tycho Brahe (p. 163) in about 1600.

d. The *regression of the nodes* is a backward motion of the nodes along the ecliptic, similar in effect to the precession of the equinoxes (p. 37). The nodes make the complete circuit of the ecliptic in about nineteen years. Because of this effect, the time taken by the moon to pass round its orbit from node to (the same) node is not the same as the time taken from a given star back to the same star. It is known as the *nodical* or *draconitic month* (27.21222 days) and differs from the sidereal month (27.32166 days) in exactly the same way as the tropical year differs from the sidereal year (p. 36). The nodical month is an exact submultiple of the interval at which eclipses can occur, hence the name "draconitic," a throwback to the time when a dragon was supposed to swallow the sun at a total eclipse.

e. The *inclination of the moon's orbit* varies cyclically between 4°59' and 5°18'. This perturbation, and the motion of the nodes, were first observed by Flamsteed in about 1670.

f. The *progression of the line of apsides* causes the whole orbit to turn on itself once in about every 9 years. Newton tried to account for this effect, first observed in his time, but his published prediction did not fit the facts. A century later the French mathematician Clairaut found that Newton had neglected some small terms in the equations, and he brought theory and fact into agreement by taking such terms into account. The correct calculation was later found among Newton's unpublished papers: he had detected his own error but had never corrected it in print!

g. The *inequality* * *of the apogee* and

h. The *inequality* * *of the nodes* were first predicted by Newton in his discussion of the consequences of the law of gravitation; they had not previously been observed, but were looked for and found.

The eight perturbations of the moon's motion just mentioned are only the most conspicuous of a far greater number. The interplay of sun, moon, and earth produces a motion so complex that the relative positions of the

* "Inequality" here denotes "variable time of arrival at."

three are never exactly repeated. The motion of the moon can be broken down into about 150 principal periodic motions along the ecliptic, and about the same number perpendicular to it: there are also about five hundred smaller terms. The moon's perturbations are not only of interest as a study in complex motion, but they can also be used as a way of measuring the sun's distance.

The moon does not always rise to the same altitude on the *meridian* (the north-south line through the celestial pole). The moon's orbit is not in the ecliptic, but makes an average angle of 5°8' with it. Neither, therefore, is it in the same plane as the earth's equator, which makes an angle of about 23°27' with the ecliptic (the inclination of the ecliptic, p. 36). The angle between the moon's path and the equator varies because of the regression of the moon's nodes. When the ascending node is at vernal equinox, the angle between the moon's path and the equator, which determines how high the moon will rise, is the *sum* of these inclinations, 5°8' + 23°27' = 28°35'. The moon's altitude can range through twice this value, or 57°10'. When the descending node is at vernal equinox, the angle

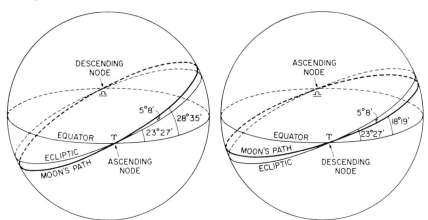

Fig. 5.12. Cause of the variation of the moon's declination and altitude. The intersection of the moon's path and the ecliptic (the *nodes*) moves around the ecliptic in about nineteen years.

In the left-hand figure, the *ascending node* (where the moon's path crosses the ecliptic from south to north) is at vernal equinox (♈). The angle between the moon's path and the equator, which determines the moon's declination, and therefore its altitude as seen from any one terrestrial latitude, is the *sum* of the angle between the moon's path and the ecliptic and the angle between the ecliptic and the equator.

In the right-hand figure, the ascending node is at autumnal equinox (♎); the *descending node* (where the moon's path crosses the ecliptic from north to south) at the vernal equinox. Therefore the angle between the moon's path and the equator is the difference of the angles made by the moon's path with the ecliptic and by the ecliptic with the equator.

The difference in time between the first and second figures is 1 9/2, or about 9 1/2 years.

between the moon's orbit and the equator is the *difference,* $23°27' - 5°8'$, of the two inclinations, or $18°19'$, and the moon's altitude has a range of $36°38'$. Thus the declination of the moon and its meridian altitude as seen from a given latitude change regularly in a cycle of nineteen years, the time taken for the nodes to make one turn around the ecliptic.

Because of its orbital motion, the moon appears to move eastward against the stellar background, and rises, on the average, 50.5 minutes later each day for observers in the usual latitudes. The interval varies somewhat because of the ellipticity of the moon's orbit, which results in nonuniform motion, and also because of the perturbations. There are, however, some occasions on which the times of successive moonrise can be much closer together than 50 minutes.

The full moon nearest the time of autumnal equinox is called the *harvest moon,* and in some latitudes may rise only about twenty minutes later each night. At autumnal equinox, the sun is at the autumnal equinox in the ecliptic, and the full moon (opposite the sun in the sky) is therefore at or near vernal equinox in the ecliptic. At this point the ecliptic has the smallest inclination to the eastern horizon, and in the Northern Hemisphere, the moon "comes up sooner" than at times when the ecliptic intersects the horizon at a larger angle, and the retardation of rising is at its smallest.

The moon's period of rotation relative to the earth is equal to its period of revolution around the earth. It turns once on its axis as it makes one circuit, and therefore always turns the same face to the earth. If the moon's orbit were circular, and if its axis of rotation were exactly perpendicular to its orbit, we should see half its surface.

Actually we see, at one time or another, 59% of the moon's surface because of the so-called *librations* (rockings), which are of several kinds and have various causes. Most of the librations can be seen to be simple results of the geometry of the moon's orbit:

a. The moon spins about an axis that is not exactly perpendicular to its orbit, but inclined to it by $6°5'$. However, the axis always points in the same direction *in space.* Therefore, the moon inclines alternate poles toward the earth at intervals of two weeks. Thus we alternately see $6°5'$ "over the top" and "under the bottom" of the moon. (In exactly the same way, the earth inclines alternate poles toward the sun at intervals of six months: the cause of the seasons.) The variations of the moon's meridian altitude cause the size of this libration to vary concurrently.

b. The ellipticity of the moon's orbit causes it to travel faster near perigee than near apogee. But its period of rotation on its axis is constant. Therefore we see $7°.75$ "round the side," first at one limb, then at the opposite one, at two-week intervals.

c. As the earth turns, the position of an observer changes (by the length of the earth's diameter) every twelve hours, and he can see about 1° around, on alternate limbs of the moon, at these intervals.

d. There are also slight variations in the moon's own rate of rotation, probably results of slight irregularity of shape, and these bring a small zone into view at one limb or the other.

The combined result of all the librations enables us to view 59% of the moon's surface at one time or another. A period of nearly thirty years must elapse before the librations have brought all possible areas of this percentage into view. Finally, 41% of the moon's face is permanently hidden.

ECLIPSES AND TIDES

Nothing is strange, nothing impossible,
Nor marvellous, since Zeus the father of gods
Brought night to midday when he hid the light
Of the shining sun. Grim fear has smitten us,
And anything can happen to mankind.
Let no man marvel if he sees the flocks
Yield up their grassy pasture to the dolphins
And seek the salty billows of the deep,
Grown dearer to them than their native meadows,
While to the fishes sweeter seem the mountains.

ARCHILOCHUS
On the solar eclipse
of April 6, 648 B.C.

Beam of the Sun!
What wilt thou be about, far-seeing one,
O mother of mine eyes, O star supreme,
In time of day
Reft from us? Why, O why hast thou perplexed
The might of man,
And wisdom's way,
Rushing forth on a darksome track?

PINDAR, Paean IX,
On the solar eclipse
of April 30, 463 B.C.

Eclipses have traditionally struck terror into mankind. But when the arrangement of the solar system is understood, they can be seen as consequences of simple geometry. The earth and moon throw long conical shadows in sunlight: the earth's shadow causes eclipses of the moon; the moon's shadow, eclipses of the sun. Similar eclipses occur throughout the solar system for most planets with moons.

Eclipses of the Moon. The shadow thrown by the earth is 1,382,000 kilometers long on the average; between the greatest and least distances from earth to sun it varies in length by 22,500 kilometers. Refraction in the lower part of our atmosphere bends some light into the shadow cone, so that its edges are not perfectly sharp; the cloud layer near the earth's surface makes the shadow about 16 kilometers wider than it would be for a cloudless earth. The shadow of the earth is more than three times as long as the distance of the moon from the earth, and therefore it must strike the moon if it is pointing in the right direction, i.e., if earth, moon, and sun are in a line. At the distance of the moon the shadow is, on the average, more than 9170 kilometers wide, over two and a half times the diameter of the moon. The width of the shadow where the moon crosses it depends on the distance from earth to sun at the time, and also on the distance of the moon from the earth. Both these distances vary with the positions of earth and moon in their respective orbits, so the width of the earth's shadow at the moon's distance may be over 300 kilometers greater or less than the average.

The earth's shadow will point toward the moon at full moon when sun, moon, and earth are in a line, at or near the *nodes* of the moon's orbit. At other times the shadow falls north or south of the moon, and no eclipse takes place.

Because the sun is larger than the earth, the earth's shadow is a cone that points away from the sun; there is another "half-shadow," a diverging cone, within which only part of the sun is cut off by the earth. The complete shadow is the *umbra;* the partial shadow, the *penumbra.* Because of atmospheric refraction, the umbra and penumbra of the earth's shadow are not sharply divided.

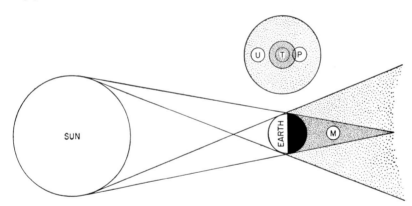

Fig. 6.1. Sun, earth, and moon at the time of a total eclipse of the moon. When the moon is inside the conical shadow of the earth, the eclipse is *total;* when it is partly inside, the eclipse is partial; when it is within the diverging partial shadow, the eclipse is *umbral.* The passage of the moon through the shadow is shown in the inset above, where total, partial, and umbral eclipses are marked *T, P,* and *U.* Dimensions and distances are not drawn to scale.

When the moon is full and near one of its nodes, the *umbra* of the earth will cover its face completely, producing a *total lunar eclipse.* Somewhat farther from the node, only part of the umbra will fall on the moon— a *partial lunar eclipse.* Still farther from the node, only the penumbra may fall on the moon (Figure 6.1). If the moon is more than about 10° from the node, there is no eclipse. As the moon and sun will *both* be near the nodes only twice a year, usually not more than two lunar eclipses occur in one year.

The moon's hourly motion in the sky is a little greater than its own diameter. As the earth's shadow at the moon's distance is more than two and a half times the moon's diameter in width, the whole moon may be in shadow (*totality*) for over an hour; shadow may cover part of the moon for about two hours (*partial phase*).

A lunar eclipse begins when the moon enters the penumbra—not a conspicuous sight, though the light is somewhat dimmed. As the moon enters the umbra, the edge of the earth's shadow passes over the moon's face, and we note that the shape of the uneclipsed crescent is different from the shape produced by the moon's phases (the former is bounded by a circle and the projection of a circle on a sphere; the latter, by a circle and an ellipse). The shadow is not completely dark, but of a coppery-red hue that differs from eclipse to eclipse. The color is due to light bent into the umbra by atmospheric refraction, and is reddish for the same reason that the setting sun is reddened. Differences in color at various eclipses are results of different conditions in the part of the atmosphere by which the light is refracted and scattered.

Eclipses of the moon are not nearly so interesting to astronomers as eclipses of the sun. Ironically enough, these less interesting eclipses require no eclipse expeditions; they are visible at the same moment from every part of the earth's surface where the moon is above the horizon at the time.

Eclipses of the Sun. Though the shadow of the earth cannot possibly miss the moon if sun, earth, and moon are exactly in line, the full shadow cone of the moon may very well miss the earth when sun, moon, and earth are lined up. The average length of the moon's shadow is 375,000 kilometers, less than the average distance (384,500 kilometers) from earth to moon. But the length of the moon's shadow differs with distance from the sun, and the distance from earth to moon differs because of the eccentricity of the moon's orbit. The longest possible shadow will occur when the sun is farthest (aphelion); the smallest distance of the moon from the earth's surface (when the moon is at the node and new) is 350,000 kilometers. Under these circumstances the moon's shadow is long enough to extend 29,300 kilometers beyond the earth's surface, and the shadow thrown on the earth by the moon is the largest possible. A *total eclipse* occurs within this shadow. The diverging cone of the penumbra will also throw a shadow on the earth, within which an observer sees part of the sun, hence a *partial eclipse* occurs. The greatest diameter that the shadow can have at the earth's surface is 269 kilometers, but if it strikes the earth obliquely, the projected shadow will be (nearly) an ellipse, with least diameter 269 kilometers, and largest diameter much greater.

More often than not, however, the shadow cone of the moon is too short to reach the earth's surface, and from places between its bounding lines, the rim of the sun is seen to form a bright ring around the obscuring moon—an *annular eclipse*. An eclipse may be total in part, if the cone *just* reaches the earth's surface at some place, and annular elsewhere. Annular eclipses are more frequent than total eclipses.

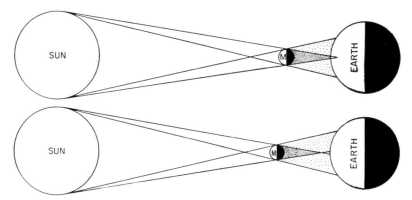

Fig. 6.2. Positions of sun, moon, and earth at the time of an eclipse of the sun. The upper figure shows a *total eclipse*, when the tip of the moon's conical shadow reached the surface of the earth; within the diverging shadow, part of the sun's disc is visible, and the eclipse is *partial*. The lower figure shows an *annular* eclipse; within the extension of the moon's conical shadow, the edge of the sun is seen all around the moon; within the surrounding diverging shadow, the eclipse is partial. Dimensions and distances are not drawn to scale.

The penumbra is much wider than the umbra, and a partial eclipse can be seen about 3000 kilometers on each side of the zone of totality.

The moon's shadow sweeps very fast across the earth's surface, and a total eclipse lasts a very short time. The speed of the shadow in space is about 3380 kilometers an hour, but this is partly compensated by the earth's rotation in the same direction, and the actual speed across the earth's surface may be 1680 kilometers an hour at the equator (much faster at higher latitudes). Totality lasts, therefore, longer at the equator than at higher latitudes. The longest possible total eclipse at the equator would last 7^m40^s; at latitude 45°, the greatest duration possible is 6½ minutes. An annular eclipse can last as much as 12^m24^s at the equator.

That the apparent sizes of the sun and moon are so nearly equal is an extraordinary, and very fortunate coincidence. At the most favorable total eclipse, the moon's apparent diameter is only 1′19″ larger than the sun's; at the least favorable annular eclipse, the sun's disc is 1′35″ larger than the moon's. Would the chromosphere and corona have been discovered if the moon did not so nearly cover the sun's face at the time of an eclipse?

The Prediction of Eclipses. Our present knowledge of the motions within the solar system makes possible the accurate prediction of the times of eclipses, the paths of the tracks on the earth's surface, and the duration of totality. But even before this detailed knowledge was available, the occurrence of eclipses was predicted empirically. The ancient Babylonians noticed

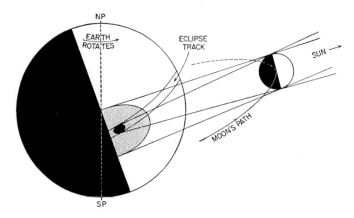

Fig. 6.3. The moon's shadow produces a total eclipse at the earth's surface. The moon catches up with and passes the sun in the sky, and the shadow moves across the earth's surface in the same direction. The earth rotates in the same sense, which causes an apparent backward displacement of the shadow. The speed of the shadow across the earth's surface is the difference between the true speed of the shadow and the speed with which the earth's surface moves under it. The shadow, which is circular in section, is shown projected obliquely on the earth's surface. The size of the shadow has been much exaggerated in the figure.

that eclipses recur near the same place at intervals of 18 years, 11⅓ days (the *Saros*), which is almost equal to 223 synodic months, to 19 "eclipse years" (the interval at which the sun passes the moon's ascending node, 346.62 days), and to 239 nodical months. At integral multiples of the synodic month the moon is *at the same phase* (it must of course be new at the time of a total solar eclipse); at integral multiples of the "eclipse year" the sun is *at the right position* for an eclipse to occur—near the moon's node. The ancient Babylonians are alleged to have predicted eclipses by means of the Saros, but this is probably a myth.

The occurrence of an eclipse of sun or moon at a given new or full moon depends on whether the shadow of earth or moon is directed exactly toward moon or earth. That is, it depends on how far the sun and moon are from the nodes of the moon's orbit. If the difference is greater than 18°31′, a solar eclipse cannot occur; if it is greater than 12°15′, a lunar eclipse cannot occur. If, on the other hand, it is less than 15°31′, a solar eclipse *must* occur; and a lunar eclipse is inevitable if the distance is less than 9°30′. For positions between these *ecliptic limits,* eclipses may or may not occur according to other circumstances. It may seem surprising that the ecliptic limits are smaller for lunar than for solar eclipses. But the shadow cone of the earth is a convergent cone. When the moon is on the rear side of the earth (lunar eclipse), the distance in its orbit within which it can suffer eclipse is 9170 kilometers; when the moon is on the near side of the earth (solar eclipse) the corresponding distance, within which it can produce an eclipse, is greater, about 15,000 kilometers (Figure 6.4).

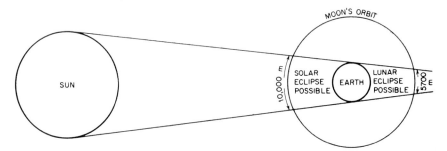

Fig. 6.4. Diagram to illustrate the reason for the greater width of the solar ecliptic limits than that of the lunar ecliptic limits. Eclipses (both solar and lunar) can take place when the moon is between the lines that bound the edges of sun and earth. The distances and dimensions are not drawn to scale.

Because the solar ecliptic limits are wider than the lunar, more solar than lunar eclipses occur *within a given time;* however, more lunar eclipses are visible in a given interval *from a given station* because lunar eclipses are visible from a whole hemisphere of the earth. The path of a total solar eclipse, on the other hand, is less than 300 kilometers wide, and even the partial phase is not visible from a whole hemisphere of the earth. There can be as many as seven eclipses in one year (five solar and two lunar, or four solar and three lunar); or there can be as few as two (both solar). Total solar eclipses are visible somewhere on earth about twice in every three years; but at any one place they happen, on the average, only once in 360 years.

The calculation of the time, place, and duration of a particular eclipse is not difficult in principle, but complicated in practice. The eclipse sweeps across the earth in a narrow band (the eclipse track), and both time of occurrence and duration must be calculated separately for each point on the track. The tracks of all total and annular eclipses between 1207 B.C. and A.D. 2162 have been calculated and mapped by Oppolzer. Early eclipses are very interesting in providing means of dating historical events.

Observation of Solar Eclipses. Unlike lunar eclipses, solar eclipses give results of great astronomical importance. Only rarely does an eclipse track cross an observatory, and many of the trivia of astronomy relate to the adventures of eclipse expeditions that have gone, often to remote parts of the earth, to make the observations.

At the best, an eclipse lasts only about seven minutes, and the work during those few minutes must be accurately planned and timed if the desired results are to be obtained.

A total solar eclipse is an impressive sight; the first sign of its incidence is the appearance of the edge of the moon's disc against the sun (*first*

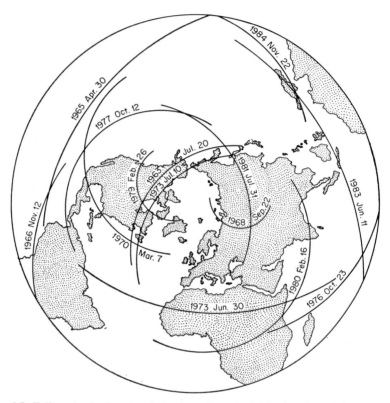

Fig. 6.5. Eclipse tracks for the sixth, seventh, and eighth decades of the twentieth century (from Oppolzer's *Canon der Finsternisse*).

contact). Gradually the sun's crescent grows smaller and the light fades into twilight. At the moment when the sun vanishes behind the moon (*second contact*) the corona flashes into view. The dark moon is rimmed with a pinkish glow—the chromosphere. All around the sun a gleaming halo—the inner corona—stands out like a crown of spun glass. And pearly streamers ring the sun in an immense aura—the outer corona. Perhaps a large prominence is poised above the sun's surface like a gigantic feather. Perhaps a comet, hitherto unseen, comes into view, or Mercury and Venus shine near the sun.

The few available minutes pass rapidly in observations. The rim of the sun reappears (*third contact*); the corona vanishes in the glare of sunlight. The sun's crescent grows, the last dent in the edge of the disc vanishes (*fourth contact*), and the eclipse is over. Throughout the eclipse, the shape of the sun's crescent is unlike that of the moon (circle and ellipse), or of the eclipsed moon (disc projected on sphere); it is the intersection of two circles.

The simplest observations at an eclipse are the determination of the exact moments of the four contacts, which give detailed information about the moon's position, and improve our knowledge of its motion. The exact width of the zone of totality, and the exact path, also give information about the moon's position and motion. At the moment of totality the disappearing crescent of the sun breaks up into a number of glowing points (*Baily's beads*), which give information about the exact shape of the mountains at the moon's edge.

Observation may be concentrated on the sky near the sun. Comets near perihelion are occasionally seen. Search has been made for a planet revolving within the orbit of Mercury, but none has ever been found.

The positions of stars near the sun may be photographed, with the purpose of testing the theory of relativity. Light is attracted toward a gravitating body, both according to the Newtonian theory of gravitation, and according to the theory of relativity. In each case the attraction of the sun on the light will pull it inward, and displace the star images apparently outward (Figure 6.6). The difference between the two theories lies not in

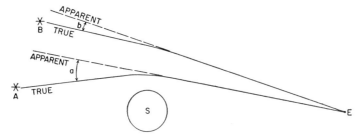

Fig. 6.6. Light rays are bent toward the sun by the sun's gravitational field. The effect is greatest for rays that pass nearest to the sun's edge. Therefore the light from star *A* is displaced through the angle *a*, greater than the angle *b* through which the light from star *B* is displaced. The angles are greatly exaggerated in the drawing.

the displacement, which both predict, but in its size. The positions of the stars, as displaced by the sun, must be compared with the positions of the same stars, measured if possible under the same conditions, when the sun is out of the way. This can be achieved by reobservation of the same region from the same spot with the same equipment several months later than the eclipse (when the sun is out of the way). The measurements are similar to those for the proper motion of stars, and require the utmost precision. All observers are not yet in agreement about the results, and probably the relativity shift will be a major eclipse problem for a long time to come. The observed displacement is too large to be in accord with Newtonian theory, and is nearer to that predicted by the theory of relativity; but the agreement is not yet considered good enough by all investigators to be quite conclusive.

The photography of corona and prominences has been a by-product of many eclipses. But the spectroscopic study of the sun during an eclipse is not only the most important eclipse problem, it is also the one that has contributed more than perhaps any one observation to our detailed understanding of the sun and, therefore, of all stars. When the flash spectrum was observed in 1874, both Lockyer and Janssen saw in it a bright line in the yellow region that had never been observed on the earth. Under the impression that it came from a substance peculiar to the sun, they gave it the name "helium" [Greek: *helios,* the sun]. Only after twenty years was the same gas found on earth, and we know now that it is the second commonest kind of atom in the universe. Other helium lines in the flash spectrum were ascribed to a hypothetical "cosmium." The bright lines in the spectrum of the corona, also, were long ascribed to "coronium," thought to be a substance otherwise unknown. Only during the past decade has this myth also been exploded, and the corona has been shown to consist of some of the commoner metals, but at a fantastically high "temperature" (p. 114). It is now known that some of the lines in the coronal spectrum occur in the spectrum of certain peculiar stars.

The spectra of chromosphere and corona can now be examined with the aid of the coronagraph, so the few minutes a year, or hours a century, previously available for the spectroscopic study of the outer layers of the sun can now be extended indefinitely. The spectra of the chromosphere and corona have been described in Chapter IV.

Other observations that can be made during an eclipse of the sun are: measurement of the changing intensity of light of various colors; study of the brightness of the corona, the heat that it reflects, and the polarization of its light; study of the effects of the eclipse on the earth in affecting the ionized layers of our atmosphere. The rippling "shadow bands" that run over the ground just as the shadow arrives give information about the effects of the churning motions of our atmosphere on the passage of the shadow edge. There have also been many interesting studies of the behavior of birds, animals, and plants as they react to the growing darkness, but these can scarcely be classed as astronomy.

Eclipses in History. If an eclipse is known to have occurred at a given place, and the date is recorded within a century or two, the event can be accurately dated from knowledge of the motions of the sun and moon. The average interval between solar eclipses at any one place is almost four hundred years.

Some authorities think that the earliest eclipse of which a record survives was the total solar eclipse of October 22, 2137 B.C., mentioned in *Historical Documents* of ancient China. From the same source it appears that eclipses were predicted in China by the seventh century B.C.; in those days astronomy had its occupational hazards:

> Being before the time, the astronomers are to be killed without respite; and being behind the time, they are to be slain without reprieve.

Confucius (551–479 B.C.) recorded thirty-six eclipses of the sun in the *Annals of Lu,* and all but four have been identified. The dates range from February 22, 720 B.C. to July 22, 495 B.C. The description of the eclipse of 720 B.C. is one of the earliest astronomical publications on record:

> In the 58th year of the 32nd cycle in the 51st year of the Emperor King-Wang of the Chou dynasty, the 3rd year of Yin-Kung, Prince of Lu, in the spring, the second moon, on the day called Kea-Tze, there was an eclipse of the sun.

We shall see how an eclipse of the moon, two weeks later, was recorded in Babylon. Early astronomical records from China and Japan have been recently found not only to confirm but to supplement Western knowledge (p. 394).

Euphratean and Greek literature contain many references to eclipses. Perhaps the passage in the *Odyssey* (xx, 356–7):

> The sun has perished out of the sky, and a thick fog spreads over all . . .

refers to an eclipse of the sun that was total in Ithaca about 1200 B.C., but the exact date is uncertain.

Ptolemy enumerated several eclipses of the sun and moon that had been recorded by Euphratean peoples. That of the sun on June 15, 763 B.C. is the famous "eclipse of Nineveh," recorded also on a surviving Assyrian tablet, and in the book of Amos. Three eclipses of the moon (March 19, 721 B.C., March 8, 720 B.C., and September 1, 720 B.C.), were recorded by Ptolemy, and one of them is described in a Babylonian fragment:

> To the king my Lord, thy servant Abil Istar: May there be peace to the king my Lord. May Nebo and Merodach to the king my Lord be favorable. Length of days health of body and joy of heart may the great gods to the king my Lord grant. Concerning the eclipse of the moon of which the king my Lord sent to me: in the cities of Akkad Borsippa and Nippur, observations they made, and in the city of Akkad we saw part.
>
> The observation was made and the eclipse took place . . . . And when for the eclipse of the sun we made an observation, the observations were made and the eclipse did not take place. That which I saw with my eyes to the king my Lord I send.

The record is of the partial eclipse of the moon, March 8, 720 B.C. It shows that the eclipses were predicted, and that an eclipse of the sun was looked for near the same date. It was visible not in Babylonia, but further east, and was recorded by Confucius, as we have seen.

The Egyptians, and probably the Greeks, are supposed to have learned the prediction of eclipses from the Babylonians. The knowledge is said to have been applied by Thales of Miletus (640–546 B.C.) on an occasion that is perhaps the earliest example of a familiar popular fiction situation (e.g., in *A Connecticut Yankee in King Arthur's Court* by Mark Twain and *King Solomon's Mines* by H. Rider Haggard). The eclipse was on May 28, 585 B.C., and the tale is told by Herodotus in his history, *The Persian Wars* (Book I, Chapter 74):

> Just as the battle was growing warm, day was on a sudden changed into night. This event had been foretold by Thales of Miletus, who forewarned the Ionians of it, fixing it for the very year in which it actually took place. The Medes and Lydians, when they observed the change, ceased fighting, and were alike anxious to have terms of peace agreed upon.

Babylonian and classical eclipses are a fascinating subject that links astronomy with history and chronology. Eclipses recorded by Thucydides and Plutarch have been used to fix or verify dates in classical history. The reference by Amos (7:9) to the eclipse of Nineveh suggests prediction:

> I will cause the sun to go down at noon, and I will darken the earth on a clear day.

It has been suggested that on the occasion when Joshua commanded the sun to stand still, the passage, "and the sun stood still and the moon stayed" (Joshua 10:13), may be rendered "and the sun ceased . . . ," and may refer to an eclipse.

Historical Eclipses and the Length of the Day. Eleven ancient eclipses can be identified with considerable certainty by the verisimilitude of the records.

ANCIENT ECLIPSES

Eclipse of Babylon:	July 31	1063 B.C.
Eclipse of Nineveh:	June 15	763 B.C.
Eclipse of Archilochus (p. 134):	April 6	648 B.C.
Eclipse of Thales:	May 28	585 B.C.
Eclipse of Pindar (p. 134):	April 30	463 B.C.
Eclipse of Thucydides:	August 3	431 B.C.
Eclipse of Agathocles:	August 15	310 B.C.
Eclipse of Hipparchus:	November 20	129 B.C.
Eclipse of Phlegon:	November 24	29 A.D.
Eclipse of Plutarch:	March 20	71 A.D.
Eclipse of Theon:	June 16	364 A.D.

When actual eclipse tracks are calculated for these eclipses on the basis of the current motions of sun and moon, they are found to deviate slightly to the eastward from the observations, and more so, the further back we go. This indicates that the moon has continually speeded up at a very slight rate, 10″.3 per century. The change is proportional to the *square* of the elapsed time, and amounts to nearly a degree over two thousand years, enough to shift the eclipse tracks appreciably. This apparent change in the moon's rate (the *secular acceleration of the moon's motion*) was first detected by Edmund Halley, the friend of Newton. It arises partly from planetary perturbations, partly by a steady increase in the earth's rotational period (the length of the day). The earth is slowing down, and the day increasing, by about $1/1000$ of a second a century. This slowing of the earth's rotation is caused by the braking effect of the tides.

9. THE TIDES

The association of the tides with the moon is obvious. In most seas there are on the average *two* high and *two* low tides in 24^h50^m, which is the average interval between successive passages of the moon across the meridian; high water occurs every 12^h25^m. The range of the tides is greatest at new and full moon (*spring tides*), least at first and third quarter (*neap tides*). There are tides in the land as well as in the ocean, but their range is much smaller, about 4½ inches, and they are not evident without measurement. With respect to land tides the earth displays a rigidity greater than that of steel (see p. 18).

The range of the tides varies widely. In mid-ocean it is about 2½ feet, even less in the Gulf of Mexico. In lakes it is an inch or less. In New York harbor it is about 5 feet; much larger tides occur on the coasts of Cornwall in England and Brittany in France, and the highest tides in the world are found in the Bay of Fundy (40 feet). These differences are largely due to the configuration of the coastlines.

The tide-raising force at a point on the earth's surface is the *difference* between the force exerted by the moon at the distance of the earth's center and the force exerted at the point. The moon pulls the near surface of the earth away from the center, and the center away from the far side of the earth. Therefore the tide-raising force is always *outward* along the line joining earth and moon; perpendicular to that line, it is inward. Thus it tends to distort the earth's shape, and raises a tide *toward* the moon, *and also on the opposite side* of the earth. The body of the earth, more rigid than steel, responds only slightly (land tides), but the fluid oceans are more affected.

Successive tides at one spot are not of equal height, and the difference is greater the smaller the moon's altitude. In some parts of the earth this difference (the *diurnal inequality*) may be so large that only one tide is appreciable. There would be little or no diurnal inequality if the earth's axis were not tipped relative to the plane of its orbit (Figure 6.7).

The sun raises tides as well as the moon. Though much more massive than the moon, the sun is so much more distant that its tide-raising force is only about one-third that of the moon. *Spring* and *neap* tides are caused by the interplay of the tide-raising forces of the sun and moon. At full and new moon, the two forces are in line, and tides are highest; at the quarters, the moon's tide-raising force is perpendicular to the sun's, and the lower neap tides represent the resulting compromise on the part of the oceans.

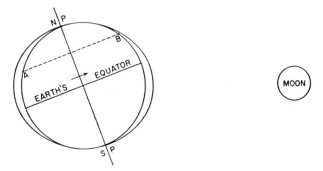

Fig. 6.7. The moon raises tides on opposite sides of the earth. Because the earth's axis is tipped, two successive tides at one place are of unequal height; after twelve hours, point *A* has moved to *B*, where the tide is lower.

Especially high spring tides occur when new or full moon coincides with perigee, for at that time the moon (because of greater proximity to the earth) exerts a tide-raising force 30% larger than when it is at apogee.

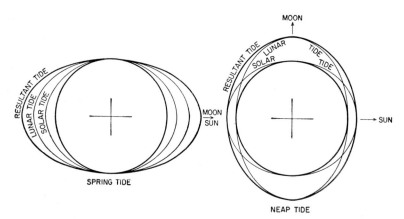

Fig. 6.8. Spring and neap tides are the result of the interplay of the tide-raising forces of sun and moon. The moon's tide-raising force is about twice the sun's. On the left, the tide-raising forces of sun and moon are in the same direction (new or full moon); the total range of the tides is greatest. On the right, the tide-raising forces of sun and moon are at right angles to one another (quarters of the moon), and the range of the tides is least. The heights of the tides are greatly exaggerated in the drawing.

At places on the coast, the time of high tide is ordinarily later than that of the moon's meridian passage by as much as six hours. This lag of the tides is a consequence of the irregular form and depth of the ocean bed, and of the time taken by the water to adjust itself to the rapid variations of the tide-raising force. Unlike the oceans, the body of the earth adjusts itself immediately: it is not merely more rigid than steel, but also more elastic (i.e., more resistant to distortion).

The friction of the tides on the ocean floor dissipates energy and robs the earth of energy of rotation. The rate of dissipation is about two billion horsepower. Most of the effect is produced in the Bering Sea, an appreciable fraction also in the Irish Sea, where the tides drive the water through narrow, shallow passages. The energy lost in a year, 4.75×10^{26} ergs, is only one ten-billionth of the earth's total rotational energy. Small as the loss is, it slows up the earth's rotation enough to lengthen the day by one-thousandth of a second a century. This change in the day causes the moon to forge ahead by $5''.8$ per century and the sun by $0''.75$ per century, which, in combination with the change in the moon's position by planetary perturbations ($6''.08$ per century) adds up very nearly to the observed secular acceleration of the moon's motion, as deduced from ancient eclipses.

As the moon raises tides on the earth, the earth must also raise tides on the moon; its tide-raising force (proportional to its mass) is eighty-two times as large as the moon's. If the rigidity of the moon were similar to the earth's, the land tide on the moon would be several feet. But very likely it is much larger. The moon always keeps the same face to the earth, which suggests that the earth raised huge tides in a partially plastic moon and forced its tidal bulge into a line directed toward the earth. If this tidal bulge remains on the solidified moon, the distortion might even be great enough for detection by radar.

As a consequence of the tides, steady, though very slow, changes must be taking place in the relationship between day and month. The moon (through the tides) is retarding the earth. Angular momentum (which results from energy of spin) cannot be destroyed, so the moon is being accelerated at the earth's expense, and its angular momentum in its orbit increased. The angular momentum is proportional to the square root of the radius of a circular orbit. Thus, as the angular momentum in the orbit increases steadily, so do the size of the orbit and the time taken to describe it. The moon must be receding from the earth, and the length of the month increasing.

Long ago (perhaps a hundred billion years) the moon is calculated to have been only nine thousand miles from the earth; the earth, with more angular momentum than now, spun faster, and the day was about five hours long. The month was a little longer. In the far future, similar calculations predict that the earth, constrained by tidal friction, will always turn the same face to the moon. Day and month will be equal, and will occupy forty-seven of our present days. But this stage cannot be reached before many millions of years have passed.

Theory looks even farther into the future, when the sun, through the tides, will in its turn begin to rob the earth-moon system of angular momentum. The moon will then draw nearer to the earth; the month will become shorter than the day; and the moon will finally be disrupted by tidal forces, and may end up as a ring like Saturn's. But the interval required for these changes to be completed is beyond the horizon of the astronomical time-scale.

CHAPTER VII

THE SOLAR SYSTEM

Around the ancient track marched, rank on rank
The army of unalterable law.
GEORGE MEREDITH

1. PLANETARY MOTIONS

That the naked-eye planets "wander" among the stars was one of the earliest astronomical observations. At first it was not understood. The ancient Greeks gradually saw its significance; indeed, Aristarchus of Samos realized that the sun is the central body of the solar system. But the tide of opinion ebbed, and a central earth held the field until Copernicus redis-covered the heliocentric system in the sixteenth century. In this chapter we shall trace the ebb and flow of ideas about the motion of the planets. For clearer understanding, we shall first look at the system from the modern standpoint.

The motion of a planet can be specified by the time it takes to make a circuit of the sky from one star back to the same star (the *sidereal period*). This is the true period of the planet around the sun. A planet's motion can also be described by the intervals at which the planet occupies the same position relative to the sun and earth (the *synodic period*).

MEAN DISTANCES AND PERIODS OF THE PLANETS

Planet	Distance from Sun (millions of kilometers)	Sidereal Period	Synodic Period
Mercury	58	88 days	116 days
Venus	108	225 days	584 days
Earth	150	365 days	...
Mars	228	687 days	780 days
Asteroids	220–850	1.76–13.7 years	394–845 days
Jupiter	779	11.9 years	399 days
Saturn	1428	29.5 years	378 days
Uranus	2872	84.0 years	370 days
Neptune	4501	164.8 years	367.4 days
Pluto	5915	248.4 years	366.7 days

When the directions (as projected on the plane of the solar system) from the earth to sun and planet coincide, the planet is in *conjunction* with

149

the sun. At conjunction, earth, sun, and planet are in a line. *Superior conjunction* occurs when the planet is on the far side of the sun; all planets can reach superior conjunction. When the planet is on the near side of the sun, between sun and earth, it is in *inferior conjunction;* only the planets with orbits smaller than the earth's (*inferior planets*) can reach inferior conjunction.

A planet seen on the opposite side of the sky from the sun is at *opposition.* Here again the projections of the three on the plane of the system are in line. Only the planets with orbits larger than the earth's can be in opposition (*superior planets*). The interval between superior conjunctions is the synodic period. Both conjunction and opposition are sometimes known as *syzygy.*

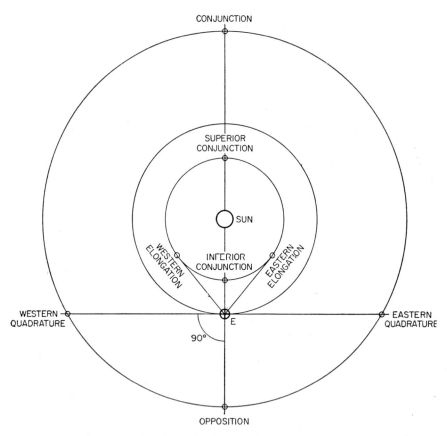

Fig. 7.1. Planetary configurations.

When the direction of a planet, as seen from the earth, makes an angle of 90° with the direction of the sun, the planet is at *quadrature* (east or west). Only superior planets can reach quadrature. Mercury can be 28°

from the sun at most, Venus, 48°. The angular distance of a planet from the sun at any time is called its elongation. Superior planets can have any elongation up to 180° east or west. The planetary configurations are illustrated in Figure 7.1.

Planets can also be said to be in conjunction with one another, or with the moon, and ancient (as well as modern) astrology ascribed great importance to these planetary conjunctions—a fact for which we should be grateful, because it led to early study of planetary positions and motions. Chaucer (versed in astronomy, as his *Treatise on the Astrolabe* shows) described a planetary conjunction with the moon in *Troilus and Creseide:*

> The bente Mone with her hornes all pale,
> Saturnus and Jove, in Cancro joyned were.

The time of this conjunction has been calculated, and used as a means of assigning a date to the poem.

When a planet is at conjunction, it is on the meridian at noon; when at opposition it crosses the meridian at midnight, and when it is at quadrature, meridian passage is at either 6 a.m. or 6 p.m.

The apparent complexity of the planetary motions is a result of the fact that as we watch them describing their orbits we are changing position in our own. When the planet has gained or lost one lap on the earth, one synodic period has been completed. Mercury catches up by one lap before the earth has covered one-third of a circuit; Neptune moves only 2° while the earth goes completely around. The earth and Mars have so nearly the same (sidereal) period that the earth requires nearly two years to gain a lap on Mars, and therefore Mars has the longest synodic period of all the planets.

The relation between the sidereal period (P) of a planet, the sidereal period (E) of the earth, and the synodic period (S) of the planet is simple:

$$1/S = 1/P - 1/E \quad \text{(inferior planet)}$$
$$1/S = 1/E - 1/P \quad \text{(superior planet)}.$$

Inferior planets, which we see illuminated by the sun at all angles, show phases like the moon, and for the same reason. Superior planets can be gibbous but not crescent.

The apparent motions of the planets against the narrow background of the zodiac are results of the extreme flatness of the solar system, which causes us to view their orbits very nearly edgewise, and of the fact that all the planets (the earth included) go around their orbits in the same direction, though with different speeds.

It is as though we were at a spot on the rim of a rotating wheel, looking at a spot on the rim of another wheel of different size, and both wheels were turning in the same direction on the same axle at different rates. If the two spots are being brought closer by the motion of the wheels, *from our standpoint* the other spot will seem to be approaching; if they are being brought farther apart, *from our standpoint* the other will seem to be receding. And it is from our standpoint that we view the motions of the planets.

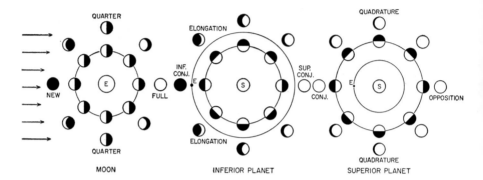

Fig. 7.2. Difference between the phases shown by the moon, and by the inferior and superior planets. The relative positions of the earth, sun, and moon or planet are shown in the interior of each figure. The phases of the moon or planet are shown in the outer circle, *as seen from the earth*.

The earth-moon system is supposed to be illuminated from the left by the sun. New moon occurs when the moon is between earth and sun, full moon when the earth is between sun and moon; half moon occurs at the quadratures.

For an inferior planet, "new" phase occurs when the planet is at inferior conjunction, "full" phase when it is at superior conjunction. The planet is *gibbous* at the quadratures, a crescent near inferior conjunction. The diagram does not show the changes of apparent size of the planet's disc as distance from the earth varies.

A superior planet is never seen at "new" phase; it is "full" at opposition and conjunction, slightly gibbous at or near quadrature.

The orbits *relative to the earth* are a series of loops. The planets spend more time apparently going in the same direction as the earth, eastward (*direct* motion), than in the opposite direction (*retrograde* motion).

At *a* in Figure 7.3, Jupiter is in superior conjunction; its (direct) motion is eastward; as it loops inward, it seems to turn west (retrograde motion) through opposition, *b,* and then turns eastward again, completing one synodic period between *a* and *c.*

For Venus, superior conjunction is at *d* in Figure 7.4, inferior conjunction (retrograde motion) at *e,* and one synodic period is completed between *d* and *f.*

We view the loops on edge. If the orbits of all the planets were exactly in the plane of the ecliptic, their positions would seem to oscillate back and forth in the ecliptic, with a preponderantly direct motion. But they are slightly inclined to the ecliptic (see p. 180): Venus by 9°, Mars by 7°, Mercury by 5°, the others even less. Asteroids may have much larger inclinations—40° for Eros, for example. In consequence we see projections of the loops, and at various times the planets seem to describe paths with flattened loops or kinks.

The superior planets always move westward *relative to the sun,* because the earth has a smaller orbit and a greater angular speed. Therefore they rise earlier each night than the previous night. They do, however, move eastward most of the time *relative to the stars,* fastest near superior conjunction; near opposition their (retrograde) motion is westward among the stars.

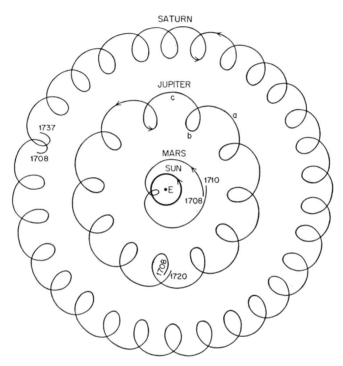

Fig. 7.3. Apparent orbits of Saturn, Jupiter, Mars, and the sun, *on the supposition that the earth is at rest.*

When a superior planet is at superior conjunction, it is on the far side of the sun. It moves *westward* of the sun, soon appears before dawn as a morning star, and passes toward quadrature, when it rises at midnight. As it moves toward opposition, it rises earlier in the night, and at opposition it rises at sunset; later it moves toward eastern quadrature, when it is on the meridian at sunset and rises at midday. Thereafter it continues to move west with respect to the sun and is an evening star until it disappears in the twilight, and it finally completes its synodic period by rising again with the sun (conjunction).

An inferior planet never reaches quadrature or opposition. It seems to oscillate about the sun, westward swing (relative to the sun) through inferior conjunction being made faster than eastward swing near superior conjunction. After superior conjunction it is an evening star; it swings away

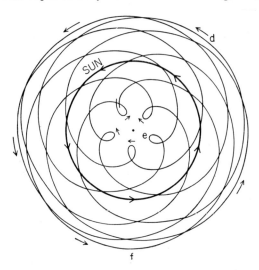

Fig. 7.4. Apparent orbits of Venus and the sun, *on the supposition that the earth is at rest.* The apparent orbit of Mercury is similar, but contains many more loops.

from the sun until it reaches the greatest eastern elongation permitted by the size of its orbit. The sun then begins to gain on it in its eastward path around the ecliptic. It speeds toward the sun in the ecliptic, passes through inferior conjunction, and thereafter becomes a morning star; it then moves rapidly to western elongation, and completes a synodic period by catching up with the sun again at superior conjunction.

It is difficult enough to describe the observed motions of the planets when we start from a knowledge of the actual arrangement of the solar system. Small wonder that the process of reconstructing the system from the observed motions was a slow one. The next few sections contain an outline of the steps by which it was effected.

2.   CLASSICAL GREECE AND THE SOLAR SYSTEM

We have seen that Thales of Miletus (640–546 B.C.) successfully predicted a solar eclipse, probably from knowledge of the *Saros* derived from the Babylonians. He probably also realized that it was caused by the moon coming between earth and sun, for he thought the sun an "earthy body" like the moon. He cannot have understood lunar eclipses correctly, for he believed the earth to be a disc or cylinder floating in water, and the heavens a hemisphere above the floating earth, that moved around, not under it, when they set. The upper hemisphere, also, was supposed to be surrounded by water. The same picture occurs in the first chapter of Genesis: "And God made the firmament, and divided the waters which were under the firmament from the waters which were above the firmament."

Thales determined the solstices and the equinoxes. He also made an excellent estimate of the apparent diameter of the sun, as recorded by Apuleius:

> The same Thales in his declining years devised a marvellous calculation about the sun, which I have not only learned but verified by experiment, showing how often the sun measures by his own size the circle which he describes. Thales is said to have communicated his discovery soon after it was made to Mandrolytus of Priene, who was greatly delighted with this new and unexpected information, and asked Thales how much by way of fee he required to be paid to him for so important a piece of knowledge. "I shall be sufficiently repaid" replied the sage, "if when you set to work to tell people what you have learned from me, you will not take credit for it yourself but will name me, rather than another, as the discoverer."

The estimate, 1/720 of the circle, or half a degree, is nearly correct. It is not only in modern times that men of science are jealous of scientific credit.

Evidence that might have suggested the sphericity of the earth to a man with sufficient imagination, was incredulously recorded by Herodotus (b. 484 B.C.), in his description of the expedition sent out by Necho (610–594 B.C.) (*The Persian Wars,* Book IV, Chapter 42):

> Libya . . . is washed on all sides by the sea, except where it is attached to Asia. This discovery was first made by Necos, the Egyptian king, who . . . sent to sea a number of ships manned by Phoenicians, with orders to make for the Pillars of Hercules [Gibraltar], and return to Egypt through them, and by the Mediterranean. The Phoenicians took their departure from Egypt by the Erythraean [Red] Sea, and so sailed into the southern Ocean. When autumn came they went ashore, wherever they might happen to be, and having sown a tract of land with corn, waited until the grain was fit to cut. Having reaped it, they again set sail; and thus it came to pass that two whole years went by, and it was not till the third year that they doubled the Pillars of Hercules, and made good their voyage home. And they said a thing that does not seem probable to me (though others may believe it), that when going round Libya they had the sun on their right hand.

The picture evolves in the hands of Anaximander (611–547 B.C.), pupil of Thales. He envisaged the earth as a flat cylinder, with its depth one-third of its breadth, suspended freely in the center of the universe. Man occupied one of the flat faces of the cylinder.

The cosmogony of Anaximander has a modern ring; he imagined a boundless First Principle, and thought that eternal motion was the cause of the generation and destruction of things. He thought the stars were a fiery sphere, seen through pinholes in the firmament. The sun, to him, was a fiery wheel twenty-eight times the diameter of the earth, borne round it on a wheel twenty-seven times the earth's diameter in length. The moon, he thought, was nineteen times the size of the earth. Eclipses took place when the opening of the wheel was stopped up.

Anaximenes (585–528 B.C.) pictured the earth as a flat table floating on air, and the sun flat like a leaf. He believed that sun, moon, and stars had evolved from the earth and that the former were made of fire. The stars, he thought, could not be hot because of their great distance; he pictured them as studs attached to a crystal sphere. He seems to have been the first to recognize that the stars are farther from the earth than the sun; he also distinguished planets from stars, and thought they were "floating" like the sun. He seems to have understood correctly the cause of a lunar eclipse.

The second giant of classical astronomy was Pythagoras (572–492 B.C.). He seems to have been the first to maintain that the earth is a sphere, perhaps because he thought the sphere the perfect figure; he maintained that other celestial bodies are also spherical. He imagined the earth as being central in the universe and at rest. The fixed stars turned from east to west on an axis through the poles of the earth, and the planets moved independently from west to east. He is said to have been the first to recognize that the morning and evening stars are identical (p. 12); the discovery was certainly made in his time.

Pythagoras made great contributions to geometry, and might be called the originator of the theory of numbers. His discovery of the relation between the frequency and pitch of musical notes, which he is said to have made from noticing the ringing tones of a blacksmith's hammer, not only began the theory of vibrations but also led to the idea of musical harmony underlying the motions of the heavenly bodies—the "music of the spheres." His influence on subsequent science was enormous.

By contrast, the views of Heraclitus, the "weeping philosopher" (544–504 B.C.), are a regression. He pictured the sun as re-created daily; he thought the sun and moon bowl-shaped, the phases of the moon being results of the turning of the bowl, and eclipses due to their turning their convex sides in our direction. He imagined the sun to be only a foot in diameter! But his famous generalization, "all things are in flux," suggests the modern statistical view of the properties of matter.

Anaxagoras (b. 500 B.C.?) was an astronomer to the core. The object of being born, he said, is "the investigation of sun, moon, and heaven." He had an extraordinary common-sense grasp of the solar system. He was perhaps the first to realize the nature of moonlight and to explain lunar eclipses correctly, but it is doubtful whether he understood the moon's phases. The moon, he said, is of an earthy nature, has plains and ravines, and shines with a false light. The sun is a red-hot mass or a stone on fire, and it is larger than the Peloponnesus! In his speech in his own defense at his trial for impiety (399 B.C.), Socrates says that he is accused of maintaining "that the sun is a stone, and the moon earth." "You think," he says, "that you are accusing Anaxagoras," but he denies any knowledge of, or interest in, such matters.

Empedocles (484–424 B.C.) correctly understood the cause of night, and the occurrence of eclipses. His most astonishing flight of imagination was, however, his speculation that light must travel with finite velocity. Actual evidence of the finite velocity of light was not obtained until over two thousand years after his time (p. 211).

Practical knowledge of the solar system (as distinct from theories) was advanced by Oenopides (soon after Anaxagoras), who detected the obliquity of the ecliptic.

Growing knowledge of the actual motions of the planets led, in the fourth century B.C., to the formation of complicated geometrical pictures for embodying them. Eudoxus (408–355 B.C.) imagined an elaborate system of concentric spheres, which by their regular motions produced the planetary configurations. Three were required for the sun, three for the moon, four for each known planet, one for the starry heaven—twenty-seven in all. Aristotle (384–322 B.C.) modified the picture of Eudoxus into a compact mechanical model which required fifty-five spheres in all to account for the planetary motions. The earth was at the center in both these models.

Heracleides of Pontus (b. 388 B.C.?) made a step toward the heliocentric idea. "The stars of Mercury and Venus," he said "make their retrograde motions and retardations about the rays of the sun, forming by their courses a wreath or crown about the sun itself as center. It is also owing to this circling that they linger at their stationary points." But though he understood the inferior planets, it is doubtful that he realized that the same is true of the superior planets. A very similar picture was adopted two thousand years later by Tycho Brahe. Heracleides recognized clearly that the earth rotates on its axis.

The third giant of Greek astronomy was Aristarchus of Samos (310–230 B.C.), who came seventy-five years later than Heracleides and was twenty-five years older than Archimedes. He stands out as an observer who saw the problems of the solar system geometrically, and who set out to solve them by measurement.

Aristarchus' method for measuring the distance of the sun was perfectly correct in principle. He reasoned that at half moon, the directions from sun to moon and earth to moon form an exact right angle. The distance from earth to sun is therefore the hypoteneuse of a right-angled triangle. The angle between the directions from earth to moon and earth to sun suffices to determine the shape of the triangle, and therefore the distance of the sun in terms of the distance of the moon. Aristarchus determined the angle sun-moon-earth to be 87°, whereas it is actually 89°51′; hence his determination (which depended on the difference between the measured angle and 90°) was greatly in error. The difficulty lies in selecting the exact time of half moon. But the principle of his method shows a precise grasp of the geometry of the problem, and his book, *On the Sizes and Distances of the Sun and Moon,* is the first important astronomical treatise. Besides measuring the ratio of the distances of the sun and moon, he used his knowledge of eclipses to make a calculation of their actual sizes. We shall quote several of the eighteen propositions in the treatise. It must be remembered that Aristarchus based his results on rough estimates. There were no trigonometrical tables in his day; trigonometry had not been invented! Even the value of π was only roughly known; Archimedes later obtained the value 22/7.

The hypotheses and some of the conclusions of Aristarchus are as follows.

HYPOTHESES:

1. *That the moon receives its light from the sun.*

2. *That the earth is in the relation of a point and center to the sphere in which the moon moves.*

3. *That, when the moon appears to us halved, the great circle which divides the dark and bright portions of the moon is in the direction of our eye.*

4. *That, when the moon appears to us halved, its distance from the sun is then less than a quadrant by one-thirtieth of a quadrant. (This is the statement of the measured 87° between sun, moon, and earth.)*

5. *That the breadth of the earth's shadow at the moon, is that of two moons.*

6. *That the moon subtends a fifteenth part of a sign of the zodiac (i.e., 2°, actually four times the true value).*

PROPOSITIONS:

7. *The distance of the sun from the earth is greater than eighteen times, and less than twenty times, the distance of the moon from the earth.*

8. *The diameter of the sun is greater than eighteen times, but less than twenty times, the diameter of the moon. (This conclusion follows from a previous one and from the fact that the moon exactly covers the sun at a total eclipse.)*

9. *The sun has to the moon a ratio greater than that which 5832 has to 1, but less than that which 8000 has to 1 [in volume].*

The result, that the sun is 6¾ times the earth's size, is of course far below the true value, 108.9; the value obtained for the moon's diameter, 0.36 of the earth's, is much nearer the true value, 0.27. Even the size for the sun is a great improvemnt on Anaxagoras' "larger than the Peloponnesus."

Aristarchus was, moreover, the first to state a heliocentric hypothesis definitely. Archimedes wrote that "Aristarchus brought out a book consisting of certain hypotheses. . . . His hypotheses are that the fixed stars and the sun remain unmoved, that the earth revolves about the sun in the circumference of a circle, the sun lying in the middle of the orbit." When Copernicus enunciated the heliocentric idea, he knew, and mentioned, that Aristarchus had suggested it.

The next giant of Greek astronomy was also primarily an observer. Hipparchus, who worked at Rhodes in 140–120 B.C., was attracted to astronomy by observing a new star, and compiled a catalogue of stars that has come down to us, through Ptolemy, in the form of the "Almagest." He discovered the precession of the equinoxes and made a very good estimate of its rate (p. 37). He also noted the evection of the moon (p. 130).

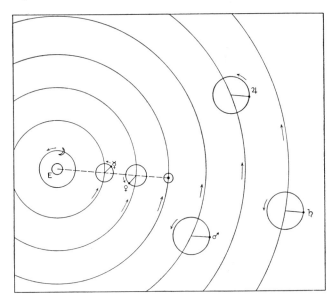

Fig. 7.5. The Ptolemaic system (not drawn to scale). Each planet, as well as the sun, was supposed to travel around the earth once a year on a circular orbit, the *deferent*. The five planets, in addition, were thought to move in circular orbits (*epicycles*) round a point that moved steadily round the deferent. Mercury and Venus were supposed to remain always in a straight line between earth and sun. For the three outer planets, the line that joined the center of the epicycle and the planet was thought always to remain parallel to the line joining earth and sun.

Classical astronomy reached its peak with Aristarchus and Hipparchus. Ptolemy (second century A.D.) revised the star catalogue of Hipparchus and collected older observations of eclipses. He did not, however, accept the idea that the sun is at the center of the solar system. He worked out the elaborate system of epicycles (circles rolling on circles) that held sway

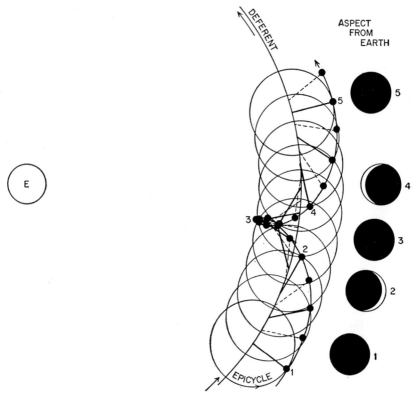

Fig. 7.6. Illustration of the way in which a combination of deferent and epicycle reproduces the apparent (geocentric) motion of a planet. The epicycle moves along the deferent, turning like a wheel. The line that joins the center of the epicycle to the position of the planet sweeps out the looped path of the latter. Only alternate positions of the epicycle are drawn.

The figure to the right shows the phases, *for an inferior planet,* that would result from motions in the Ptolemaic system. The sun's orbit is *outside* the orbit on which the deferent of an inferior planet moves (see Figure 7.5), and the center of the epicycle moves so that it is always in a straight line between sun and earth. When the line from the center of the epicycle to the planet—the spoke of the wheel—is parallel to the line earth-sun (positions 1, 3, 5) the planet will be at "new" phase. At positions 2, 4 the line sun-earth is slightly oblique to the line sun-planet, and the planet will therefore appear as a thin crescent. Notice that, on the Ptolemaic system, *an inferior planet can never be seen except as a thin crescent,* as its deferent always keeps it between sun and earth. Galileo's observation, that Venus shows phases *like the moon* (i.e., from thin crescent to full) at different points of its orbit, therefore showed that the Ptolemaic system was incorrect.

over the astronomical world for 1200 years under the name of the "Ptolemaic system." It resembled the systems of Eudoxus and Aristotle in having the earth at the center, but it departed from them in using circles instead of spheres.

It must be remembered that (with proper refinements, to take account of perturbations) a geocentric system can give just as good a *formal* picture of the motions within the solar system as a heliocentric one. As a schematic picture, it has the disadvantage of being far more complicated. It is only when the motions are being not merely described, but also understood in terms of gravitation, that it becomes natural to consider the most massive body in the system as the central one. The Ptolemaic system, however, did not give a correct formulation of the geocentric motions of the planets.

After the time of Ptolemy, the tide of scientific thought ebbed in the world of Europe, and throughout the Middle Ages it was kept flowing by the Moslems. Through them the catalogue of Hipparchus and Ptolemy was preserved, and received the name "Almagest." Few European scholars contributed to astronomy during this time. Soon after 1200 A.D. Roger Bacon began to urge the importance of observation and experiment. He called attention to the need for correcting the Julian calendar, and he is even thought to have had the idea of a telescope. However, astronomy did not flower again in Europe until the Renaissance.

3. COPERNICUS

At the close of the Middle Ages, the accepted European idea of the solar system was geocentric. The tremendous influence of Aristotle and the Ptolemaic system (which, like the Aristotelian, was centered on the earth) were transmitted and consolidated by the authority of Saint Thomas Aquinas (1225–1274). The frescoes of Daddi and Gaddi (fourteenth century) in the Spanish Chapel of Santa Maria Novella in Florence show the temper of the times by setting Thomas Aquinas enthroned above the sciences. The scope of astronomy at the time is typified by the figure of the astronomer (Frontispiece) sculptured on the wall of Giotto's Campanile by Andrea Pisano (1270–1348); he holds a little quadrant, and reminds us that sighting instruments were the only means of precise study of the stars, the telescope being two hundred years away.

The *Divine Comedy* shows the astronomical picture as it appeared to an educated man at the beginning of the fourteenth century. Dante's view is essentially Thomist, and the spheres of the sun, moon, and planets in the *Paradiso* are those of the Ptolemaic system. Dante loved the stars; each of the three books of the *Divine Comedy* ends with the word *stelle* [stars].

He takes pains to get the details correct for the date on which he begins
his journey (Good Friday, April 8, 1300):

> The fair planet that incites to love
> Was making all the east to laugh,
> Veiling the Fishes that were in her train.

Apart from the sheer beauty of its poetry, the *Divine Comedy* attracts us
because it reveals the contemporary view of the universe: the earth is
stationary and central; man resides in the northern hemisphere, of which
Jerusalem is the center; and purgatory is at the antipodes of a spherical
earth.

Leonardo da Vinci was not only a great artist but also a practical scientist. "Those sciences are vain and full of errors," he wrote, "which are not
born from experience, the mother of all certainty, and which do not end
with one clear experiment." He foreshadowed the laws of motion and
formulated the principle of inertia: "Nothing perceptible by the senses is
able to move itself . . . every body has a weight in the direction of its
movement." This is in contrast to the statement of Thomas Aquinas:
Movetur igitur corpus celeste a substantia intellectuale; * compare Dante's
Amor che muove il sole e l'altre stelle.† Leonardo knew that falling bodies
are accelerated, but not the correct law of acceleration. He thought of the
heavens as a celestial machine governed by definite laws and realized that
the earth is a heavenly body that would reflect light like the moon.

Shakespeare, always a touchstone for contemporary ideas, speaks in
Pythagorean terms:

> There's not the smallest orb that thou behold'st
> But in his motion like an angel sings

refers to the "music of the spheres." He makes Hamlet a student at Witten-
berg, where Tycho Brahe actually studied in 1570, and he must have seen
Tycho's nova of 1572, which was, in a sense, the event that dates the beginning of modern astronomical observation. But the modern theory of the
solar system had already been formulated by Copernicus.

Nicolaus Koppernik (Copernicus) was born at Thorn in Polish Prus-
sia, in 1473. A Doctor of Arts and Medicine, he spent some time in Rome
as professor of mathematics, and returned to become a canon at Frauenburg
on the Vistula.

Copernicus compiled tables of the planetary motions that remained
useful until they were superseded by Tycho Brahe's careful measures. In
about 1507 he became convinced of the heliocentric picture of the solar
system. He was aware that the planets traverse different parts of their orbits
at different speeds and ascribed the effect to the orbits being circular but

---

* Therefore the body of heaven is moved by an intellect.
† Love, that moves the sun and the other stars.

not quite concentric. (That it is the result of ellipticity was first realized by Kepler.) He also knew that precession is a conical motion of the earth's axis, after the existence of precession had been known for almost two millenniums. He seems even to have thought of universal gravitation: he suggested "that gravity is not an influence of the whole earth, but is a property of its substance, which, it is thinkable, may extend to the sun, moon, and other stars."

Copernicus' first treatise was written about 1530. The contemporary pope (the unfortunate Clement VII, Giulio de' Medici, d. 1534) approved of the work and asked for a complete presentation, which was finished in 1540 and printed in 1543; Copernicus received it on his deathbed. He dedicated it to Pope Paul III (1534–50), and a cardinal paid for the printing. He did not fear religious criticism: "If there be some babblers," he wrote, "who, though ignorant of mathematics, take upon them to judge of these things, and dare to blame and cavil at my work, because of some passage of Scripture which they have wrested to their own purpose, I regard them not, and will not scruple to hold their judgment in contempt." Indeed, during the lifetime of Copernicus his work received the approval of the Church. Not until 1616 was it declared "false, and altogether opposed to Holy Scripture," and placed on the Syllabus Errorum.

At the end of the sixteenth century the theory of Copernicus was warmly, if not hotly, upheld by Giordano Bruno, who welcomed its aid in his attack on Aristotle. Bruno was imprisoned, excommunicated, and burned at the stake in 1600, and scientific men of succeeding generations cannot have been unmindful of his fate.

4. TYCHO BRAHE

The first modern observer, Tycho Brahe, was born in 1546, three years after the death of Copernicus. He was a Danish nobleman, a fantastic person scornful of tradition. When only fourteen he saw an eclipse of the sun, became interested in astronomical instruments, and began to construct quadrants and sextants. At that time the measurement of position was the only branch of observational astronomy. The new star of 1572 stimulated him to build an observatory—the first in modern Europe. Uraniborg [the city of the heavens] was built on the island of Hven, off the Swedish coast, in 1576; its ruins still stand there.

Tycho undertook the study of planetary motions, and constructed the most accurate instruments that were possible at the time. He took great precautions in making observations and was thoroughly modern in his attempts to avoid and evaluate errors. His magnificent series of planetary observations made possible Kepler's study of the laws of planetary motion, the basis of our modern picture of the solar system.

Although Tycho was aware of the heliocentric ideas of Copernicus, he did not accept them. He believed that the earth was the stationary center of the solar system; its motion, he argued, is not felt and is difficult to picture. Moreover, he saw that the stars should show annual parallax if the earth has orbital motion; he did not realize that his failure to detect the parallax was a result of the great distance of the stars and the consequent extreme smallness of the effect. He pointed out (as Copernicus had done) that Mercury and Venus should show *changing phases* if the sun is at the center of the system; we now know that they display such phases, but Tycho could not detect them. He argued that if the earth is in motion, a stone should not fall vertically; we know now that it does not (p. 34). Lastly, Tycho was an orthodox religious man, reluctant to depart completely from the Thomist formulation of science. His picture of the solar system (the Tychonic system) showed the sun, moon, and superior planets carried around the earth, but with Mercury and Venus going in orbits around the sun (Heracleides of Pontus had held the same view two millenniums before).

Tycho's imperious character caused offense on the island of Hven, and he was forced to abandon his observatories, Uraniborg and Stjärneborg [the city of the stars]. He was called to Prague by Rudolph II of Bohemia, and there spent the remainder of his life in the compilation of tables of planetary motion (called the "Rudolphine Tables" after his patron). In his later years he was assisted by Johann Kepler, who thus absorbed the facts of planetary motions and prepared himself to formulate their theory.

The published works of Tycho fill many volumes, which contain astronomical observations, mathematical studies, and a generous sprinkling of horoscopes. Astrological predictions were a reputable part of the astronomer's work in the sixteenth century, and accounted for much of his revenue.

5.   KEPLER

Johannes Kepler (1571–1630) was twenty-five years younger than Tycho, and seven years younger than Galileo. He was the son of a poor man, and attended a charity school. In his twenties he was interested in the planetary motions, and published a theory based on the forms of the regular solids. Fuller knowledge of the facts destroyed this formal picture after he began to work with Tycho in 1599.

Kepler worked primarily with Tycho's observations of Mars, and he devoted enormous labor to improving the Copernican picture of circular motion in an orbit not quite centered on the sun. In 1609 he announced the first two of his three laws of planetary motion, and the third in 1618. Kepler's Laws are one of the landmarks of astronomy.

1. The planets move in ellipses, with the sun at one focus.

2. The radius vector sweeps out equal areas in equal times.

3. The square of the time of revolution is proportional to the cube of the mean distance.

The first law describes the *shapes* of the orbits; the second, the *varying speed* of motion in an orbit; and the third, the relation of the *size* of the orbit to the time of revolution. Kepler's formulation was *empirical;* his laws describe the orbital motion of the planets perfectly, but they do not account for it—that was left to Newton, fifty years later. Kepler's Laws have been completely verified and still form the basis of our study of orbital motion, not only within the solar system, but also among stars and stellar systems. We should remember that without the careful observations of Tycho Brahe they would not have been formulated.

Kepler, like Tycho, derived much of his income from computing horoscopes and astronomical almanacs. Another idea of the times in which he lived is given by the fact that his mother was tried for witchcraft (1615–21) and ably defended by her son. The redoubtable old woman (who was 69 at the beginning of her ordeal) was finally acquitted after a harrowing examination.

6. GALILEO

The son of a noble Florentine, Galileo Galilei was born in 1564, the year that Michelangelo died; he died in 1642, the year that Newton was born. At seventeen he was a student of medicine, but the works of Euclid and Archimedes induced him to abandon medicine for mathematics; at twenty-six he was a professor of mathematics at a salary small compared to that of a professor of medicine.

Galileo was a good Catholic all his life, and his first great discovery was made in church! He noticed the regular swing of a lamp in Pisa Cathedral, timed it by his pulse beats, and discovered the isochronism of the pendulum. The first use to which he put the discovery was the timing of pulses; it was Huyghens who applied it to the regulation of clocks. The lamp, a masterpiece of Benvenuto Cellini, is still shown suspended in the Cathedral at Pisa.

The laws of falling bodies were first experimentally studied by Galileo; the Aristotelian view that the rate of fall depends on the weight of the body had not before been experimentally tested. Galileo, in the famous experiment that he made from the top of the Leaning Tower of Pisa, showed that a 100-lb weight and a 1-lb weight, if dropped simultaneously, reach the ground at the same moment. He extended his study of falling bodies by means of inclined planes, which slow the fall so that it can be measured, and reached the important conclusions:

1. The final velocity acquired is independent of the angle of slope, but depends on the vertical height through which the body falls.

2. The height through which the body falls is proportional to the square of the time.

3. All bodies fall at the same rate.

Galileo's laws of falling bodies, like Kepler's laws of planetary motion, are *empirical;* Newton extended them in his discussion of the laws of motion.

Galileo was interested in the planetary motions, and received with pleasure Kepler's present of a copy of the latter's *Mysterium Cosmographicum.* "I have been for many years an adherent of the Copernican system," he wrote. "I have collected many arguments for the purpose of refuting [the commonly accepted hypothesis], but I do not venture to bring them to the light of publicity, for fear of sharing the fate of our master Copernicus, who . . . has become the object of ridicule and scorn . . . I should certainly venture to publish my speculations if there were more people like you. But this not being the case, I refrain from such an undertaking." Even with Kepler's encouragement he was reluctant to publish; caution and humility marked him through life.

A new star appeared in 1604, and Galileo lectured on it (in Italian, not in the scholars' Latin); but he was impatient of the greater public interest in an ephemeral novelty than in the fundamental facts of nature. He realized, however, that the new star had dealt a final blow to the Aristotelian doctrine of the immutability of the starry heaven. He now began to speak and write publicly of the Copernican system.

The greatest contribution of Galileo to astronomy, however, was his application of the newly discovered telescope to observation of planets and stars. He made a little refractor (with a convex objective and a concave eyepiece, which gave an erect image) that magnified three times. We recall that Milton, who visited the aging Galileo at Arcetri, called it the "glazed optic tube." Galileo had to make his own lenses; his largest telescope magnified thirty times.

Galileo published fascinating illustrated accounts of his observations with the telescope. His study of the moon revealed the familiar surface features, and he recognized the lunar mountains and made rough measures of their heights. He understood earth-light correctly: the earth, he said, is brighter than the moon because its surface is cloudy. This is perfectly correct—as we have seen, the earth's albedo is about five times that of the moon.

To his amazement he observed the four brightest moons of Jupiter, and he showed that they were in orbital motion around the planet. Even Kepler, when he first heard of it, was incredulous: "we roared with laughter without stopping," he wrote, but the facts soon convinced him. Public reaction was less reasonable: many disbelieved; some people even refused to look and see for themselves. Galileo is said to have expressed the hope that one skeptic had seen the moons on his way to heaven!

The little telescope was not perfect enough to reveal the true nature of the rings of Saturn, but Galileo noted that the image of the planet was out of the ordinary: *Ultimam planetam tergeminum observavi,* he wrote [I have observed the farthest planet to be triple]. In 1610 he described in more detail what he saw:

> I have observed with great admiration that Saturn is not a single star but three together, which, as it were, touch each other. They have no relative motion . . . the middle being much larger than the lateral ones . . . Saturn has an oblong appearance, almost like an olive, but by employing a glass that multiplies the superficies by more than 1000 times, the three globes will be seen very distinctly and almost touching . . . I have already discovered a court for Jupiter and now there are two attendants for this old man, who aid his steps and never leave his side.

But two years later he received a shock; the ring was presented on edge and was invisible to him. He wrote with characteristic humility and caution:

> Looking at Saturn within these last few days, I found it solitary and without its accustomed stars, and in short perfectly round and defined like Jupiter. Now what can be said of so strange a metamorphosis? Are perhaps the two smaller stars consumed like spots on the sun? Have they suddenly vanished and fled? Or has Saturn devoured his own children? Or was the appearance indeed fraud and delusion? . . . Now perhaps the time is come to revive the withering hopes of those who . . . have fathomed all the fallacies of the new observations and recognized their impossibility . . . The shortness of time, the unexampled occurrence, the weakness of my intellect, the terror of being mistaken, have greatly confounded me.

We know now that Saturn presents its rings to us at different angles as it goes around its orbit, and that in 1612 the rings were presented edgewise toward the earth. Five years later Galileo was relieved to see Saturn "triple" again. The ring was once more presented with considerable inclination.

Other early observers drew bizarre pictures of Saturn. Huyghens (1629–1695) first correctly described the rings: *Annulo cingitur, tenui, plano, nusquam cohaerente, ad eclipticam inclinato* [it is girdled by a ring, thin, flat, nowhere touching, inclined to the ecliptic].

The phases of Venus were observed by Galileo for the first time, and he announced them characteristically in the form of an anagram: *Haec immatura a me iam frustra leguntur o.y.* [*Cynthiae figuras aemulatur Mater Amorum:* the mother of the Loves (Venus) imitates the phases of Cynthia (the moon)]. Thus one of the objections to the heliocentric hypothesis was removed. The important part of the discovery was that Venus shows phases *like the moon,* i.e., from crescent to full.

Slight phase changes were even correctly observed for Mars; as one of the superior planets, Mars is only seen either full or slightly gibbous.

The discovery of sunspots was made by Galileo, but other observers had also seen them at the same time or earlier. Finally, he turned the tele-

scope to the Milky Way and, for the first time, showed that the hazy band
of light consists of large numbers of faint stars.

In addition to inaugurating the great era of telescopic observation,
Galileo's data seemed to point decisively to the heliocentric hypothesis:
the moons of Jupiter were a solar system in little, and full phase was demonstrated for an inferior planet, which, on the Ptolemaic system, could
only appear as a crescent (see Figure 7.2).

All his life Galileo was a sincerely religious man. He was well aware
of the edict against the Copernican books, and when the Church decided
that they should be corrected, he offered his services and stayed in Rome.
He wrote an apparently dispassionate dialogue that compared the Ptolemaic
and Copernican systems; however, the astronomical arguments for the
latter that it contained were cogent and unanswerable, and the book was
suppressed.

The story of Galileo's trial and humiliation is a sad commentary on the
temper of the times. At the age of seventy he was called to Rome to answer
criticism of his books, which he had already twice revised. He pleaded age
and infirmity, but was not spared the ordeal. Although he was severely
humiliated, he was neither imprisoned nor ill-treated. In the little Church
of Santa Maria Sopra Minerva, which stands in Rome near the Pantheon,
on the site of the temple of the goddess of wisdom, he made public sub-
mission: "I do not hold, and have not held this opinion of Copernicus since
the command was intimated to me that I must abandon it . . . I swear
that in future I will never say or assert, verbally or in writing, anything
that might furnish occasion for a similar suspicion against me." They are
the words of the humble and cautious man that Galileo always was. The
legend that he murmured "it does move, nevertheless" under his breath
is probably apocryphal.

Galileo retired to Arcetri, near Florence, where his house still stands
near the modern astrophysical observatory. He continued his scientific
studies, though he kept away from the Copernican theory. His work on the
laws of motion paved the way for Newton; he wrote also on optics and
on the librations of the moon. Here the young poet Milton visited the blind
old man, who solaced himself by playing the lute as well as by scientific
studies.

7. NEWTON

Qui humanum genus ingenio superavit
[who surpassed the human race in genius]

*   *   *

Nature and nature's law lay hid in night
God said, Let Newton be, and all was light

POPE *

* In our own day, a pendant was added to this celebrated couplet by J. C. Squire.
"It did not last. The devil howling: Ho!/Let Einstein be! restored the status quo."

Isaac Newton (1642–1727) was one of the brilliant group of English-men who met in the early days of the Royal Society of London; he was the friend of Halley (p. 241), Hooke (famous for his study of the stresses in rigid materials), and Wren (best known as an architect, though also a contributor to pure science). The inscription from his monument (quoted above), and the words of Leibnitz: "Taking mathematics from the begin-ning of the world to the time when Newton lived, what he had done was much the better half," assess the greatness of his contribution. His life was the externally uneventful one of a retiring scholar. Most of his important scientific work was done in his earlier years; in his old age he wrote at great length on theological problems. Much of this latter work remains un-published; Newton himself thought it his most important production. Even in earlier years he was reluctant to go to the trouble of publishing his results, and his great work the *Principia* would probably never have appeared if his friend Halley had not paid for it and seen it through the press.

The *Principia,* which contains the derivations of the laws of planetary motion, is a formulation of celestial mechanics that has not been essentially improved up to the present. It strikes the modern reader as difficult because all the proofs are geometrical, and we are more accustomed to analytical formulation. Nonetheless, it is fascinating reading, as a glance at some of its contents will show.

The book begins with the formulation of the three laws of motion:

1. Every body persists in its state of rest, or of uniform motion, except insofar as it is compelled by impressed force to change that state.

2. Change of motion (i.e., *acceleration*) is proportional to impressed force and inversely as the mass, and takes place in the direction in which the force acts.

3. Every action has an equal and opposite reaction.

The familiar symbolical expression of the second law is

$$f = ma$$

where f is the impressed force, m the mass, and a the acceleration. The third law states that for two bodies designated by E and S:

$$f_S = f_E; \quad m_S a_S = m_E a_E; \quad m_S/m_E = a_E/a_S$$

Newton goes on to prove that for central forces (i.e., forces acting along the radius vector), the law of areas (Kepler's second law) must be obeyed, whatever the nature of the force, and that if the law of areas is obeyed, the force must be central.

The next problem to be solved is how the force between bodies varies with the distance between them. He examines the law of force that will produce motion in an elliptical orbit round the *center* of the ellipse (the mid-point of the major and minor axes, not the focus) and shows that in this case the force will vary *directly as the distance* from the center. Thus all periodic times about a given mass at the center would be equal, whatever the size of the ellipse. This is not in accordance with Kepler's third law (which states that the squares of the periodic times are proportional to the cubes of the mean distances), and therefore the force does not vary directly as the distance.

Newton then investigates motion under a central force in an ellipse about the *focus,* and shows that it implies an *inverse square law.* He goes on to prove that motion in confocal ellipses under an inverse square law of attraction will result in periodic times that are as the 3/2 powers of the major axes (Kepler's third law). Thus he shows that the observed planetary motions are compatible with (and also require) a law of attraction that is proportional to the inverse square of the distance between the attracting bodies (the *law of gravitation*). He extends the proof to the other conic sections (p. 31), the hyperbola ("because of the dignity of the problem") and the parabola.

In order to apply the law to bodies of finite size, Newton had to prove (and successfully did so) that a sphere attracts as though its mass were concentrated at its center.

The complete statement of the law of motion of two bodies under their mutual gravitation is that *they revolve in similar conics about each other and their common center of gravity.*

Newton then considers how to determine orbits. A conic is fixed by five points through which it passes; however, the determination of the motion of a planet in a conic requires only three, because the law of motion supplies the equivalent of the other two.

The *Principia* ranges over a vast variety of subjects, including the motion of very small bodies (with applications to the reflection and refraction of light, considered as corpuscular); motion in a resisting medium; hydrostatics; and hydrodynamics. Newton discusses the actual motion of the (known) moons of Jupiter and Saturn, and that of the planets. He gives the theory of the figure of the earth (p. 25) and of the tides (p. 145). He makes a dynamical study of precession (p. 35). The astonishing passage on the motion of the moon refers severally to the "propositions" in which the various motions of the moon are described or predicted.

> Our moon moves faster, and, by a radius drawn to the earth, describes an area greater for the time, and has an orbit less curved, and therefore approaches nearer to the earth in the syzygies than in the quadratures, excepting so far as these effects are hindered by the motion of eccentricity: for the eccentricity is greatest when the apogee of the moon is in the syzygies, and least when the same is in the quadratures; and upon this account the perigaean moon is swifter, and nearer to us, but the apogaean moon slower and farther from us, in the syzygies than in the quadratures. Moreover, the apogee goes forwards, and the nodes backwards; and this is done not with a regular but an unequal motion. For the apogee goes more swiftly forwards in its syzygies, more slowly backwards in its quadratures; and by excess of its progress over its regress, advances yearly forwards. But the nodes, on the

contrary, are quiescent in their syzygies, and go fastest back in their quadratures. Further, the greatest latitude of the moon is greater in the quadratures of the moon than in its syzygies. And the mean motion of the moon is slower in the perihelion of the earth than in its aphelion. And these are the principal inequalities taken notice of by astronomers. [See p. 130.]

But there are yet other inequalities not observed by former astronomers, by which the motions of the moon are so disturbed that to this day we have not been able to bring them under any certain rule. For the velocities or hourly motions of the apogee and nodes of the moon, and their equations, as well as the difference between the greatest eccentricity in the syzygies and the least eccentricity in the quadratures, and that inequality which we call the variation, are in the course of the year augmented and diminished as the cube of the sun's apparent diameter. And besides the variation is augmented and diminished nearly as the square of the time between the quadratures.

The reader may be interested in identifying from this description the eight principal perturbations of the moon's motion mentioned in Chapter V. The advance represented by Newton's work over that of his predecessors could hardly be more strikingly illustrated.

The story of how the control of the moon by gravitation was suggested to Newton by the fall of an apple is well known. He applied the idea to calculating how far the moon should fall toward the earth in a minute under an inverse square law of gravitation. The distance fallen was calculated to be $d^2/120R$, where R is the earth's radius and d the linear distance traveled by the moon in one minute. He knew the moon's distance to be 60 earth radii, and he believed the earth's radius to be given by the rough nautical estimate of 60 miles to $1°$ of latitude. These numbers led to a fall of less than 14 feet per minute; but under gravity, it should be 16 feet a minute, and Newton laid aside his work in disappointment. It seemed that theory and observation did not check. Six years later, however, Picard made a better measure of the length of a degree (p. 25), and Newton found that with the new data, theory and observation were in agreement. From this beginning grew the tremendous work that is condensed in the *Principia,* begun in 1672 by a youth of twenty-four, and first confirmed by observation six years later.

In order to carry out the theory of the changing motions of planets, Newton was compelled to invent a new mathematical technique, which he called the "Method of Fluxions"; it is essentially our modern *calculus.*

He did not publish his results at once. It was not until 1684, when Wren offered a small prize for the first who should prove that a body under the inverse square law would describe an ellipse, that Halley spoke to Newton of the problem, and found that he had solved it sixteen years earlier! Halley persuaded Newton to publish, and the *Principia* appeared in 1687.

In a chapter dealing with the planetary motions, we cannot do more than mention Newton's epoch-making experiments in optics and his demonstration of the production of the spectrum and the nature of white light.

Equally fruitful was his reflecting telescope (p. 81). His second great work was the *Optics,* but much of his writing, including his theological speculations, has never been printed.

Celestial mechanics, the study of the motion of planets (and stars) under gravity, has not been essentially changed since Newton formulated it, but the application and extension of his principles continues, and the subject is one of the most difficult and fruitful in modern astronomy. The theory of relativity has shown, during the past thirty years, that the Newtonian laws are applicable only at comparatively small velocities (small, that is, compared with the velocity of light). The success of Newtonian mechanics is a result of the fact that most of the velocities of the heavenly bodies are (in this sense) small; in fact, the astronomical phenomena that can be used to verify the predictions of relativity theory are few (see p. 141, p. 182, and p. 291).

With Newton we may consider that the process of formulating and understanding the motions within the solar system is complete. We have traced the growth of understanding through five hundred years of Greek science, the long interregnum between Ptolemy and Leonardo da Vinci, and the rapid flowering in the two centuries between Copernicus and Newton. Such another burgeoning of astronomy is hardly to be seen before the opening of the present century, which has expanded our view outside the solar system and its immediate neighborhood as far as the remote galaxies. But the astronomy of galaxies is still in the Copernican stage; it has yet to find its Newton.

8.   THE SCALE OF THE SOLAR SYSTEM

Before Kepler could formulate his laws of planetary motion, he had to map out the forms of the planetary orbits with great care. This he did, first for the orbit of Mars, later for the other planets, from comparing the direction in which a given planet was seen, at intervals of exactly one sidereal period, from the earth. Tycho's precise observations provided the material, and Kepler was able to construct a picture of the known parts of the solar system all on the same scale. It was from such a plan that the "harmonic law," relating period to mean distance, was derived. But although the plan is all on the same scale, the actual scale is not determined unless at least one distance is known with accuracy. The distances of all the other planets are usually expressed in terms of the mean distance of the earth from the sun—the average of major and minor axes of the earth's orbit. This distance is defined by the *solar parallax.*

The solar parallax is the angular size of the earth's radius as seen from the sun. Since the earth is not quite spherical, the equatorial radius is the one used, and because its distance from the sun varies, the mean radius, corresponding to the mean distance, is employed. The quantity measured is called, strictly, the sun's *mean equatorial horizontal parallax: mean,* because it is referred to mean distance; *equatorial,* because referred to equatorial radius; *horizontal,* because it is the angle between the direction of the sun on the horizon and the direction it would have if viewed from the earth's center. It is thus a *geocentric parallax,* on the baseline of the earth's radius; it must not be confused with *heliocentric parallax,* whose baseline is the earth's orbit.

The mean equatorial horizontal parallax of the moon is fairly easy to measure because the moon is so near; its value is 57'2".7. The greater distance of the sun makes its geocentric parallax harder to measure, the value being less than 9". Aristarchus tried to measure it directly, as we have seen (p. 157), but obtained far too small a value for the distance. Ptolemy, Kepler, and Huyghens also attempted a direct determination, but with only moderate success. Even today the *direct* measurement of the solar parallax is not attempted.

A better approach is to measure the distance from the earth of another member of the solar system whose orbit has been accurately placed in the scale drawing of the whole. There are a number that approach the earth nearer than the distance of the sun, and, since their distances are smaller, their parallaxes are easier to measure accurately.

The first such parallax measure for another member of the solar system was made by means of the planet Mars, the nearest of the planets to us. Every fifteen or seventeen years, Mars comes to the point where it is nearest to us, because opposition coincides with the perihelion of its eccentric orbit. At its nearest, Mars' distance from us is little more than one-third of our distance from the sun, and its geocentric parallax is over 23".

Cassini obtained measures of the parallax of Mars. The planet was observed at the same time from two stations (Paris and Cayenne, South America), widely separated on the earth's surface, and a value of 9".5 was obtained for the solar parallax. In 1862, values of 8".96, 8".94, and 8".85 were determined by different observers. Measures in 1877 and 1892 gave parallaxes of 8".78 and 8".80 for the sun. Over a span of fifty years, the measures of solar parallax became more accurate, and the values obtained were a little smaller. The problem, though easier than for the sun, is still difficult, for Mars has a fairly large disc.

Minor planets are better subjects than Mars; they have smaller, almost stellar images, and some of them approach nearer than Mars to the earth. Five different asteroids, between 1873 and 1889, led to values from 8".80 to 8".87 for the solar parallax. In 1897 the asteroid Eros was discovered; it can approach the earth within 0.15 of the earth's distance from the sun, and its parallax can be as great as 60". Unhappily, it was discovered a little too late to use the closest opposition (which occurred in 1894), but even so, it was by far the best means yet found for measuring the solar parallax geometrically. In 1900–1901, Eros yielded a value of 8".806; and the opposition of 1931 was attacked by a number of co-operating astronomers and observatories, and gave the best geometrical value hitherto obtained, 8″.790.

From all the geometrical measures of the solar parallax, the mean value is calculated to be 8″.803 ± 0.001. In combination with the best value for the earth's equatorial radius, this gives for our mean distance from the sun the value 149,450,000 ± 17,000 kilometers, or 92,870,000 miles.

The direct geometrical method is not, however, the only one for determining our distance from the sun. One of the smaller periodic terms in the moon's motion is the *parallactic inequality,* which causes the moon to be behind at the first quarter and ahead at the third, by an amount that depends on the distance of the "mean moon" from the mean sun, and also on the solar parallax. The size of the parallactic inequality is determined from a careful study of the timing of occultations of stars by the moon (which give the same kind of information as the times of contact for a solar eclipse, as described on p. 141). The best determination of the parallactic inequality that has hitherto been made gives a value of 8″.805 for the solar parallax.

Another method depends on measuring the earth's velocity relative to some distant star (p. 31). From a study of the earth's changing orbital speed relative to Arcturus, the solar parallax has been found to be 8″.805 ± 0.007. The geometrical, dynamical, and spectroscopic determinations of the solar parallax are in close agreement.

The solar parallax defines the size of the *astronomical unit,* the mean distance of the earth from the sun. This distance is not only fundamental to distances measured within the solar system. It also enters our determination of the sizes of the orbits of double stars, which are measured by a comparison of their orbital motions with our own, and is basic to our scale of distances for the stellar universe.

CHAPTER VIII

THE PLANETS

The strong love of thy son, Hyperion
    I there sustained, and saw how in a ring
    Maia and Dione close beside him shone.
Next there appeared to me Jove's tempering
    Between his father and son; and I could note
    Their places and their stations' varying.
And all the seven were shown me, and I thought:
    How swift they are in moving and how great,
    And each one from the other how remote!
                  DANTE
       Paradiso, Canto XXII, 142–150 *

The sun is attended by an enormous number of lesser bodies, the members of the solar system. Most conspicuous are the nine planets—Mercury, Venus, Earth, Mars, Jupiter, Saturn, Uranus, Neptune, and Pluto. They are the subject of the present chapter. Between Mars and Jupiter circulate the *minor planets,* or *asteroids,* nearly all too small to be seen with the naked eye, and represented by bodies of a great variety of sizes; the *meteorites* are probably closely related to them. The *comets,* some of which pass near the sun, are spread much more widely throughout the system, and are near relatives of the *showers of meteors.* These families of smaller bodies will be described in the next chapter.

A general picture of the solar system is shown in Figure 8.1. The distances of the principal planets from the sun show a simple relationship, which is known as Bode's Law after the man who formulated it in 1772. Write down the series 0, 3, 6, 12. . . . Add 4 to each member of the series, and divide each result by 10. The numbers thus obtained will be found in the third column of the table. The fourth column gives the actual distances of the planets from the sun, in terms of the earth's distance (the *Astronomical Unit*). The "law," as may be seen, predicts fairly well the distances of all the planets known to Bode.

* Dante, as we have seen, thought of the solar system in Ptolemaic terms. "The seven" are, therefore, the moon, the sun (Hyperion's child), Mercury (son of Maia), Venus (daughter of Dione), Jupiter, placed between his father (Saturn) and his son (Mars).

BODE'S LAW

Planet	Series (+4)		Sum/10	Distance from Sun (A.U.)	Known Moons	Notes on Discovery of Planet
Mercury	4	0	0.4	0.39	0	Prehistoric
Venus	4	3	0.7	0.72	0	Prehistoric
Earth	4	6	1.0	1.00	1	. . .
Mars	4	12	1.6	1.52	2	Prehistoric
Asteroids	4	24	2.8	2.65 (average)	. . .	First in 1801
Jupiter	4	48	5.2	5.20	12	Prehistoric
Saturn	4	96	10.0	9.54	9	Prehistoric
Uranus	4	192	19.6	19.19	5	1781
Neptune	4	384	38.8	30.07	2	1846
Pluto	4	768	77.2	39.52	. . .	1930

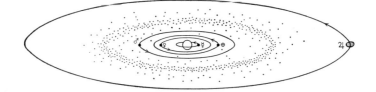

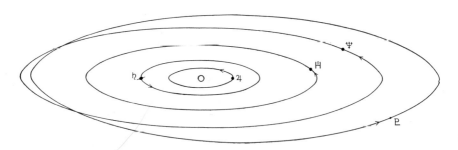

Fig. 8.1. Perspective view of the solar system. The upper figure shows the orbits of the planets and asteroids out to Jupiter. The lower figure, on a much smaller scale, shows the orbits of the planets from Jupiter to Pluto.

Only the "prehistoric" planets were known to Bode. The gap between Jupiter and Mars is occupied by the asteroids, of which there are at least a thousand; their average distance fits the "law" rather closely. Uranus, discovered later than 1772, also fits the prediction of the "law," but the next prediction would fit Pluto much better than Neptune. Many relationships that are noted in the inner parts of the solar system are followed less rigidly near its bounds. Bode's "law" is probably an empirical accident, but some modern theories of the origin of the solar system have attempted to take account of it.

The nine planets display a rich variety of size and surface conditions. Mercury is small, arid, hot, and devoid of atmosphere; Jupiter is enormous, intensely cold, and wrapped in a blanket of gases. The differences between the planets stem from their sizes, their positions in the system, and also as we shall see, from the conditions under which they were formed.

The larger planets fall into two natural groups. The *major planets* (Jupiter, Saturn, Uranus, Neptune) are all large, of low density, and surrounded by huge atmospheres. The *terrestrial planets* (Mercury, Venus, Earth, Mars, Pluto) are more like the earth. All but Venus and the earth appear to be solid globes with comparatively little or no atmosphere, and all are far smaller than the major planets.

The more important physical properties of the planets are summarized in the Table. To understand how this information has been obtained, we must grasp the mechanics of the system, and visualize the way in which all the planets and lesser bodies move about the sun. The preceding chapter described how the laws of motion were formulated by Kepler and used by Newton to derive the law of gravitation. Orbits that are described under gravitation are *conic sections* (almost precisely). Most of the members of the solar system travel very nearly in *ellipses* with the sun at one focus. The sun, by virtue of its greatest mass, plays the major role in determining the motions, but the small effects of the other bodies in the system produce small disturbances, or *perturbations,* in the simple elliptical motion.

The motion of a body in an orbit around the sun can be completely described by seven numbers (the *orbital elements*): five describe the size, shape, and orientation of the orbit; and two are needed to fix the position at which the body lies at a given moment. If the orbital elements are known for a planet, *we can calculate its exact position for any time either in the past or in the future,* insofar as it depends only on the sun's attraction. In refined calculations, the effect of other bodies of known mass must also be evaluated, as we shall see.

The orbital elements of a member of the solar system, as shown in Figure 8.2, are:

   a. *Semi-major axis,* which gives the *size;*

   b. *Eccentricity,* which tells the *shape;*

   c. *Inclination,* which describes *how the orbit is tilted* relative to the plane of the ecliptic;

   d. *Longitude of ascending node,* Ω (measured from the First Point of Aries in the direction of the earth's motion), which tells the *position at which the orbit cuts the plane of the ecliptic* from south to north;

   e. *Longitude of perihelion* (reckoned from the ascending node), which is the sum of Ω and the angle ω between ascending node and perihelion and

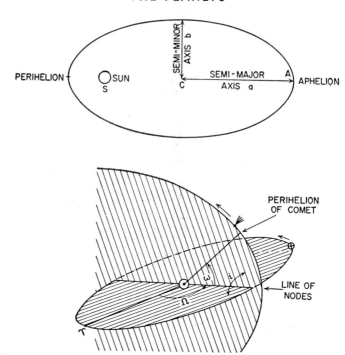

Fig. 8.2. Elements of a planet's or comet's orbit. The upper figure shows an ellipse with the sun at the focus. The eccentricity of the ellipse is equal to the ratio of the sun's distance from the center to the semi-major axis: $e = SC/AC$. In terms of semi-major and semi-minor axis, $e = (1/a) \sqrt{a^2 - b^2}$. The lower figure shows the inclination, i, of a comet's orbit relative to the ecliptic (defined by the plane of the earth's orbit); the *longitude of the ascending node*, Ω, where the orbit cuts the ecliptic from south to north; and the *longitude of perihelion*. The longitude of the ascending node is measured from the vernal equinox, ♈, in the direction of the earth's motion; longitude of perihelion is measured by the sum of Ω (in the plane of the ecliptic) and ω, the angle between the ascending node and the perihelion point (in the plane of the orbit, measured in the direction of the comet's or planet's motion).

tells the *position in the orbit at which the planet is nearest to the sun,* and therefore the direction in which the long axis of the orbit points; and

f. *Time of perihelion passage,* which tells an actual *time at which the planet is at perihelion,* and locates the position of the planet in the orbit at any other time with the aid of the period.

g. The *period* is the *time taken by the planet to go around the sun,* and is related to the *mean distance* (fixed by the semi-major axis and the eccentricity) by Kepler's third law.

1. MERCURY

The planet Mercury, whose place
Is nearest to the sun in space
Is my allotted sphere.
LONGFELLOW

The planet nearest to the sun appears in Babylonian records as far back as the fourth century B.C. The Greeks associated it with Mercury, the swift-footed messenger of the gods, and called it *Stilbon,* "the twinkler." As Mercury can never be more than 28° from the sun, it is most readily visible just before sunrise or just after sunset, near the horizon. The obscuring and reddening effects of the thick layer of atmosphere near the horizon give it a rosy color and make it twinkle, in spite of the disc that causes most planets to shine with a steady light, which distinguishes them from stars. Mercury at best is as bright as Sirius, and is easy to find if looked for near a favorable elongation. Copernicus apparently never saw it, and in a murky urban sky it is more rarely noticed than it was in the clear air of Babylonia.

Mercury is unique among the planets. It is the smallest in size and mass and has the shortest period and the greatest linear speed. Its orbit is more eccentric than that of any planet except Pluto, and it receives by far the most heat and light from the sun.

The sidereal period of Mercury is 88 days, its synodic period 116 days. The large eccentricity of the orbit, 0.2056, causes its distance from the sun to vary from 47 million kilometers at perihelion to 69 million kilometers at aphelion, so that the sun is over 11 million kilometers from the center of the orbit. The speed in the orbit, which is governed by the law of areas, accordingly varies from 56 km/sec at perihelion to 37 km/sec at aphelion. The mean distance from the sun is 58 million kilometers, a little more than a third of the earth's.

As earth and Mercury go around their respective orbits, they are closest to each other at inferior conjunction, furthest apart at superior conjunction. The actual distances on these occasions vary because both orbits are elliptical; at the most "favorable" inferior conjunction (earth near perihelion, Mercury near aphelion) they are 80 million kilometers apart, and at the most remote superior conjunction the distance is 219 million kilometers, nearly three times as great.

Mercury shows phases, *like those of the moon and Venus;* we see its illuminated hemisphere from all angles. At superior conjunction it is "full," but (owing to greater distance) the disc is smallest; at inferior conjunction it is "new," and the disc is largest. The apparent brightness depends both on phase and distance; like Venus, and unlike the moon, it is brightest when in the crescent phase, not when "full."

The Planets

	Mercury	Venus	Earth	Mars	Jupiter	Saturn	Uranus	Neptune	Pluto
Sidereal period:	87.97^d	224.7^d	365.26^d	687.0^d	11.86^y	29.46^y	84.02^y	164.8^y	247.7^y
Synodic period:	115.88^d	583.9^d	...	779.9^d	1.092^y	1.035^y	1.012^y	1.006^y	1.004^y
Mean distance, 10^6km:	57.94	108.27	149.68	228.06	778.73	1,427.7	2,872.4	4,500.8	5,914.8
Astron-units:	0.387	0.723	1.000	1.524	5.203	9.539	19.19	30.07	39.46
Orbital speed, km/sec:	47.9	35.0	29.8	24.1	13.6	9.6	6.8	5.4	4.8
Orbital eccentricity:	0.2056	0.0068	0.017	0.093	0.048	0.056	0.047	0.0086	0.249
Orbital inclination:	7°0'	3°24'	...	1°51'	1°18'	2°29'	0°46'	1°47'	17°19'
Mean diameter, km:	5,000	12,400	12,742	6,870	139,760	115,100	51,000	50,000	12,700?
Earth diameters:	0.39	0.973	1.000	0.532	10.97	9.03	4.00	3.90	0.46
Volume (earth volumes):	0.06	0.92	1.00	0.15	1,318	736.	64.	39.	0.10
Mass (earth masses):	0.04	0.82	1.00	0.11	318.3	95.3	14.7	17.3	1.0?
Density (earth densities):	0.69	0.89	1.00	0.70	0.24	0.13	0.23	0.29	?
in g/cc:	3.8	4.86	5.52	3.96	1.33	0.71	1.26	1.6	?
Surface gravity (earth's):	0.27	0.86	1.00	0.37	2.64	1.17	0.92	1.44	?
Velocity of escape, km/sec:	3.6	10.2	11.2	5.0	60.	36.	21.	23.	11?
Maximum surface temperature, °F:	770.	140.	140.	86.	−216.	−243.	−300?	−330?	−348?
Length of day:	88^d	30^d?	1^d	1^{d}37^{m}23^s	9^{h}55^m	10^{h}38^m	10^h.7	15^h.8	?
Inclination of equator to orbit:	...	0°?	23°27'	25°12'	3°7'	26°45'	98°.0	29°	?
Oblateness:	0.00	0.00	1/296	1/192	1/15.4	1/9.5	1/14	1/45	?
Albedo:	0.07	0.59	0.29	0.15	0.44	0.42	0.45	0.52	0.04?
Atmosphere:	none	CO_2	(see p. 21)	H_2O?	CH_4,NH_3	CH_4,NH_3	CH_4,NH_3	CH_4,NH_3	none
Moons (known):	none	none	1	2	12	9	5	2	none

The elongation that can be reached by Mercury depends on the part of the orbit occupied by the planet at the time. When it is farthest from the sun (aphelion) at elongation, it can be as much as 28° from the sun; but if elongation coincides with perihelion of Mercury, the angle is only 18°. The orbit of the planet has the large inclination to the ecliptic of 7°.

The size of Mercury can be determined from its apparent diameter and known distance. The apparent diameter varies from 5″ at superior conjunction to 13″ at inferior, which gives the diameter as 5000 kilometers, or 3100 miles. Thus Mercury has 0.06 of the volume, and 0.15 of the surface, of the earth.

The best way of measuring the mass of a planet depends on the perturbations of its motion caused by other planets, or on a study of the motions of its moons. Mercury has no moons, and it is so near to the sun that the latter is the chief influence on its motions, the planetary perturbations being very small. The mass is therefore not accurately known. The best determination makes Mercury 1/8,000,000 of the mass of the sun, 1/24 of the earth's. Thus Mercury's density is 0.7 of the earth's, gravity at the planet's surface has 0.27 of the terrestrial value, and the velocity of escape is 3.6 kilometers a second. The specific gravity of Mercury is 3.8, half-way between the values for earth and moon.

The sun beats fiercely on the face of Mercury; on the average the planet receives seven times as much sunlight per unit area as the earth. But Mercury always turns the same face to the sun (just as the moon does to the earth), and so, apart from librations, one side of the planet is in continual sunlight, the other in perpetual shadow. Under the sun's rays the surface of Mercury is kept at a temperature near 350°C, or 660°F; when nearest to the sun at perihelion, the planet has a surface temperature as high as 770°F, hot enough to melt lead! But the other side of the planet is in eternal darkness, and its temperature cannot be far from absolute zero, −273°C, even colder than the surface of the remote Pluto. Mercury presents the paradox of being at once the hottest and the coldest planet in the solar system.

Mercury is an outsize edition of our own moon, about 50% larger in diameter, and (if we make allowance for the difficulty of seeing detail on the planet's surface) it looks very like our satellite. Its specific gravity is 3.8, against the moon's 3.3. Its reflecting power (*albedo*) is 0.07, the same as the moon's, and the colors are similar (darkish brown). Even the difference of reflecting power with angle (*phase effect*) is about the same and shows that the planet has a roughish surface. There is little doubt that similar forces have played their part in the formation of the surfaces of Mercury and our moon.

Like the moon, Mercury gives no observable evidence of an atmosphere and, indeed, would be incapable of retaining one at the enormous subsolar temperature because of its low velocity of escape.

A planet whose orbit lay in the plane of the ecliptic would pass between us and the face of the sun at every inferior conjunction. However, the orbit of Mercury is so much inclined to the ecliptic that we see its motion across the sun's face (*transit*) rather rarely, at intervals of seven or thirteen years. These transits can, of course, take place only when inferior conjunction occurs near the nodes. Mercury is so small that its transits can only be seen with a telescope. Their chief importance is that their timing adds to information about the planet's motion in its orbit. The last favorable transit of Mercury visible in the United States was on November 14, 1953.

Mercury provides one of the few observational tests of the theory of relativity. Gravitational theory predicts (as for the moon) that the line of apsides will revolve slowly at the rate of 532" per century. The observed rate of the advance of perihelion is, however, sensibly greater, being 574" per century. An attempt was made by Leverrier to ascribe this unexplained excess to perturbations caused by another planet inside the orbit of Mercury, and the hypothetical planet even received the name Vulcan. Such a planet has never been seen, though it should be observable at some solar eclipses. The discrepancy is no longer attributed to a planet inside Mercury's orbit. Another suggestion was that the orbital motion is influenced by a "cloud" of meteoric matter; however, if such a cloud were present it should be detected by the light it reflects, and this idea also has been abandoned.

The general theory of relativity interprets the unexplained advance of perihelion in other terms. At the close of the last chapter, it was mentioned that departures from Newtonian motion are to be expected at large relative velocities. The velocity of Mercury in its orbit is just large enough to produce such a deviation. An otherwise undisturbed planet going round the sun should experience an advance of perihelion and, therefore, an apsidal revolution proportional to $3v^2/c^2$ per period, where v is the planet's velocity and c the velocity of light. On this basis, an advance of 42".9 per century is predicted for Mercury's perihelion. The observed excess is 42".84 ± 0".41, in very close agreement, and is therefore satisfactorily explained.

Mercury affords a good test of the theoretical advance of perihelion required by the theory of relativity, because the planet's period is small and its orbital velocity therefore large; moreover, the orbit is quite eccentric, so that perihelion is easily located. Venus, the next nearest planet to the sun, has a large enough velocity for the test, but its orbit is so nearly circular that perihelion is difficult to fix. Mars has an orbit that is satisfactorily elliptical, but its distance from the sun is so great that the orbital velocity is rather small; the advance of perihelion predicted by relativity theory is 1".3 per century, whereas the observed unexplained advance is 5". These quantities are too small and too hard to measure to furnish a reliable test, and no other planet is suitable for studying the problem.

2. VENUS

For a breeze of morning moves,
   And the planet of love is on high,
Beginning to faint in the light that she loves
   On a bed of daffodil sky,
To faint in the light of the sun she loves,
   To faint in his light, and to die.
                              TENNYSON, *Maud*

The second planet from the sun, and the brightest object in the sky after sun and moon, formed one of the heavenly trinity in Babylonia. Observations of her heliacal risings are some of the earliest on record and were carried out in the third millennium B.C.

Venus has the most nearly circular orbit of any planet, with eccentricity 0.007, inclined by 3°24' to the ecliptic. Her mean distance from the sun is 108,270,000 kilometers, and varies with position in the orbit by less than 2 million kilometers. Mercury, with a much smaller orbit, but a much more eccentric one, varies in distance from the sun by over 22 million kilometers. The mean velocity of Venus in her orbit is 35 kilometers a second, again much less variable than Mercury's. At a distance of a little less than three-quarters of our own from the sun, Venus receives almost twice as much light from the sun as we do.

The sidereal and synodic periods of Venus are 225 and 584 days, respectively. She moves from superior conjunction (on the far side of the sun) to greatest elongation in 220 days, but the interval from greatest elongation to inferior conjunction is only 72 days, less than a third of the time. At elongation the distance from the sun may be 48° to 47°, depending on position in the orbit; that the range of elongation is so much smaller than for Mercury is another illustration of the difference of eccentricity of the orbits.

The distance between earth and Venus varies between 257 million kilometers at superior conjunction and 42 million at inferior conjunction. Thus Venus comes nearer to the earth than Mars at opposition—nearer, indeed, than anything in the solar system except the moon, some comets, and a few asteroids.

The great difference in the distances from which we see Venus makes her apparent size vary more than sixfold: when farthest she shows a disc of 10″ diameter, when nearest, her diameter is 64″. It might be expected that the large apparent diameter would make the size of Venus easy to measure accurately, but this is not the case. The brilliant surface of the planet produces an "irradiation effect" that causes astronomers to overestimate the diameter. When Venus, at one of her rare transits, is silhouetted against the brilliant sun, irradiation works the other way and makes the measured size a little too small. Direct measures give a diameter of about 12,550 kilometers; transits, about 12,230 kilometers. The value usually adopted is 12,390 kilometers, with an uncertainty of about 1%. Thus Venus has a diameter 0.973 of the earth's: she and the earth are nearly twins in size.

Fig. 8.3. Photographs of Venus at different phases and displaying the true relative sizes of its disc at different points in the orbit. (Photograph by E. C. Slipher, Lowell Observatory.)

Venus has a mass a little less than that of the earth. She is far enough from the sun for her motion to be appreciably perturbed by earth and Mars, so even though she has no moons, her mass is well determined. The mass is 1/408,000 of the sun's and 82% of the earth's; the density is 89% and the surface gravity 86% of our own planet's. A weight of 160 pounds would weigh 138 pounds on Venus, and the velocity of escape from her surface is about 10 kilometers a second. All this suggests surface conditions very like those of earth.

The suggestion, however, is not borne out by the appearance of Venus. An outside observer looking at the earth would see the oceans and continents, overlaid by changing cloud-banks, through a bluish atmospheric haze. On Venus we see nothing of the kind. To the eye and on an ordinary photograph she presents a uniform, brilliantly white surface. Ill-defined markings show on photographs taken in ultraviolet light, and their rapid changes (within a day) show that they are probably clouds, not permanent features.

That Venus is covered by thick clouds is also shown by her great reflecting coefficient. Her albedo is 0.59, the largest in the solar system, approached only by the major planets, which are also cloud-covered. Moreover the phase effect shows that the surface of Venus is not rough, like those of Mercury and the moon. She reflects sunlight without changing its color, unlike Mercury, the moon, or Mars, whose rocky or sandy surfaces return the sunlight to us much reddened. Evidently Venus is covered by a thick layer of opaque cloud: we cannot see her actual surface, and hypothetical beings on Venus could presumably not see the heavens.

Besides the evidence for an atmosphere provided by the high albedo, the smoothness of the visible surface, and the changing markings, the appearance of the cusps at twilight also indicates its presence. The cusps of the moon are perfectly sharp, but those of Venus in the crescent phase are seen to extend round the dark edge of the disc. The effect must be caused by the scattering of sunlight in an atmosphere.

Probably the atmosphere of Venus below the clouds is not very deep. If the solid body of Venus has the same density as the earth, the cloudy layer has 4% of the observed radius, or about 500 kilometers. The earth is (with the possible exception of Pluto) the densest known body in the solar system, and Venus is unlikely to be denser. Certainly Venus is not built like Jupiter and Saturn, whose gaseous envelopes are at least as large as the planets themselves.

The quality of the light reflected from the surface of Venus tells a great deal about the nature of its atmosphere. The whiteness of the reflected light rules out scattering by gas or dust, which would redden it. Sunlight must be reflected by surface particles, solid or liquid globules.

The spectrum of Venus comes from light that has gone through a layer of her atmosphere, been reflected from the globules, and come back through the atmosphere again. The atoms in the atmosphere above the reflecting globules have left their imprint on the light and robbed it of the frequencies peculiar to them (p. 72). The absorption spectrum of Venus therefore contains some lines not present in the original sunlight. They show that the principal constituent of the atmosphere of our twin planet is carbon dioxide (CO_2), two hundred and fifty times as much (per unit volume) as is found at the earth's surface. Strangely enough, there is no evidence of oxygen or water vapor; if any such atoms are present in the atmosphere of Venus, they are too scarce to make a visible impression on the spectrum —less than one thousandth of the number found at the surface of the earth. Nitrogen, the commonest of our atmospheric gases, has not been detected, but the relevant parts of its spectrum are inaccessible; nitrogen is very likely present on Venus.

No acceptable suggestion has been made about the nature of the globules that give the atmosphere of Venus its brilliant and opaque surface. The clouds may perhaps be similar to our own; but the absence of water vapor from the spectrum of Venus would in that case be puzzling.

Where there is so little free oxygen, animal life as we know it could not exist. Plant life is conceivable, but if the temperature at the surface of Venus is as high as the boiling-point of water, which seems just possible from the greater intensity of sunlight and the atmospheric blanket, anything like our higher plants could not exist, and we must restrict our speculations to the kind of simple, undifferentiated organism that is found in hot springs.

The spectrum of Venus conveys another piece of important information: the rotation of the planet is too slow to produce an appreciable Doppler effect (p. 290) between opposite limbs. In other words, Venus takes at least a month to turn on her axis, and unlike Mercury, which always turns the same face to the sun, she is therefore illuminated successively on all sides. The cloud markings move slowly; Jupiter, where they are seen to whirl around once in ten hours, presents a great contrast.

The measured temperature on Venus at the subsolar point is about 60°C: if anything, the surface of the planet is likely to be hotter. On the dark side, the temperature is −25°C, well below the freezing point, but nowhere near absolute zero, as it would be if that part of the planet never received any sunlight. This is one of the reasons for believing that the sidereal day and year of Venus are not equal (as they are for Mercury), but that Venus turns so slowly that the night side has a chance to cool off greatly, which it would not do if the period of rotation were very short.

When Galileo turned his telescope on Venus, he saw that she has phases *like those of the moon,* and this was a crucial test of the heliocentric idea. Even by the Ptolemaic system, Mercury and Venus should show phases, but *not like the moon:* they would always be seen as crescents, because they were pictured as always between earth and sun. Galileo found that Venus is *gibbous* [Latin: *gibbosus,* hump-backed] near superior conjunction.

The changing brightness of Venus, like that of Mercury, is a joint consequence of differences of distance and of phase. When the planet is "full," at superior conjunction, she is actually 2½ times fainter than she is at her brightest. Greatest brilliance occurs when the crescent is similar to that of a 5-day old moon, at elongation 39°, 36 days before and after inferior conjunction (at which latter time she presents her dark face to us, and is invisible). The change in brightness with advancing phase (after allowance is made for changing distance) is much less rapid than that of the moon— the *phase effect.* The roughness of the moon's surface causes it to be a less efficient reflector in an oblique, than in a vertical, direction; Venus reflects as though the surface is smooth.

Transits of Venus are of rare occurrence; they must take place within

1°45′ of the nodes. Five synodic revolutions of Venus occupy 8 years (to within a day), and 152 synodic revolutions, 243 years. Therefore transits can take place 8 years apart, and no other transit can occur at the same node for 235 or 243 years. Usually a transit, or pair of transits, will take place at the other node near the half-way point.

The first record of a transit of Venus is of the one in 1639 that was predicted and observed by Horrox and Crabtree in England. Since then, there have been transits in 1761 and 1769, 1874 and 1882. The next pair will occur in 2004 and 2012.

The nodes of the orbit of Venus are reached by the sun on June 7 and December 8, so transits can occur only at those times. The transits of 1639, 1874, and 1882 were December transits; those of 1761 and 1769 were June transits, as will be those of 2004 and 2012. If Venus in transit passes more than 12′ from the sun's center, it must pass over the sun's disc again 8 years earlier or later; but transits nearer to the center are "solitary." We are now in an interval of paired transits, which will be succeeded by an interval of solitary transits.

Fig. 8.4. Venus—photograph taken when the planet was passing between the earth and sun, showing a complete ring of light around the rim of the then dark disc of the planet. This ring of light is produced by the planet's atmosphere bending the sun's rays around the edge of the disc, thus providing objective proof that Venus possesses an extensive atmosphere. Likewise, a similar bright ring would appear around the night-side of the earth if we could observe it from Mars when it passed between that planet and the sun. The photograph was taken when Venus was about 2¾° from sun. (Photograph by E. C. Slipher and J. B. Edson, Lowell Observatory.)

Transits of Venus are among the rarest of astronomical phenomena; many astronomers cannot possibly see one during their lifetimes. They have, however, only a limited scientific interest. They can be used (as Halley first suggested) for determining the solar parallax; but better methods are now available. The most interesting fact that they have revealed is the bright ring seen round the edge of the planet just before it comes completely in front of the sun—the first recognized evidence for the atmosphere of Venus that refracts and scatters the sunlight.

3. EARTH

My sight through each and all of the seven spheres
Turned back; and seeing this globe there manifest,
I smiled to see how sorry it appears;
And I approve that judgment as the best
Which least accounts it, and that man esteem
Most worthy, who elsewhere brings his thoughts to rest.

DANTE
Paradiso, Canto XXII, 133–138

The earth has been described in some detail in Chapter II. It is the third planet out from the sun, and the first that has a satellite. Our moon, with a little over a quarter of the earth's diameter, is by far the largest satellite in the solar system relative to its primary. In fact the earth-moon system, viewed from outside, would suggest a pair of dissimilar planets rather than a planet attended by a moon.

The period of rotation of the earth (the day) is much shorter than those of the two planets nearer to the sun (Mercury, 88 days; Venus, perhaps several months) or than that of the sun itself (about 25 days). Periods almost equally short, or shorter, are found for all the planets farther out, whose rotation can be detected.

The comparatively rapid rotation of the earth causes it to be slightly oblate (1/296). Sun, Mercury, and Venus are not measurably oblate, and the moon is only slightly so, at least on the profile that we see (0.0006). All the planets outside the earth's orbit (except, conceivably, the unmeasurable Pluto) are more oblate, in harmony with their rapid rotations.

The eccentricity of the earth's orbit is 0.01674; the only planets with less eccentric orbits are Venus and Neptune. The orbit is, of course, not inclined to the ecliptic: *it defines the ecliptic.* The axis of rotation, however, is inclined to the orbit by 66°31'01", which produces the obliquity of the ecliptic relative to the equator. As with the moon, and indeed with all the other planets, the eccentricity of the orbit and the inclination of the ecliptic are changing very slightly as a result of planetary perturbations.

The earth, with a density of 5.52 gm/cc, is the densest body known in the solar system; Pluto may be as dense or denser, but the value for it is uncertain, as it is for the numerous smaller bodies whose sizes cannot be directly measured. The velocity of escape from the earth's surface is 11.2 kilometers a second, a little greater than for Venus, and greater than that for any other terrestrial planet (again with the possible exception of Pluto). But the velocities of escape from the major planets are very much higher.

4. MARS

Mars
As he glow'd like a ruddy shield on the Lion's breast
TENNYSON, *Maud* *

The red planet Mars has been associated with carnage and disaster since the Babylonians called it the Star of Death. It has cut a great figure in the popular eye because of the speculations about the possibility of life on its surface, and is a standing boon to science fiction. We shall soon see the tenuity of the case on which these fantasies are built.

The mean distance of Mars from the sun is 228 million kilometers. The eccentricity of the orbit, 0.093, is quite large for a planet, and in consequence, the radius vector varies by more than 40 million kilometers. The distance from Mars to the earth varies a great deal more. When the two planets are at conjunction, Mars is about 380 million kilometers from us, and appears a little fainter than Regulus. At an average opposition, the distance between the two planets is only 78 million kilometers. If opposition coincides with Mars' arrival at aphelion, the distance is as much as 100 million kilometers, and Mars appears a little brighter than Canopus. But if opposition coincides in time with the arrival of Mars at perihelion, where the orbits of Mars and earth are at their closest, the distance is only about 50 million kilometers, and the red planet appears three times as bright as Sirius. Favorable oppositions of Mars occur in the latter part of August, every fifteen to seventeen years.

The sidereal and synodic periods of Mars are 687 days (the Martian year) and 780 days, of which the planet advances for 710 days and retrogrades for 70. The synodic period, the longest for any of the planets, is a consequence of the comparatively small difference between the sidereal periods of earth and Mars: the former takes over two of our years to catch up by one lap.

The varying distance from the earth produces a sevenfold variation in the apparent diameter, from 3".5 at conjunction to 25".1 at favorable opposition. The diameter, unlike that of Venus, is easy to measure accurately and is 6780 kilometers. Therefore, Mars has a little more than half the earth's diameter, 0.291 of its surface, and 0.153 of its volume.

Mars, like the earth, is slightly flattened at the poles. Direct measures of the oblateness give a value $1/190$, and the perturbations of the equatorial bulge on the two satellites give $1/192$, in good agreement. Thus, even though Mars' day is almost the same as the earth's, indeed a little longer, the rotational distortion is greater. Mars must be less condensed toward the center than the earth is, and its density also is rather smaller.

* The poem can be dated in the spring of 1854 by the fact that the poet places Mars in the constellation Leo.

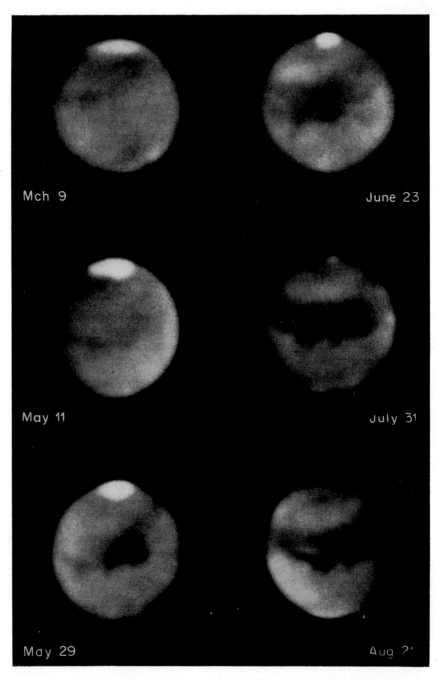

Fig. 8.5. Seasonal changes on Mars—the dates given are Martian seasonal dates, taken to correspond to those on the earth. (E. C. Slipher, Lowell Observatory)

To measure the density directly the mass must be determined, and this is best done from the two satellites. The mass of Mars is 1/3,093,500 of the sun and 0.1078 of the earth. Therefore, the mean density is 0.70 of the earth's; the surface gravity, 0.37 of the terrestrial value; and the velocity of escape from the surface, about five kilometers a second.

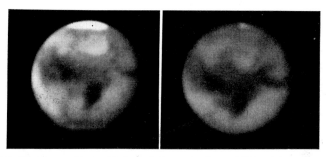

Fig. 8.6. The great Syrtis Major—left, springtime, 1939, right, summer, 1941. Photographs of the same face of the planet displaying change in the axial tilt, reduction of the southern snow cap and the darkening of various areas with the advance of summer in the south of Mars. (E. C. Slipher, Lowell Observatory)

The surface of Mars reflects light poorly, about twice as well as the moon's and one-third as well as the earth's. Red light is reflected better than blue; the photographic albedo is 0.090 and the visual albedo, 0.154: the planet reflects less blue light, and so it looks red. Probably the reflecting surface resembles dark rocks. The phase effect, however, is more like that of Venus than of the moon, so Mars cannot have a very rough surface, and probably there are no considerable mountains.

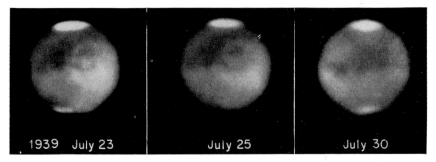

1939   July 23   July 25   July 30

Fig. 8.7. Mars (1939). Photographs displaying rapid variations about the northern pole (lower) in Martian autumn when the forming cap consists largely of cloud before the more permanent snow cap appears. (E. C. Slipher, Lowell Observatory)

Mars rotates on its axis in $24^h37^m22.58^s$. The earliest drawings of the surface, made in the seventeenth century by Hooke and Huyghens, permit a very accurate determination of the rotational period. Thus the Martian day is about half an hour longer than our own. The planet's brightness varies by about 15% as it turns, which shows that different parts of its surface have different reflecting properties.

The rotation of Mars is more nearly equal to that of the earth than is the case for any other planet. While they both turn on their axes it is pos-

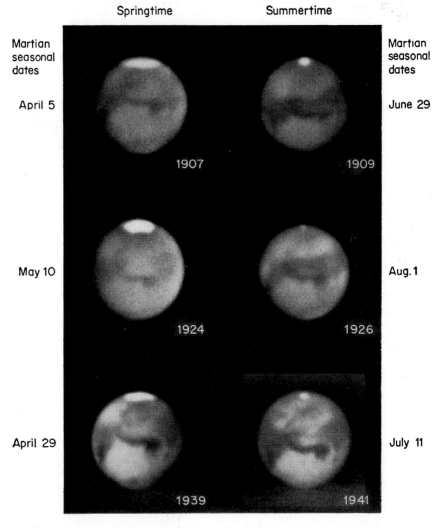

Springtime   Summertime

Martian seasonal dates

April 5

1907

Martian seasonal dates

June 29

1909

May 10

1924

Aug. 1

1926

April 29

1939

July 11

1941

Fig. 8.8. Mars. Photographs of same face of the planet in its springtime and summer (south) showing the darkening of the Deucalion region in summer. (E. C. Slipher, Lowell Observatory)

sible, from different parts of the earth, to see the whole surface of Mars in the course of twenty-four hours. From any one place on the earth, the same face is seen at intervals of forty days.

The orbit of Mars is very nearly in the same plane as our own; it is inclined at 1°51′ to the ecliptic. The inclination of its rotational axis to the ecliptic is also very like our own, about 75°. Thus the Martian seasons

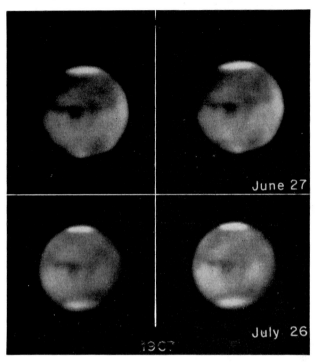

Fig. 8.9. Early photographs of Mars, made June 27 and July 26, 1907, showing same longitude, Dawes Forked Bay, and the south cap in its springtime. Note the great expansion of the northern cap in its autumn. The long series of photographs secured at that opposition recorded the seasonal decline of the south snow, and rapid fluctuations in the north cap in autumn which, with other aspects of its behavior, denoted it to consist mainly of clouds. (E. C. Slipher, Lowell Observatory)

should be similar to ours. Mars is in perihelion when its south pole is inclined toward the sun, and therefore the southern hemisphere has a hotter (though shorter) summer than the northern, just as the earth has.

Because Mars moves in an orbit outside the earth's, it can never be seen as a crescent; at the quadratures it is appreciably gibbous (as was observed by Galileo).

The surface of Mars has been intensively studied with both the eye and the photograph. Some features of the landscape are permanent, some appear to change, and some are regular seasonal phenomena. The general color is reddish-orange scattered with bluish-gray regions, and the white polar caps vary with season. The reddish surfaces are probably a Martian "Sahara." The bluish-gray regions cannot be bodies of water, or they would reflect sunlight at certain angles. They vary with season in a regular manner and are very likely associated with the growth of vegetation in areas of changing dampness.

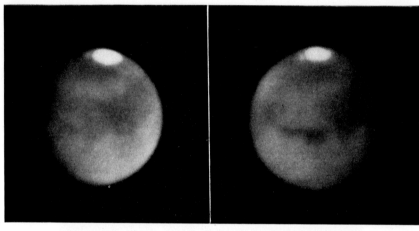

August 21, 1909   September 8, 1924
(Martian date—June 3)   (Martian date—June 3)

Fig. 8.10. Snow on the mountains of Mitchell (left, August 21, 1909, Martian date June 3) (right, September 8, 1924, Martian date June 3). Photographs showing a lingering patch of snow completely detached from the main snow cap. These photographs, although taken 15 years apart, show the planet on the same Martian seasonal date. Discovered by Mitchell in 1845, this event has been observed to occur at the same time each Martian summer that circumstances have permitted us to witness it. (E. C. Slipher, Lowell Observatory)

The most controversial features of the Martian landscape are the markings that were originally called *canali* [channels] by the Italian observer Schiaparelli, who began to observe them in 1877. An unfortunate mistranslation of the word as "canals" has conveyed the impression that these markings are constructed by intelligent beings, perhaps for irrigation. The most careful observations, since the days of Schiaparelli, have not produced any unanimity as to the detailed appearance of the so-called canals, let alone their nature.

Percival Lowell founded the Lowell Observatory, in Flagstaff, Arizona, for the study of the planets, and especially of Mars. Both climate and instruments were the best available at the time. Lowell saw the canals as straight and narrow (25 to 30 kilometers wide), uniform along their length, and covering the surface of the planet with a network of great circles. They were observed to meet (sometimes as many as fourteen to a point) in "oases," 120 to 160 kilometers across. Four hundred canals were mapped, and nearly 200 oases; 50 of the canals were recorded as double. Lowell saw the canals faint in spring; with the melting of the polar caps he noted an

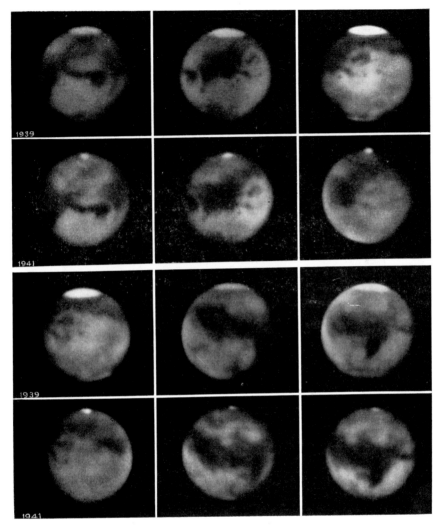

Fig. 8.11. Mars. Two series of photographs around the planet showing comparable views in 1939 and 1941. The 1939 photographs were made in South Africa, the other at Flagstaff. (E. C. Slipher, Lowell Observatory)

increase in their intensity, a "wave of quickening" flowing south at about 80 kilometers a day.

William Pickering, though he also saw canals, thought them curved and not great circles, and never saw double ones. Barnard, perhaps the keenest visual observer of modern times, never saw canals at all, only hazy lines connecting dark spots.

It is very likely that some of the details that have appeared on Martian maps are the result of unconscious amplification by the eye of markings faintly seen. The presence of real markings is undoubted; but of the details it is more difficult to be certain. And the conclusions drawn by Lowell are hardly justified: that the straightness and doubling of the canals point to intelligent agency; that the progressive darkening shows artificial propulsion

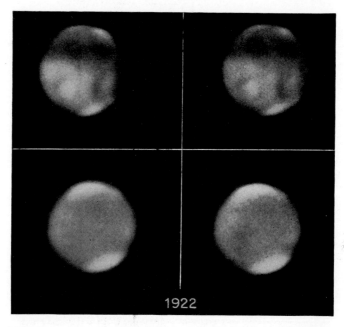

Fig. 8.12. Mars, June 17, 1922 (photographs in yellow and blue light at same hour). When early blue-light images failed to reveal the familiar surface markings of Mars it was thought at first by some to be due to selective reflection, the balancing out of the dark bluish markings and the bright reddish areas in blue light. But later the detection of occurrences of temporary "clearing" when the markings appeared distinctly in such photographs demonstrated the unexpectedly high opacity of Mars' atmosphere to blue light. (E. C. Slipher, Lowell Observatory)

of water; and that Mars must have a "world civilization," since the whole surface seems covered with a uniform network! The appeal of such ideas to the popular imagination is rarely tempered by the recollection that all observers have not agreed about the facts, and that even if established, the facts may admit of other interpretations (such, for example, as progressive change of atmospheric haziness with season, which could explain the varying visibility of the markings).

Photographs of the surface of Mars have not shown as much detail as visual observers see. Pictures made in violet and infrared light show very different features; the former no doubt reach only the outer bounds of a scattering atmosphere, while the latter reach the surface.

The polar caps show the seasonal changes, very similar to our own. The white areas near the poles are probably rather hoar-frost than thick layers of ice like our own polar caps, and they have been found to reflect light in the way characteristic of ice crystals. The south polar cap (where summer is hotter than at the north pole) disappears entirely toward the end of its

MARS 1937

May 20   Red   April 20   Blue

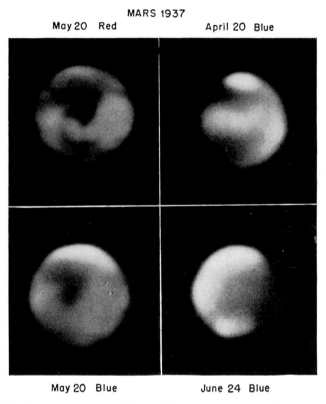

May 20   Blue   June 24   Blue

Fig. 8.13. The atmosphere of Mars, 1937 (upper left, May 20, red; upper right, April 20, blue; lower left, May 20, blue; lower right, June 24, blue). Red and blue photographs of the same face of Mars displaying extraordinary changes in the transparency of the Martian atmosphere to blue light. (E. C. Slipher, Lowell Observatory)

summer. An isolated white patch, nearby, is the last to disappear, and may come from a high plateau; however, the plateau cannot be more than 2500 feet high or it would be visible as an irregularity at the limb of Mars.

That Mars has some atmosphere is shown by Figure 8.9. The polar caps can only be produced by precipitation, which requires an atmosphere. Moreover, there is a slight twilight arc when Mars is gibbous. Toward the edge of Mars the surface markings are obscured, as they should be by a greater atmospheric thickness there; moreover, this obscuration is greatest in blue light, just as it would be for an atmosphere like the earth's. Occasional fogs and haze are seen, and yellow clouds sometimes whirl across the deserts. But the atmosphere is tenuous compared with ours; spectro-scopic studies show that it contains not more than 5% as much water as our atmosphere does, and not more than 15% as much oxygen. Nitrogen is not detectable, but probably present. The atmospheric pressure cannot

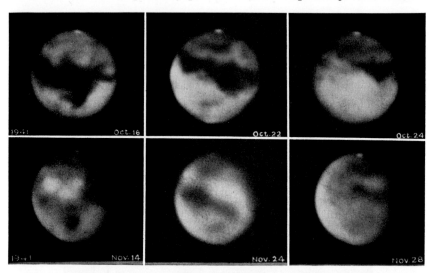

Fig. 8.14. Clouds on Mars (1941) in yellow light. The striking difference between the upper and lower rows of photographs is solely due to clouds in the atmosphere of Mars. The large dark areas so clearly visible in the upper photographs are hardly recognizable in the lower ones when they were heavily obscured by Martian clouds. Such Martian storms are rather rare and extremely transitory, lasting but a day or two. (Alternately the upper and the lower photographs are of the same faces of the planet.) (E. C. Slipher, Lowell Observatory)

be more than one-fifth of ours. Such conditions would be unfriendly, but not prohibitive, to life of our kind.

On the equator at noon, the temperature of Mars is about 50°F, and about −90°F during the polar night. The complete melting of the polar caps in summer shows that they must be very thin, and it is estimated that the whole surface of Mars has only as much water as is contained in Lake Erie. The planet must therefore be essentially a desert.

Positive information about living things on Mars is practically unattain-able. The green areas seem not to be grassy, since they do not show the characteristic reflection by chlorophyll of infrared light. Perhaps they are covered with something like the terrestrial lichens, fungi and algae that have solved the problem of living under very adverse conditions by a sym-biotic partnership: the alga undertaking the task of nutrition and the fungus, of reproduction. Such lichens as *Lecanora calcarea* can live on such un-promising sites as chalk and mortar, and some even bury themselves. Lichens, which can live at low temperatures and resist desiccation, seem the most likely living things for the surface of Mars. The fantastic creatures so dear to science fiction have no basis in observed astronomical fact.

The most fantastic things about Mars are its two moons, Phobos and Deimos (names of the attendants of Mars, culled from the *Iliad*), which

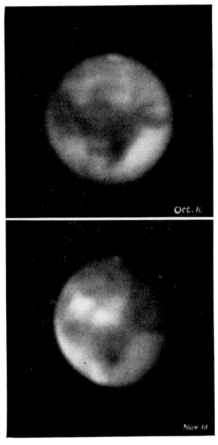

Fig. 8.15. Clouds on Mars (1941). Upper photograph shows the face of the planet under a normally clear Martian sky; the lower photograph shows same face some weeks later, widely obscured by bright yellowish clouds. (E. C. Slipher, Lowell Observatory)

were discovered by Asaph Hall in 1877. They travel around the planet in nearly circular orbits, inclined respectively 1¾° and 1° to the equator of the planet. The equatorial bulge of Mars acts on these satellites as the bulge of the earth acts on the moon: the lines of apsides advance, and the nodes regress, with periods of 56 years for Deimos, 2¼ years for Phobos. Phobos, the nearer to the planet, has a diameter of about 16 kilometers; Deimos is about half that size.

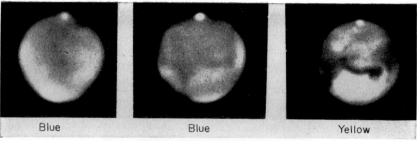

Blue   Blue   Yellow

Fig. 8.16. Blue clouds of Mars. These images show about the same face of the planet. From left to right they are: (1) in blue light showing the normally hazy and obscured appearance of the planet in short wave lengths; (2) in blue showing the occasional widespread clearing which permits photographing surface features in the blue; (3) in yellow for comparison showing the normally clear aspect of the planet in longer wave lengths, and delineating the surface markings quite clearly. (E. C. Slipher, Lowell Observatory)

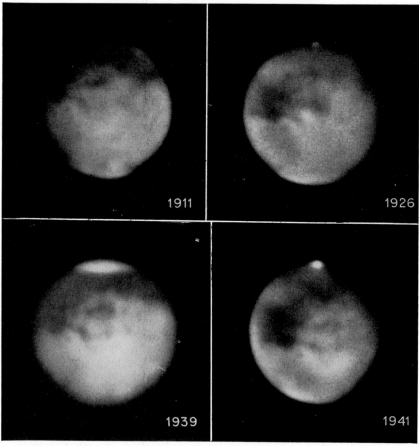

Fig. 8.17. Changes in the Solis Lacus. Photographs in 1911, 1926, 1939, and 1941, recording changes in the size, shape and structure of Solis Lacus of Mars. (E. C. Slipher, Lowell Observatory)

200

Both moons go around Mars very close to the surface in the same direction as the planet rotates and revolves. The orbit of Deimos has a radius of 23,500 kilometers; that of Phobos, 9370 kilometers. Deimos, the outer one, has a period of 30^h18^m, only a little longer than the Martian

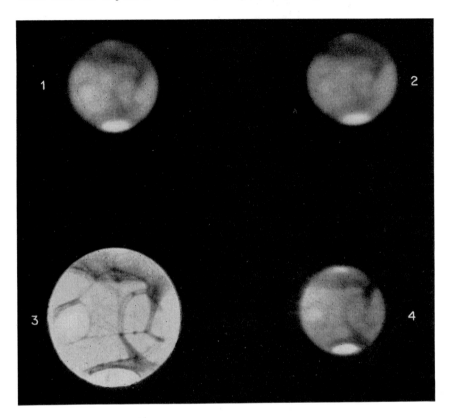

Fig. 8.18. Mars (February, 1916). Direct photographs (1, 2), an independent drawing (3), and a telescopic photograph of the drawing (4). Comparison of these telescopic photographs of the planet with the drawing and the telescopic photograph of it, shows a fair agreement between them and confirms the existence of the "canals." (E. C. Slipher, Lowell Observatory)

day. As seen from Mars, it would rise (like our moon) in the east, but because its month is but little longer than Mars' day, it would not be seen to rise again for 132 hours. Phobos, on the other hand, has a period of only 7^h39^m, and goes around Mars more than three times while Mars itself turns only once! No other satellite has a period less than that of its primary. Phobos accordingly rises in the west and sets in the east, to rise again eleven hours later, so it passes across the sky twice every (Martian) night, apparently going the opposite way from Deimos.

Small as it is, Phobos is so near to the surface of Mars that it would look to the hypothetical Martians one-third the size of our moon, though only 4% as bright. Deimos, larger but more distant, would be less than

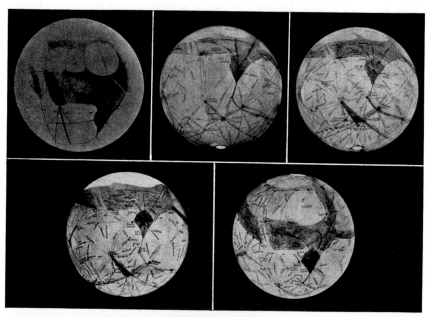

Fig. 8.19. Lowell's globes of Mars, from left to right, for the years 1894, 1901, 1903, 1905, and 1907. Drawn on globes and photographed. (Lowell Observatory)

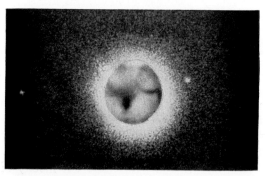

Fig. 8.20. Satellites of Mars photographed August 17, 1924. Phobos is the planet at the right. Deimos is considerably farther from the planet at the left. (Image of Mars has been superimposed.) (E. C. Slipher, Lowell Observatory)

one-quarter the size of Phobos in the sky, and about one-fortieth as bright. The moons of Mars, if not the oddest in the solar system, are unique in their way.

5. JUPITER

Turning, I perceived
The whiteness round me of the temperate star
The sixth, whereinto I had been received.
And in that torch of Jove, I was aware
Of sparkles from the love within it warm . . . .
O sweet star, with how many and rare a gem
Didst thou prove that the justice we obey
Proceedeth from the heaven thou dost begem!
DANTE, *Paradiso,*
Canto XVIII, 67–71, 115–117 *

Mercury, Venus, earth, and Mars complete the tally of the inner terrestrial planets. Outside the orbit of Mars, where the Bode Law anticipates another planet, we find not one, but over a thousand—the asteroids, or minor planets. They are so different from the nine principal planets that we shall consider them, with the other small bodies of the system, in Chapter IX. Outside the asteroids lie the four major planets, Jupiter, Saturn, Uranus, and Neptune; and the tiny Pluto. They form a natural group; in fact they resemble one another, in many respects, more than the terrestrial planets do. The first and largest of the major planets is Jupiter, rightly named after the ruler of the Olympians.

Fig. 8.21. Jupiter, photographed August 10, 1938, showing the region of the Great Red Spot. At the time the red spot itself was almost white, but its location is marked by the dark elliptical ring lying just above the center of the disc. One of Jupiter's moons, Satellite I, is seen here just beginning to cross the face of the planet (the small white spot to the right). The black spot near it is its shadow on Jupiter and therefore represents a total solar eclipse on that part of Jupiter. (E. C. Slipher, Lowell Observatory)

* Dante, in accordance with the Ptolemaic system, makes Jupiter the sixth "star" of the solar system.

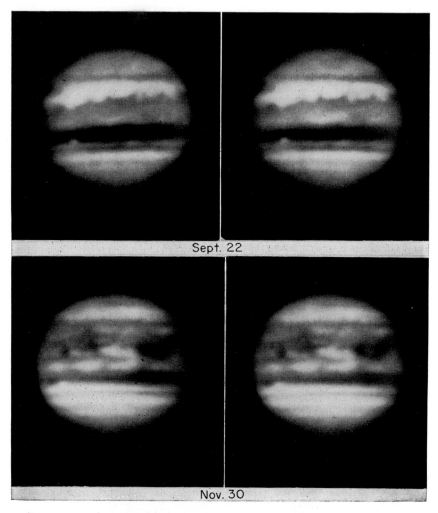

Sept. 22

Nov. 30

Fig. 8.22. Jupiter (1928). Photographs of the same face of the planet taken September 22 and November 30, showing a rapid transformation of the surface features. (E. C. Slipher, Lowell Observatory)

Jupiter has a mean distance of 5.20 astronomical units from the sun, 778.73 million kilometers. With a rather small orbital eccentricity, 1/20, its distance from the focus of its orbit still varies by 76 million kilometers. It can be as far from the earth as 965 million kilometers, and as near as 590 million; at its faintest it is a little fainter than Sirius, and at its brightest, more than twice as bright. Excepting Venus, and occasionally Mars, it is the brightest of all the planets in the sky.

The apparent diameter varies between 50″ at a favorable opposition and 32″ at conjunction. The surface is so bright that, as with Venus, the measured size is slightly falsified by irradiation. The best value of Jupiter's diameter is obtained from the width of its shadow when one of its numerous satellites is eclipsed. The resulting equatorial diameter is 142,880 kilometers.

Jupiter is so oblate that the flattening is readily visible to the eye through a suitable telescope. The actual oblateness, like the diameter, is hard to measure correctly, and the best value is obtained from studying the perturbations produced by the equatorial bulge on the motions of the nearest satellites; the resulting value is 1/15.4. Thus the diameter through the poles is 133,600 kilometers, and the mean diameter, 139,760 kilometers. Jupiter is truly a giant in dimensions, nearly 11 times the earth in diameter, 120 times in surface area, 1318 times in bulk. In both volume and mass, it is larger than all the other planets put together.

The mass of Jupiter is best determined from the motions of its inner satellites and the perturbations it produces on the motions of asteroids. It is 1/1047 of the mass of the sun and 318 times that of the earth. From mass and dimensions, the density of Jupiter is found to be less than a quarter of the earth's, and the density 1.33 gm/cc, a little smaller than that of the sun! Small densities are typical of all the major planets, and are one of the most significant differences between them and the terrestrial planets.

In spite of its large size, Jupiter has so great a mass that the mean gravity at the surface has 2.64 times the value it has on the earth, and the velocity of escape, the greatest for any planet, is 60 kilometers a second. The equatorial diameter of Jupiter is so much greater than the polar, and the planet is spinning so fast, that gravity is 15% smaller at the equator than at the poles. The corresponding difference on the earth is only 1/189. Nevertheless, the distortion produced by the rapid spin of Jupiter is less *in proportion to the centrifugal force* than that of the earth; we may infer from this that Jupiter is even more centrally condensed than the earth is.

Jupiter is an excellent reflector, almost as good as Venus, and has an albedo of 0.44. The phase effect also is small, and both these facts point to a cloudy surface. The disc is, moreover, greatly darkened at the limb, where it is only one-eighth as bright as in the center, and this effect is certainly the result of absorption in a thick layer of atmosphere.

The orbital period of Jupiter about the sun (sidereal period) is 11.86 years; the synodic period only 399 days, the time taken by the earth to catch up with Jupiter by one lap.

Jupiter's day is, surprisingly, the shortest found for any of the planets, only 9^h55^m; it is measured by the rate at which we see the cloud features turning and is different in different latitudes, most rapid near the equator (compare the sun). The rotational differences show no simple relationship to the latitude and are not symmetrical about the equator of the planet; a given zone or feature may change its rotational period appreciably in

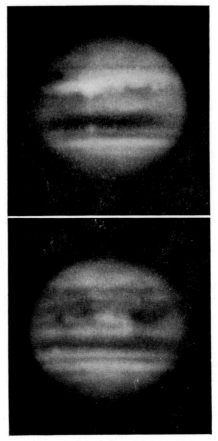

Fig. 8.23. Jupiter. Two photographs of the same face of the planet showing rapid changes in surface markings. Great Red Spot is visible near the left edge. (E. C. Slipher, Lowell Observatory)

either direction. Clearly we are not seeing the solid surface of a planet with permanent markings on it.

The rich orange, red, brown, and occasionally green, bands that run roughly parallel to the equator are probably weather zones, and to a much smaller extent our earth would show a similar banded appearance to an observer outside it. The bands are certainly belts of clouds, and may show slow, or even rapid, changes.

Although we do not see the actual surface of Jupiter, some features of the surface that we do see have a permanency that suggests the presence of something less evanescent than clouds, below the visible surface. Such a feature is the Great Red Spot, which appeared in 1831 as a sort of hollow among the cloud belts, and was seen in 1878 as a deep red spot, 50 thousand kilometers long and 11 thousand kilometers wide. The

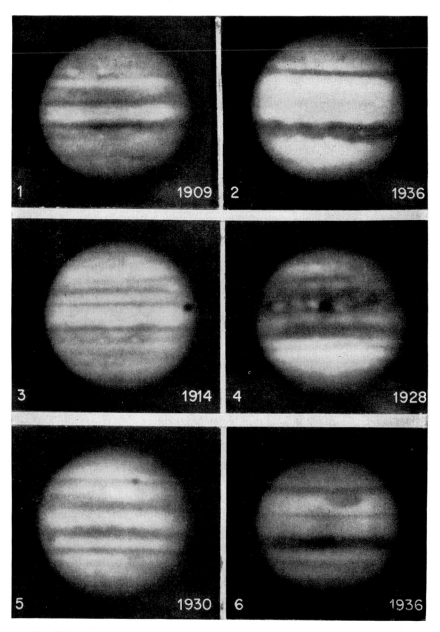

Fig. 8.24. Jupiter. Examples of photographs illustrating the type and magnitude of changes which occur on this planet. Sometimes a certain zone may be very dark, at another time practically white, as in Nos. 1, 2, 3, and 4. At one time the Great Red Spot is nearly pure white as in No. 5, at other times dark red as in No. 6. Black dots on Nos. 3 and 5 are the shadows of Satellites III and I. (E. C. Slipher, Lowell Observatory)

redness later disappeared, but a notch was still visible in the cloud bands; still later the Red Spot reappeared. Here is a feature that has been visible in some form for over a hundred years, though it has not gone around with quite a constant period, but has drifted erratically about a fixed

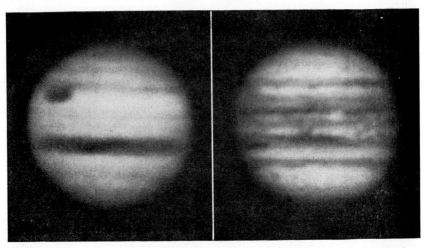

Fig. 8.25. The giant planet Jupiter (left 1936, right 1938) showing characteristic changes in his cloud banded surface. The famous Great Red Spot is the dark oval in left-hand photograph. (E. C. Slipher, Lowell Observatory)

position. It is not at all unlikely that the Red Spot represents the result of a volcanic eruption on the planet's surface, which has thrown up metallic vapors into the clouds. Jupiter's constitution is so unlike the earth's that when we speak of a volcano we do not necessarily mean an erupting mountain; a volcano on Jupiter might resemble a gigantic bubble bursting up from a semi-fluid interior.

The spectrum of Jupiter shows some striking features that are not found for the terrestrial planets, but are common to a greater or less extent to all the major planets. Dark bands in these spectra, long a mystery, have been identified as molecular spectra of two well known gases, methane (CH_4) and ammonia (NH_3). In the spectrum of Jupiter we see the equivalent of a thirty-foot column of ammonia at atmospheric pressure, and of half a mile of methane. Thus the atmosphere of Jupiter is very rich in hydrogen, which has combined with all the available nitrogen and carbon. If water vapor is present it would be frozen, for the surface temperature is $-100°C$. At this temperature, most of the ammonia (which boils at $-2°$ and freezes at $-42°$) would be solid, and only a little vapor present. Methane, however, boils at $-126°$ and freezes at $-150°$, so all the methane present is probably gaseous.

Although the spectrum of gaseous hydrogen is not seen in the light of the major planets, hydrogen may be present as such in large quantities; the lines of its spectrum would not appear at the temperature of Jupiter's surface. The low average density of Jupiter, and the fact that hydrogen is cosmically by far the most abundant element, suggest that most of the outer envelope consists of gaseous hydrogen, mixed with as much methane and ammonia as the available carbon and nitrogen can produce, and contaminated with metallic vapors, which give the cloud banks their reddish and yellowish tinge.

The structure of Jupiter (as suggested by Wildt) is very different from that of any terrestrial planet; the latter are solid planets with a thin envelope of atmosphere, a few per cent of the diameter of the planet in depth, or even altogether absent. Most of the volume of Jupiter, on the contrary, seems to be atmosphere. The actual planet, thought to consist of a metallic-rocky core of density 6 gm/cc, occupies only about 43% of the observed radius, and 8% of the observed volume. The core is imagined to be overlaid by a thick layer of ice (frozen water and other light materials) of density of about 1.5 gm/cc, that occupies about 39% of the radius. The outer 18% or so of the radius is thought to be an atmosphere of hydrogen, contaminated with other gases and some metallic vapors, and with a mean density about 0.35 gm/cc. We cannot picture a sharp boundary between atmosphere and icy layer, but rather a gradual transition from gaseous envelope to a slushy surface and a frozen zone.

The outer temperature of Jupiter is measured by the quality of the light it reflects. We do not know the inner temperatures, but they may be somewhat higher. The rocky surface can hardly be more than a few tens of degrees hotter than the exterior. Certainly Jupiter is not self-luminous, as was once believed; if it were, we should see the moons faintly illuminated when they are eclipsed, whereas they disappear completely in Jupiter's shadow. Also the quality of Jupiter's light, as recorded by the thermocouple, shows that all is reflected sunlight.

Perhaps the most interesting thing about Jupiter, as about Mars, is his system of twelve known moons. The four brightest were seen by Galileo; he dubbed them the "Medicean stars" in honor of his patron, but they have since received mythological names connected in legend with Jupiter's amours. The other eight known moons have, oddly enough, remained unnamed.

Only the diameters of the four largest moons are directly measurable; the values are slightly affected by irradiation. Better determinations can be made from the timing of their eclipses. The masses, in terms of Jupiter's mass, are determined from their considerable mutual perturbations. Densities are calculated from the observed masses and sizes.

All four bright satellites are comparable to the moon in size, and the two largest are actually bigger than Mercury. If they were not so near to

The Moons of Jupiter

No.	Name	Discovery	Mean Distance (km)	Sidereal Period	Inclination to Jupiter's Orbit	Orbital Eccentricity	Diameter (km)	Apparent Magnitude*
5	...	Barnard 1892	181,500	$11^h57^m23^s$	3°6.9'	0.0028	150?	13.0
1	Io	Galileo 1610	422,000	$1^d18^h27^m34^s$	3°6.7'	0.0000	3,730	5.5
2	Europa	Galileo 1610	671,400	$3^d13^h13^m42^s$	3°5.3'	0.0003	3,150	5.7
3	Ganymede	Galileo 1610	1,071,000	$7^d\ 3^h42^m33^s$	3°2.3'	0.0015	5,150	5.1
4	Callisto	Galileo 1610	1,884,000	$16^d16^h32^m11^s$	2°42.7'	0.0075	5,180	6.3
6	...	Perrine 1904	11,500,000	250.7^d	28°45'	0.155	120?	13.7
7	...	Perrine 1905	11,750,000	260.0^d	27°58'	0.207	50?	16.
10	...	Nicholson 1938	11,750,000	260.0^d	28°	0.08	20?	17.8
12	...	Nicholson 1951	21,200,000	625.419^d	147°	0.155	22?	18.9?
11	...	Nicholson 1938	22,500,000	692^d	163°	0.21	25?	17.4
8	...	Melotte 1908	23,500,000	738.9^d	148.1°	0.378	50?	16.
9	...	Nicholson 1914	23,700,000	745.0^d	157°	0.27	22?	16.

* At mean opposition of Jupiter

Satellite:	Io	Europa	Ganymede	Callisto
Diameter:				
direct (in km):	3,960	3,220	5,700	5,390
eclipses:	3,730	3,150	5,150	5,180
Mass:				
(in terms of moon):	1.09	0.65	2.10	0.58
Density:				
in g/cc:	2.9	2.9	2.2	0.6

Jupiter we could see at least three of them easily with the naked eye, and they are a beautiful sight even with field glasses.

The density of Io and Europa is not very different from that of our moon, and they are probably rocky bodies. Callisto, however, has a density lower than that of the least dense planet (Saturn), and is very likely a chunk of ice. Perhaps Ganymede is partly rock, partly ice. All four satellites have albedos comparable to that of Jupiter; if our moon resembled them it would be three to six times as bright as it is. All four show a phase effect, so their surfaces are probably rough. They can hardly have atmospheres; surface gravity and velocity of escape are too low.

The satellites are eclipsed when they pass into Jupiter's shadow, and both they and their shadows are seen passing across the planet's disc when they come in front of it. On almost any night one can observe one or more eclipses or transits of the large satellites. The timing of the eclipses is the most accurate means of studying the motion of the satellites in their orbits and determining the perturbations.

The timing of the eclipses of Jupiter's satellites led to the first detection and actual measurement of the velocity of light. Empedocles had realized that light travels with a finite velocity in the fifth century B.C., but the effect was not observed until Ole Roemer detected and measured it in 1675. He noticed that when eclipses of Jupiter are viewed across the earth's orbit, they seem to occur relatively later than when Jupiter and earth are on the same side of the sun. The difference represents the time taken by light to cross the orbit of the earth. Roemer found this time to be 16½ minutes, and thus made the first calculation of the speed of light. In modern times this speed has been more accurately measured experimentally, by Michelson and others, than Roemer was able to do, and with the modern value (299,796 km/sec) the calculation can be reversed, and the diameter of the earth's orbit found from the "light equation" for Jupiter's moons. This value of the solar parallax cannot, however, compete in accuracy with other methods.

The table of Jupiter's moons, and the picture of their orbits, shows that they fall into three groups. The first group contains the five inner moons, which have very small orbital eccentricities, and orbits only slightly inclined to the planet's equator. Next comes a group of three moons with rather eccentric orbits, inclined to the planet's equator by almost 30°;

they are all nearly at the same distance, between 11 and 12 million kilometers from it. The third group of satellites is very different; these are distant between 20 and 24 million kilometers from the planet, have quite eccentric orbits, and the motion is *retrograde* (as is shown by the fact that

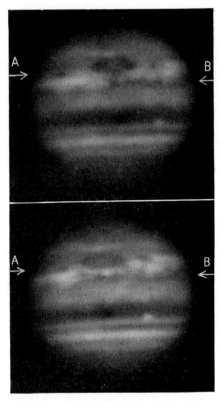

Fig. 8.26. Jupiter (1928). Two photographs of the Great Red Spot region taken 49½ hours apart showing change in shape of the Red Spot and the rapid drift of dark spots past it. Spots A drifting to right, spot B to left, relative motion 8000 miles per day. This mobility reveals that his surface consists of clouds. (E. C. Slipher, Lowell Observatory)

the tabulated orbital inclinations are greater than 90°). The objects that we have hitherto described in the solar system have *direct* motions; that is, they all go round in the same direction, and they rotate in the same direction too. The four outer moons of Jupiter are going around in the opposite direction in highly inclined orbits.

This remarkable group of moons possibly consists of asteroids that have been "captured" by Jupiter's large gravitational pull. Almost all the asteroids have orbits smaller than that of Jupiter, and therefore their linear speeds are larger, so that relative to Jupiter they are traveling from west to east, whereas the planet's own satellites go from east to west. An asteroid that came near enough to Jupiter to be deviated into a path round the planet would therefore be going the opposite way from the other moons. Very likely direct motion in such large orbits as those of the outer four would be unstable; a direct moon would be in danger of being perturbed out of its orbit around Jupiter and might become an asteroid. Even No. 8, which, because of its large orbital eccentricity, moves to the greatest distance from the planet, may be lost in this manner at some future time. On the long view, Jupiter may have a floating population of distant moons. As we shall see, this most massive of the planets has a far-reaching effect on the motions of the smaller bodies of the system, both asteroids and comets. One of the most recent theories of the origin of planets suggests the possibility of the formation of these retrograde satellites as part of the normal evolutionary process.

6. SATURN

While Saturn whirls, his steadfast shade
Sleeps on his luminous ring.

TENNYSON

The most beautiful, and in many ways the most interesting of the planets is Saturn, the remotest known in antiquity. Its distance from the sun is 9½ astronomical units, 1,427,700,000 kilometers, and varies by 160 million kilometers between aphelion and perihelion on account of the considerable eccentricity of the orbit, 0.056. The distance of Saturn from the earth varies between 1200 million and 1650 million kilometers. Despite its remoteness, the planet is always as bright as the first magnitude; its variations of brightness are caused more by the varying presentation of the ring system than by changes of distance.

Fig. 8.27. Saturn, 1941, December 8. (Lowell Observatory)

Saturn is the most oblate of all the planets: the best method of determining the oblateness, as with Jupiter, is from the perturbations of the satellites' orbits by the equatorial bulge of the planet. The result is 1/9.5, and the value from direct measures is 1/9.2. From the apparent diameter (20″ to 14″, depending on the distance from us) the mean diameter is 115,100 kilo-

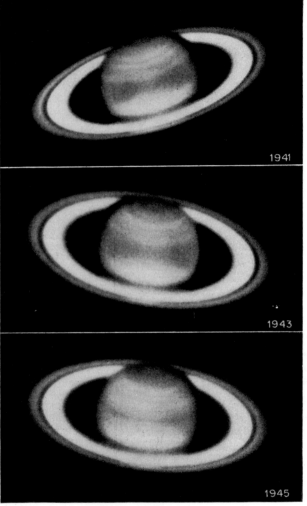

1941

1943

1945

Fig. 8.28 Saturn, representative photographs in 1941, 1943, 1945, showing moderate change in the belts. Note that the ball shines through Cassini's division and ring *A* where they cross in front of it. (Photograph by E. C. Slipher, Lowell Observatory.)

meters, or about nine times that of the earth. The equatorial and polar diameters are 119,400 and 106,900 kilometers respectively.

The mass of Saturn, as determined from observations of the inner moons and the perturbations of Jupiter by Saturn, is 1/3499 of the sun's and 95.3 times the earth's. Hence Saturn has the surprisingly low density of 13% of the earth's, or 0.715 gm/cc. Thus Saturn is the only planet that would float in water; no other planet is of such low density. The surface

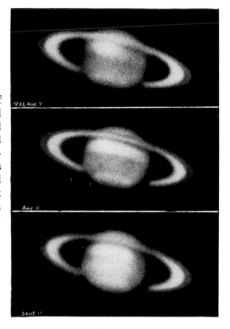

Fig. 8.29. The Great White Spot of 1933. This spot remained visible until October and like its counterpart, the Hall Spot of 1876, expanded in a forward direction (i.e., direction of planet's rotation) until it extended over three-fifths of the planet's diameter. Its north and south limits remained within the bright equatorial zone of Saturn. (Photograph by E. C. Slipher, Lowell Observatory.)

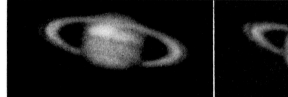

Fig. 8.30. Striking changes on Saturn. Photograph at left was taken August 9, 1933, showing the Great White Spot of that year. The photograph of September, 1934, shows marked changes over the whole Northern Hemisphere as compared to its appearance in 1936. (Photograph by E. C. Slipher, Lowell Observatory.)

gravity is 1.17 that at the earth's surface, and it varies by 30% between equator and pole because of oblateness and rapid rotation.

The low density and oblateness of Jupiter made plausible the picture of its constitution as involving a rocky core overlaid with ice and an atmosphere of hydrogen. Saturn, with an even lower density, suggests an even more extreme structure. Wildt has proposed a structure that involves a 35% rocky core, a 20% ice layer, and a 45% hydrogen atmosphere to account for the low density.

The spectrum of Saturn is in harmony with this idea. It shows stronger bands of methane and weaker bands of ammonia than Jupiter, and the lower surface temperature of Saturn is considered responsible for keeping

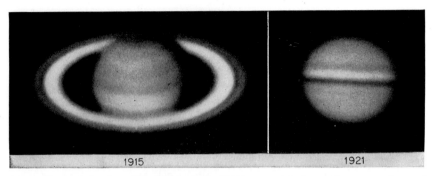

Fig. 8.31. Saturn's rings (left, 1915; right, 1921) seen at their maximum opening and edge on. (Photograph by E. C. Slipher, Lowell Observatory.)

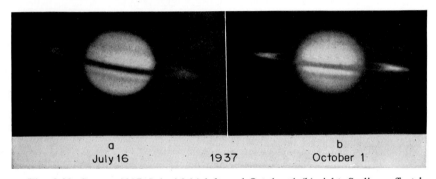

Fig. 8.32. Saturn, 1937, July 16 (a) left, and October 1 (b) right. Seeliger effect in Saturn's rings demonstrated photographically. (a) Earth is higher than sun above rings, and each particle of ring shadows the next behind causing rings to appear dark compared to ball. (b) The sun is higher than the earth above rings, and the particles as we see them are in full sunlight and the rings appear much brighter as compared to the ball. (Photograph by E. C. Slipher, Lowell Observatory.)

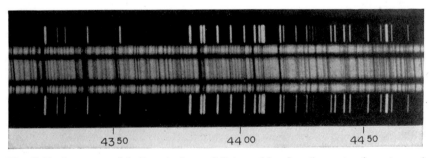

Fig. 8.33. Spectrum of ball and rings of Saturn (showing the meteoric nature of rings). (Lowell Observatory)

even more of the ammonia in a frozen state. Moreover, the methane layer is denser than on Jupiter.

The surface of Saturn is like that of Jupiter in the well-marked belts

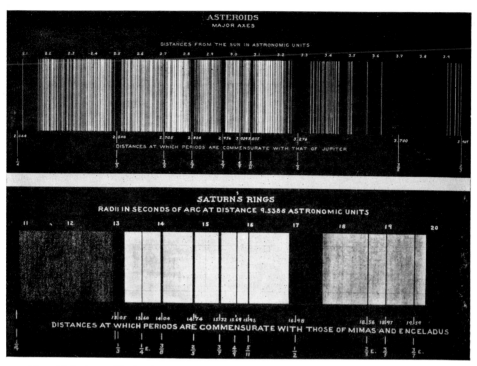

Fig. 8.34. Asteroids (major axes) and Saturn's rings (radii in seconds of arc at distance 9.5388 astronomical units). (Lowell Observatory)

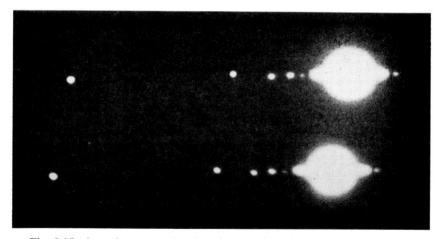

Fig. 8.35. Saturn's moons, the six brightest of its family of nine. Two photographs taken about two minutes apart, showing the six moons nearest to Saturn, namely, from left to right: Titan, Rhea, Dione, Tethys, Mimas, and Enceladus. Titan, the largest, is larger than our moon and is the only moon known to possess an atmosphere; the other five are now thought to be composed of snow and ice. (Photograph by E. C. Slipher, Lowell Observatory, 1921.)

217

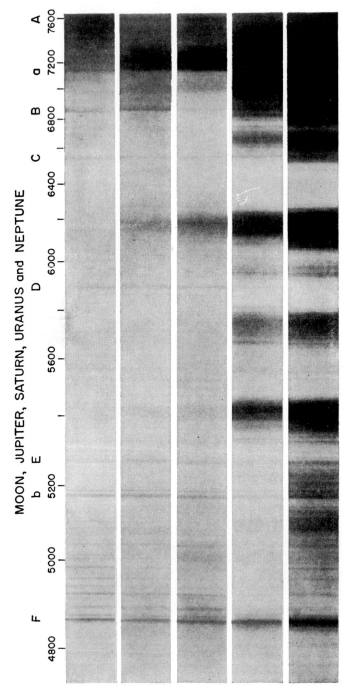

MOON, JUPITER, SATURN, URANUS and NEPTUNE

Fig. 8.36. Spectrograms of the moon and the giant planets. (Photograph by V. M. Slipher, Lowell Observatory.)

that run parallel to the equator, but the belts are less conspicuous and the colors less bright. Saturn has not displayed anything as striking as Jupiter's Red Spot, but white spots have been seen from time to time. Like Jupiter's, and for the same reason, the planet's disc is darkened at the limb.

The albedo is 0.42, a little less than Jupiter's, and the phase effect shows that the surface is smooth. The rings, however, show by their phase effect that they consist of bodies that have not smooth surfaces.

The sidereal period of Saturn is 29½ years, and the synodic period 378 days. The planet has a day nearly as short as Jupiter's, about $10^h 38^m$, and it rotates with its equator inclined $26°45'$ to the planet's orbit; since the latter is inclined $2°.5$ to the ecliptic, Saturn's equator is inclined $28°6'$ to the ecliptic. For this reason the rings show varying presentation effects, are sometimes seen at an angle of $28°$, sometimes on edge (when, as Galileo discovered, they are virtually invisible).

Saturn has nine known satellites, which are in many ways like those of Jupiter, though on the whole they are larger. They have all been given names.

Mimas, the nearest moon to the planet, is 48,200 kilometers from the outer boundary of the rings, and has, as we shall see, a great influence on them. Enceladus, Tethys, Dione, and Rhea have albedos similar to that of Saturn and low densities; perhaps they have icy surfaces. Titan is the second largest, and most massive, moon in the solar system, with almost twice the mass of our moon and more than twice its bulk. It is, moreover, the only moon known to have an atmosphere; its spectrum shows the same molecular bands that are characteristic of the planet itself. Iapetus always keeps the same face to Saturn, and its brightness varies by a factor of five. Possibly it is of irregular shape, or its surface is variegated. The motion of the outermost of the satellites, Phoebe, is *retrograde* like that of the three outer moons of Jupiter. Possibly it, too, may be a captured asteroid.

The unique feature of Saturn, the one that makes it the most beautiful of all telescopic objects, is the system of rings. Early observers depicted them in bizarre forms, and their annular nature was first recognized in 1655 by Huyghens. Twenty years later Cassini observed that the bright ring is double. In 1850 the third, or crape, ring was first seen by Bond.

The outer ring has an external diameter of 275,000 kilometers and is 16,000 kilometers wide. The Cassini division between the outer and inner parts of the bright ring is about 4800 kilometers wide. The inner portion of the bright ring is the brightest of the three rings. It is 233,000 kilometers across and 26,000 kilometers wide; its outer edge is the most brilliant, and rivals in brightness the surface of Saturn itself.

Between the inner portion of the bright ring and the third, or crape, ring is a division 1600 kilometers wide. The crape ring itself is 18,500 kilometers in width. The whole ring system is thus about 67,000 kilometers

Satellites of Saturn

Name	Discovery	Mean Distance (km)	Sidereal Period	Inclination to Orbit of Saturn	Eccentricity of Orbit	Diameter (km)	Mass (moons)	Apparent Magnitude*
Mimas	Herschel 1789	185,700	22h37m5s	26°44.7'	0.0201	450?	1/2120	12.1
Enceladus	Herschel 1789	238,200	1d8h53m6s	26°44.7'	0.0044	500?	1/520	11.6
Tethys	Cassini 1684	294,800	1d21h18m26s	26°44.7'	0.0000	1,100?	1/119	10.5
Dione	Cassini 1684	377,700	2d17h41m10s	26°44.7'	0.0022	1,100?	1/69	10.7
Rhea	Cassini 1672	527,500	4d12h25m12s	26°41.9'	0.0010	1,600?	1/30	10.0
Titan	Huyghens 1655	1,223,000	15d22h41m24s	26°7.1'	0.0289	4,600	1.86	8.3
Hyperion	Bond 1848	1,484,000	21d6h38m24s	26°0.0'	0.1043	400?	<1/600	13.0
Iapetus	Cassini 1671	3,563,000	79d7h55m32s	16°18.1'	0.0283	1,300?	<1/13	10.1–11.9
Phoebe	W. H. Pickering 1898	12,950,000	550.44d	174.7°	0.1659	300?	?	14.5

* At mean opposition of Saturn.

wide, and its edge is 11,250 kilometers from the equator of the planet. A fine rift (Encke's division) is sometimes visible in the outer ring, and Percival Lowell noted two more, even fainter divisions in it. Enormous as the ring system is, its thickness is *under 16 kilometers.*

The rings are exactly in the plane of the planet's equator; they show phases as the planet goes around its orbit, and disappear twice in the sidereal period, at intervals of nearly fifteen years, when the plane of the rings passes through the earth.

The body of the planet casts shadows on the rings, and the rings also cast shadows on the planet; the brightest ring is opaque, the outer ring nearly so, and the crape ring is semi-transparent. At some presentations, the earth and sun are on opposite sides of the rings, and the main ring system is silhouetted against Saturn as a dark line. But sunlight shows through the Cassini division and the crape ring. When a satellite of Saturn moves into the ring shadow, it is quite obscured by the central ring and almost so by the outer ring, but is easily seen in the shadow of the crape ring; the Cassini division dims it not at all. The only possible explanation is that the rings are composed of many particles, not a continuous surface. The particles in the crape ring only occupy one-eighth of the surface of the ring.

The changes in ring brightness with the angle at which they reflect light show that the particles occupy about one-sixteenth of the whole volume of the rings. When the earth is in a direct line with the sun and Saturn, each particle hides its own shadow.

If the rings consist of discrete particles, each particle must be describing its own orbit like a tiny satellite. Just as, in the solar system the planets nearest to the sun have both the smallest periods and the largest orbital velocities, the particles in the ring system must have the shortest periods near to the planet, and the greatest speeds. The speed in different parts of the ring can be measured by the spectroscopic Doppler effect, and the results are in exact agreement with expectation for satellites at the corresponding distances. Saturn, which rotates like a solid body, has an equatorial speed of 10 km/sec; the inside of the ring system gives a speed of 20 km/sec, and the outside a speed of 16 km/sec; moreover the rings show advance and recession of opposite sides and are rotating in the same direction as the planet.

These observations put the particulate nature of the rings of Saturn beyond doubt. They are probably made of rocky fragments of irregular shape, small pebbles and rough dust.

One of the most beautiful contributions ever made to theoretical astronomy was the theory of the motion of Saturn's rings, formulated by Clerk Maxwell (more famous, perhaps, for his formulation of the electromagnetic theory of light). Maxwell showed that a system of rings could be stable only if it consisted of discrete particles. A solid or liquid ring would be broken up, but a ring of particles was shown to be stable, provided its mass was small enough. The mass of the rings cannot be greater than 1/27,000 of the mass of Saturn, which is within the limit necessary for stability.

The divisions in the rings have been shown to be due to perturbations of the tiny satellite particles by Saturn's larger satellites, principally Mimas, which is the nearest to the ring system. A satellite will "sweep clean" a zone at a distance corresponding to a simple fraction of its period. The Cassini division is at a distance that corresponds to half the period of Mimas, one-third of that of Enceladus, and one-quarter of that of Tethys. The boundary between the inner ring and the crape ring corresponds to a period one-third of that of Mimas; Encke's division to three-fifths of Mimas' period, and Lowell's divisions to two-fifths and three-eighths of Mimas' period. The simpler the fraction, the more effective the perturbation, and the more conspicuous the division in the rings. Similar gaps among the asteroids ("Kirkwood's gaps") are found to correspond to fractions of the period of Jupiter.

The rings are very likely the remains of a satellite that was broken up within the tidal "danger zone" very near the planet's surface. A similar fate has been predicted for our own moon (p. 148). Exact calculations have been carried out only for a liquid satellite, of equal density with its primary; in this case the satellite could not withstand tidal disruption if it were at a distance less than 2.44 times the planet's radius (the "Roche limit"). Although the conditions are not fulfilled, and the disrupted moon was evidently solid, it is interesting to note that the distance of Mimas is 3.11 times the planet's radius, and that of the outer edge of the ring system, 2.30 times its radius, so Mimas is outside the "danger zone," and the whole ring system is within it.

7. URANUS

> Then, as I turned it on the Gemini,
> And the deep stillness of those constant lights,
> Castor and Pollux, lucid pilot-stars,
> Began to calm the fever of my blood,
> I saw, O first of all mankind, I saw
> The disc of my new planet gliding there
> Beyond our tumults, in that realm of peace.
>
> ALFRED NOYES

The six inner planets have been known since remote antiquity. The seventh, though a faint naked-eye object, was not known until 1781. It was discovered by William Herschel during his systematic search of the sky. Of German birth, Herschel settled in England and lived for more than twenty years as an organist and conductor. His symphonies, still occasionally performed, abound in cheerful vitality, a quality that distinguished him and the astronomical family that he founded. He devoted his leisure time to figuring astronomical mirrors and making reflecting telescopes (p. 84) and proudly described the images of stars in his eyepiece as "round as a button," a testimony to the skill with which the mirrors were figured.

"It has generally been supposed," he wrote of the discovery of Uranus, "that it was a lucky accident that brought this star into my view; this is an evident mistake. In the regular manner I examined every star of the heavens, and it was that night *its turn* to be discovered. . . . I perceived

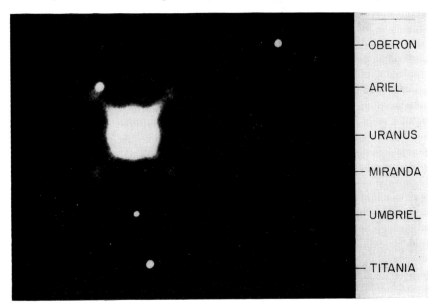

Fig. 8.37. Uranus and satellite system of five known moons. The main interest of the photo centers on the fifth satellite, Miranda, the minute object within the hala-tion ring. Its apparent magnitude is 17, period about 34 hours, and mean distance from Uranus 81,000 miles. Discovery of this moon was made by Dr. G. P. Kuiper on a negative taken by him on February 15, 1948, with the 82-inch McDonald Reflector. The photo above was taken March 1, 1948, with an exposure of 3½ minutes, at the 82-inch reflector. (McDonald Observatory)

the visible planetary disc as soon as I looked at it." He measured the diameter, and he verified the motion after a few nights. The planet had been seen, but not recognized, by earlier observers, and the records, when found, went back for almost a hundred years, and permitted an approximate determination of the orbit of the new member of the planetary system. Herschel called it *Georgium Sidus* in honor of George III of England (who gave him a post as his own astronomer in honor of the discovery), but the name Uranus, suggested by Bode, was generally adopted.

Uranus has a mean distance from the sun of 19.19 astronomical units, 2872.4 million kilometers. The orbit is very slightly inclined, by 46', to the ecliptic, and has the rather large ecentricity 0.047, so that the actual distance from Uranus to the sun varies by 270 million kilometers. The sidereal and synodic periods are 84.02 years and 369.16 days. Because of varying distance from the earth it changes in brightness by 26% between conjunction and opposition. It also varies regularly in brightness as it turns on its axis.

The planet shows a greenish disc of angular diameter 3".75, which gives a mean diameter 51,000 kilometers, about four times the diameter of the earth. It is highly flattened, more than Jupiter and rather less than Saturn, with an oblateness of 1/14. The mass of Uranus, measured from the motions of its satellites and the perturbations that it produces on the motion of Saturn, is found to be 14.68 times the earth's. Thus, although it does not compare with the giants, Jupiter and Saturn, in size and mass, it is still a major planet in comparison with the inner members of the system. The specific gravity is 1.27 and the mean density less than a quarter of the earth's—comparable to that of Jupiter.

Uranus has a high albedo, about 0.45, comparable to those of Saturn and Jupiter. Its spectrum shows very strong bands of methane. Shape, density, albedo, and atmospheric properties all indicate that it resembles the two giant planets also in structure, with a small rocky core, a layer of ice, and an extensive atmosphere of slightly contaminated hydrogen. It is evidently covered with clouds, but there are no distinct markings from which the rotation can be measured.

The length of Uranus' day has, however, been measured, from the Doppler shift of the lines coming from opposite limbs, to be 10¾ hours, near those of Saturn and Jupiter. The planet's regular variation of brightness leads to the same value, and the large oblateness confirms a rapid spin. The remarkable things about the rotation of Uranus is that the axis is nearly in the plane of the orbit: the inclination of the equator of Uranus to the plane of its orbit is 98°.0, so that its *axial* rotation is retrograde. The *orbital* motion is, of course, direct. The surface bands run parallel to its equator. An odd result of the direction of its axis is that Uranus would have no seasons of our kind: it alternately presents its poles and equator to the sun; each pole experiences a 42-year "summer" and a 42-year "winter."

Uranus has five known moons. Herschel believed that he had discovered six, but only two of his discoveries (Titania and Oberon) proved to be true moons, the other four having probably been faint stars. Two more (Ariel and Umbriel) were discovered by Lassell in 1851, and the fifth (Miranda) by Kuiper in 1948. The names of the planet and its moons are a mythological hodge-podge. Uranus was the most ancient of the gods in classical mythology, the father of Saturn; Herschel's two satellites were named from the

Midsummer Night's Dream after the fairy king and queen; Lassell took the names of sylphs from *The Rape of the Lock:*

> For, that sad moment when the sylphs withdrew
> And Ariel weeping from Belinda flew,
> Umbriel, a dusky melancholy sprite
> As ever sullied the fair face of night
> Down to the central earth, his proper scene
> Repair'd to search the gloomy cave of Spleen.
>
> POPE

The name of Miranda was taken from *The Tempest.*

MOONS OF URANUS

Moon	Distance from Uranus (km)	Period	Inclination to Orbit of Uranus	Eccentricity	Diameter (km)	Magnitude
Miranda *	130,360	1^{d}9^{h}54.41^m	97°59′	0.000	...	...
Ariel	191,800	2^{d}12^{h}29^m	97°59′	0.007	500?	15.2?
Umbriel	267,300	4^{d}3^{h}28^m	97°59′	0.008	400?	15.8?
Titania	438,700	8^{d}16^{h}56^m	97°59′	0.023	1000?	14.0
Oberon	586,600	13^{d}11^{h}7^m	97°59′	0.010	900?	14.2

* Unpublished elements by Harris.

The moons move exactly in the equatorial plane of the planet (as can be seen from the fact that they do not suffer perturbations, which they would if their orbits were inclined); they revolve in the same direction as the planet rotates. The system of moons presents itself on edge every 42 years, for example in 1882 and 1924, whereas in 1861, 1903, and 1945 the orbits appeared to us as circles.

Herschel's contribution to astronomy was so great that an epitaph says of him: *Coelorum perrupit claustra* (he burst open the gates of the heavens). He discovered the orbital motions of double stars, began to survey the nebulae, and to surmise the form of the galaxy. He was ably assisted by his sister Caroline, who devoted most of the ninety-eight years of her life to astronomical observations, and insisted that she had "no time for niceties." The cheerful temper of the Herschel family is typified by the occasion when they all assembled inside the 40-foot tube of the largest reflector, and sang the song composed for the occasion:

> Merrily, merrily let us all sing,
> And make the old telescope rattle and ring.

8.   NEPTUNE

To know why Uranus, uttermost planet known,
Moved in a rhythm delicately astray
From all the golden harmonies ordained
By those known measures of its sister-worlds.
Was there an unknown planet, far beyond,
Sailing through unimaginable deeps
And drawing it from its path?

                                        ALFRED NOYES

When Uranus was discovered in 1781, earlier observations of the unrecognized planet were sought and found. Flamsteed, the contemporary of Newton, and Astronomer Royal of England under Charles II, had recorded it five times, the first having been in 1690. Lemonnier actually saw it eight times in one month, and recorded it as a star on each occasion; however, so casual were his observations that some of them were written on a bag of hair powder! Bradley, the discoverer of aberration, saw it; altogether twenty pre-discovery observations were found. These observations carried the planet through more than one sidereal period and permitted its orbit to be fairly well determined. Later observations were used to improve the orbital elements.

For about twenty years the motions of Uranus seemed to be regular, but by 1800 it had begun to deviate from the predicted position, and by 1820 it was appreciably off its course. At first an attempt was made to fit the orbit by the rejection of the early observations, but the deviations continued: by 1830 the error was 20″, and by 1840, the planet was 1.5′ from the expected position.

The anomalous motion of Uranus exercised all the astronomers of the time. Some were inclined to attribute it to departures from the inverse square law of gravitation at large distances. But the most interesting suggestion was that the deviations were due to perturbations by another planet, probably with an orbit outside that of Uranus.

The problems of perturbations previously studied dealt with known planets, and calculated their effects upon each others' orbits. This problem is complicated enough, but it does not compare in difficulty with determining the position of an unknown planet that is causing observed perturbations. In 1841 the idea occurred to John Couch Adams, still an undergraduate at the University of Cambridge, to attempt to solve this unprecedented problem. After four years of work he obtained a solution, calculated the position of the unknown planet, and sent his results to Airy, the Astronomer Royal of England. Unfortunately, Airy was one of those who believed that departures from the law of gravitation were the cause of the aberrant motion of Uranus, and after making a superficial criticism, he laid the prediction aside, and did not apply the crucial test of looking for the planet in the position Adams had predicted.

In 1845 the problem was attacked by the French astronomer Leverrier, who did not know of Adams' work. His treatment was most systematic; first he discussed and evaluated all the early observations; then he calculated the perturbations of Uranus by Jupiter and Saturn, and showed that they were not large enough to account for the observed motion. Then he discussed possible explanations of the departure of Uranus from schedule: a failure of the law of gravitation; a resisting medium; an unseen satellite; and a collision with a comet. He rejected them all, and decided that an unknown planet must be responsible. The unknown could not be within the orbit of Uranus, for in that case it would also perturb Saturn and Jupiter. He proceeded to locate the disturber in an outer orbit, and in June, 1846, he published a prediction of the position, which agreed almost exactly with that of Adams, still unknown to him.

Even though Airy was not interested in the predictions, Sir John Herschel (son of William Herschel) wrote, late in 1846: "The past year . . . has given us the probable prospect of another [planet]. We see it as Columbus saw America from the shores of Spain. Its movements have been felt trembling along the far-reaching line of our analysis with a certainty hardly inferior to ocular demonstration." Finally, Airy suggested to an astronomer at Cambridge to look in the predicted position, and the latter, after a "leisurely and dignified search," actually saw it twice, but because he did not (until later) calculate its position, and compare it with those of known stars, he did not recognize it. Meanwhile, Leverrier had written of his prediction to the astronomer Galle in Berlin, and the latter looked for the planet the same night, found and recognized it. Although the observation made at Leverrier's request revealed the planet first, Adams had been the first to make the prediction. The glory of the discovery is equally shared by the two men.

Neptune is almost the twin of Uranus. Its mean distance from the sun is 30.07 astronomical units, 4,500,800,000 kilometers; it is the first planet to deviate greatly from the prediction of Bode's Law. The orbit is almost circular, with an eccentricity of 0.0086, and it is inclined at $1°47'$ to the ecliptic. The sidereal period is 164.8 years, the synodic period 376½ days. It is of the seventh magnitude and not visible to the naked eye.

In appearance Neptune is bluish-green, with a disc of 2.04" diameter, which at the planet's distance indicates a diameter of 50,000 kilometers, or 31,070 miles, almost four times that of the earth. Markings on the disc are not visible, but the spectrum shows strongly the methane bands like those of the other major planets. Undoubtedly it, too, is cloud-covered and possesses a similar structure. Its high albedo, 0.52, also puts it in the same category.

The mass of the planet is determined, from the motion of its moons and from perturbations of Uranus, to be 17.3 times that of the earth. It has therefore a density about 0.3 that of the earth's, and a specific gravity of 1.6, larger than that of Jupiter.

Neptune, like the other giants, rotates rapidly. As with Uranus, the rotation was discovered by Doppler effect. The length of the "day" is about sixteen hours, and therefore the planet must be decidedly oblate. The spectrum shows that it rotates in the same direction as its orbital motion, which is of course direct.

Fig. 8.38. Neptune and its two satellites. Left, prime focus, May 29, 1949, 5:05 UT, exp. 30 min. 103a-F, showing both Triton and Nereid (arrow). Fuzz spots are galaxies. Right, Cassegrain focus, February 24, 1949, 8:55 and 9:06 UT, showing Triton. Note absence of other close satellites. (Photograph by G. P. Kuiper, 82-inch telescope, McDonald Observatory.)

Neptune has two known satellites, one discovered in 1846 by Lassell, and known as Triton, and the other (Nereid) first seen by Kuiper in 1949. The motion of Triton is remarkable; its *retrograde* orbit is inclined by 40° to the planet's orbit but apparently only by 20° to the planet's equator, and it precesses with a period of about 580 years. This must be the effect of the planet's equatorial bulge (just as the earth's bulge causes the regression of the moon's nodes), and is an additional reason for our certainty that Neptune is oblate. Nereid is remarkable for its orbital eccentricity, the largest known for any satellite; its distance from Neptune varies from 1,335,758.2 to 9,817,018 kilometers, and its motion, unlike that of Triton, is direct.

The discovery of Neptune was one of the most remarkable achievements of celestial mechanics. The first test of a theory is the requirement that it shall take account of known phenomena. A severer test is the correct prediction of previously unknown ones. We have seen how, in Newton's hands, the law of gravitation predicted the inequalities of the lunar apogee

MOONS OF NEPTUNE

Moon	Mean Distance from Neptune (km)	Period	Inclination to Equator of Neptune	Eccentricity	Diameter (km)	Magnitude
Triton	354,000	5^{d}21^{h}2^{m}39^s	140°	0.000	4500	13.6
Nereid	5,391,000	359.4^d	20°	0.76	. . .	. . .

and of the lunar nodes. The discovery of Neptune was an even more remarkable and more spectacular test. Newton was working with known quantities in the motion of the moon; Adams and Leverrier were predicting the properties of a body of unknown position and mass. The law of gravitation could hardly have been more severely or more successfully tested.

9. PLUTO

> Then felt I like some watcher of the skies
> When a new planet swims into his ken. . . .
> KEATS *

The outermost planet is something of an enigma. Unlike the four giants whose orbits are almost wholly within its own, it is small, apparently without atmosphere, and much more like the terrestrial planets. It is so remote that our knowledge of its condition is meager.

Irregularities in the motion of Neptune suggested that perturbations were being produced by a planet still further out. Percival Lowell and William Pickering made exhaustive calculations of the probable position of the suspected planet, but many years passed before their predictions bore fruit. The final discovery of Pluto was the direct result of these efforts, and of the telescope that was especially built by the Lowell Observatory to search for it. The ecliptic was systematically photographed, and a year later, 1931, the image of the planet was found on one of them by Clyde Tombaugh. Pluto (as the new member of the solar system was shortly named) was less massive and fainter than had been expected, but it was found only a few degrees from the spot predicted by both Lowell and Pickering.

Pluto's orbit has the largest eccentricity among the planets, 0.249, and its mean distance from the sun is 39.46 astronomical units. Because of the large eccentricity, part of its orbit lies within that of Neptune, but the large inclination of the orbit to the ecliptic (17.1°) makes a collision of the planets practically impossible. The sidereal period of Pluto is 247.7 years, and the synodic period, just over a year.

---

* Keats probably had the discovery of Uranus in mind when he wrote these words.

The mass of Pluto, which must be determined from perturbations, is measured to be 1.0 ± 0.23 of the earth's. The size is difficult to measure, for Pluto does not show a disc even with the largest telescope. By an

January 29

1930

January 23

Fig. 8.39. The actual discovery positions of Pluto on small sections from the 14 × 17 inch discovery plates made with the Lawrence Lowell telescope, Lowell Observatory.

ingenious comparison with an "artificial planet" at the 200-inch telescope, Kuiper has determined the diameter to be 46% of that of the earth, so that the volume is a little less than one-tenth. With a mass equal to the earth's, Pluto would then have about ten times the earth's density—an enormous value which is difficult to accept. The adopted albedo is low. We must conclude that the physical dimensions of Pluto are still uncertain; probably the least reliable value is that of the mass. If the diameter has been correctly measured, a mass as small as one-tenth of the earth's would be required to give the planet a plausible density.

The measured diameter of Pluto is not very different from that of Triton, the larger satellite of Neptune. The suggestion has been made that Pluto is a "lost moon" of Neptune. This idea might account for its moon-like structure, the fact that its orbit passes within Neptune's, and the high orbital inclination. The remarkable arrangement of Neptune's present moons lends color to the suggestion of some exceptional event or condition in Neptune's past history. Certainly Pluto is an anomalous body among the planets, but the suggestion that it was once a moon is highly speculative.

CHAPTER IX

THE LESSER BODIES
OF THE SOLAR SYSTEM

> . . . like a comet burned
> That fires the length of Ophiuchus huge
> In th'arctic sky, and from his horrid hair
> Shakes pestilence and war.
>
> MILTON, *Paradise Lost,*
> II, 708–711

The nine bright planets form a sort of backbone for the solar system. Their orbits are very nearly circular and very nearly in one plane. They all go around in the same direction, and they all rotate in the same direction as that in which they revolve about the sun. These are the salient facts about the nine planets—major clues to the solution of the problem of the solar system's origin.

The smaller bodies that form the substratum of the solar system share the four dynamical characteristics of the nine planets to some extent, but they adhere to them less strictly. Some have eccentric orbits, some have orbits of high inclination, and a few have retrograde motion. Many of them are not spherical, and many are loosely organized compound bodies.

Physical and dynamical properties divide the smaller bodies of the system into two main groups. One contains the *asteroids* and the *meteorites* and probably includes also the *zodiacal light;* within the group we find a range of sizes from about five hundred miles across to a few inches or less. The second group includes the *comets* and the *shower meteors,* which are composed of very small solid bodies, but may be surrounded by huge gaseous envelopes. Smaller than the members of either group—in fact of atomic dimensions—are the enigmatic *cosmic rays* that pervade the solar system. The origin of the cosmic rays is still a matter of debate; there is not general agreement as to whether these particles originate in the solar system or enter it from interstellar space.

1. THE ASTEROIDS

Bode's empirical law of planetary distances gives a fair account of the positions of all the planets out to Uranus; but it requires a planet at a distance of 2.8 astronomical units, between Mars and Jupiter, where no bright planet is known. Strangely enough, this distance is near to the average for the many thousand small bodies that are known as the *asteroids.* They are sometimes called *minor planets* or *planetoids,* but we shall adhere to the general practice of calling them asteroids.

The first asteroid was discovered on the first day of the nineteenth century (January 1, 1801) by Piazzi; he named it Ceres after the tutelary deity of his native Sicily. Three more bright asteroids were discovered soon after: Pallas * in 1802, Juno in 1804, and Vesta in 1807. These

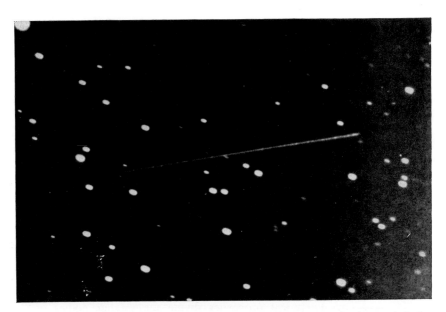

Fig. 9.1. Trail of an asteriod. The asteroid moves against the star background during the exposure of the photograph.

are the brightest of the asteroids; Vesta is indeed a faint naked-eye object. Not for thirty-seven years was another asteroid discovered, the

* Most asteroids are given feminine names, or names with a feminine ending, such as Princetonia. Those that pass within the earth's orbit, such as Hermes and Icarus, are conventionally given masculine names, and the exceptional "Trojan" group, associated with Jupiter, are called Priam, Patroclus, etc., after the heroes of the Trojan War.

much fainter Astraea. Thereafter discoveries became more frequent. At the present time, several thousand asteroids have been recorded, and 1500 good orbits had been determined by 1940.

Photography has been responsible for the rich haul of asteroids in recent years. Like planets, they move against the background of the stars, and on a long-exposure photograph, with a telescope driven to keep pace with the daily motion of the stars, an asteroid appears as a trailed image. It is also possible to make an exposure on which the asteroid image is round, and the images of stars are trailed by suitably guiding the telescope.

Most of the asteroids travel between the orbits of Mars and Jupiter; seven-eighths of them have mean distances between 2.3 and 3.3 astronomical units; the average, 2.805, fits Bode's Law very closely. The largest known orbit is that of Hidalgo, mean distance 5.71 astronomical

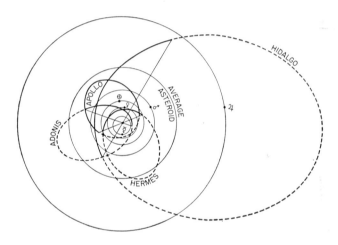

Fig. 9.2. The orbits of an average asteroid, of Adonis, Hermes, and Apollo, which all pass within the orbit of Venus, and of Hidalgo, the asteroid with the largest orbit yet discovered. The orbits of Jupiter, Mars, the earth, Venus, and Mercury are shown. Broken lines denote the parts of the asteroid orbits that lie below the average plane of the planets; straight lines join the nodes. (After Watson: *Between the Planets,* "Harvard Books on Astronomy.")

units. Some asteroids have orbits so small that they come well inside the earth's at perihelion; Adonis, Apollo, and Hermes all come within the orbit of Venus, and Icarus, with the shortest period and smallest orbit known, passes within that of Mercury.

The sizes of the orbits are not distributed uniformly: Jupiter, the most massive of the planets, has an influence on the asteroids that recalls the action of Mimas in subdividing the rings of Saturn. Gaps in the asteroid zone, known as "Kirkwood's gaps," correspond to periods ½, ⅓, ⅖, and

⅗ of Jupiter's period; and the orbits tend to cluster around distances that correspond to Jupiter's period, or ⅔ or ¾ of it (Figure 8.34).

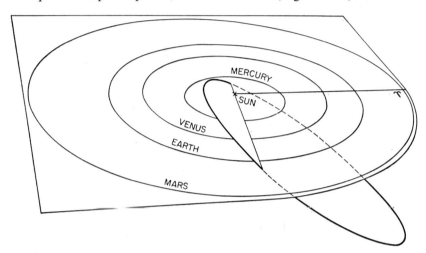

Fig. 9.3. Orbit of Icarus, the asteroid with the smallest known orbit.

The orbits of the asteroids have, on the average, larger eccentricities and larger inclinations than those of the larger planets, and very small eccentricities and inclinations are uncommon. The *average* eccentricity is 0.15 (larger than that of any planet except Mercury and Pluto) and some individuals have very large ones: Alinda, 0.53; Albert, 0.54; and Hidalgo, 0.65; 30% of all asteroids have eccentricities larger than 0.30. The *average* inclination is 9°30', larger than that of any large planet; Hidalgo has an inclination of 43°, and Pallas, of 35°.

The influence of Jupiter on the motions of the asteroids appears in the tendency of their perihelia all to lie in one direction: the orbits tend to lie parallel to that of Jupiter. The fact that the largest eccentricities and inclinations are found for the asteroids with periods between four and six years (nearly half that of Jupiter) is ascribed to the same cause.

The twelve known asteroids with periods nearly equal to Jupiter's are called the "Trojan asteroids"; they move near the corners of equilateral triangles with the sun and Jupiter at the other corners; seven precede Jupiter, five follow. Their motion is a complicated oscillation about the apexes of the triangles (the "Lagrangian points"). The theory of the "Trojans" is a beautiful special application of the dynamical "Problem of Three Bodies."

Even the largest asteroids are very small compared with the planets: Ceres is about 770 kilometers in diameter; Pallas, 488; Vesta, 386; and Juno, 190; all the others are much smaller. They have a surprising variety of albedos: Ceres resembles the moon in reflecting power, Pallas is like Mars, and the surface of Vesta is as bright as that of Venus! Vesta is consequently the brightest of the asteroids, although far from the largest. The surfaces of the asteroids are even rougher than that of the moon; they do not reflect light well in an oblique direction.

Most asteroids are too small to show a visible disc, and we can estimate their sizes from their brightness only by guessing a reasonable albedo. Probably a dozen are over 150 kilometers in diameter, and most of them are between 15 and 80 kilometers with some even smaller.

Many asteroids are evidently spotted or irregular in shape. Their common variations of brightness, in periods of a few hours, are probably results of the rotation of variegated or unsymmetrical bodies. The most thoroughly investigated variation is that of Eros, which has two maxima and two minima of brightness in 5¼ hours. This asteroid has actually been seen to be of irregular shape, and it turns on itself in a period of 5¼ hours. The rotation is *direct,* in the same sense as the revolution. The changes in the brightness of Eros are compatible with a body of dimensions 35 × 16 × 8 kilometers that rotates about its shortest axis. Probably it is an irregular lump of rock that approximates to these dimensions. Many other asteroids are no doubt similar to Eros in being rough chunks of stony material.

The asteroids are of very small mass. Ceres, the largest, weighs perhaps 1/8000 of the earth, and the thousand brightest amount altogether to no more than 1/3000 of the earth's mass. Probably the aggregate mass of all the asteroids is less than 1% of our planet's.

Innumerable asteroids have been noted on celestial photographs. Thirty thousand are probably accessible on plates made with the 100-inch telescope, and thousands can be found on the photographs of the ecliptic that were made at the Lowell Observatory in the search for Pluto. Unless an asteroid is speedily followed up on discovery it may be lost again, and the couple of thousand known orbits represent only a fraction of the asteroids that have been observed.

2. THE ZODIACAL LIGHT

Very small asteroids are not observable individually, but evidence of their presence is furnished by the *zodiacal light* and the *counterglow,* a zone of faint illumination that runs around the ecliptic, with a maximum in the direction of the sun, and a fainter one on the opposite side of the ecliptic.

The zodiacal light is responsible for 30% of the light of a moonless sky. Its spectrum is like that of the sun, so *it consists of reflecting bodies.* The light is only slightly polarized and not reddened, as it would be if the bodies responsible were primarily atoms, molecules, or even very fine dust. The observed brightness of the zodiacal light could be produced by a cloud of small bodies of the same albedo as the moon, 1 millimeter in diameter and 5 miles apart; or 10 feet in diameter and 1000 miles apart. But very small bodies would tend to spiral gradually into the sun (the Poynting-Robertson effect). Even in twenty million years (a short time compared to the probable age of the solar system), all bodies less than one centimeter in diameter would have been cleaned up. Therefore the zodiacal light probably comes from bodies larger than this, perhaps as big as baseballs.

The asteroids and the zodiacal light seem to be parts of a substratum in the solar system that consists of small rocky bodies, from several hundred miles down to several inches in diameter. The system is somewhat flattened in the plane of the ecliptic, but not so closely confined to it as the larger planets. The bodies in it all seem to have direct motions. This interplanetary material is densest in the neighborhood of the sun, and comparatively little of it seems to extend beyond the orbit of Jupiter. Fragments of it enter the earth's atmosphere continually in the form of scattered *meteorites*.

3. METEORITES

The occasional bodies that strike the earth's surface from interplanetary space are known as *meteorites*. Reports of such falls go back many centuries; the earliest on record was noted in China in 644 B.C. The one that fell at Aegospotami in 466 B.C. made a great impression on Greek thought, and probably influenced Anaxagoras, then a man of thirty-two. "Along with the visible stars," wrote Diogenes of Apollonia, "there move also invisible dark stony masses, which remain unknown for that reason. Sometimes they fall on the earth and perish, as happened with the *stony star* that fell at Aegospotami." These stones that came from heaven were often revered as divine; such, probably, was the sacred stone in the Temple of Diana at Ephesus, "which fell from Jupiter," and the holy rock that is enclosed in the Kaaba of the Moslems.

Nonetheless, scientific men were reluctant for more than two millenniums to believe that meteorites really come from outside the earth. Only in 1803 did the French Academy officially accept their extraterrestrial origin. Now we recognize their great interest, as the only bodies we can actually handle that originated outside the earth itself.

A few meteorite falls (*fireballs*) have been accompanied by brilliant light, and some (*bolides*) by loud explosive sounds. However, many of the undoubted meteorites that have been picked up must have fallen before there was any record.

Meteorites may be classed roughly in two groups: *stones* and *irons*. The stones are not very different in composition from the crust of the earth; the irons are about 90% iron, about 8% nickel, and contain traces of other elements. The irons have a very remarkable crystal structure, and display the so-called Widmanstätten figures. Their properties indicate that they must once have been liquid and have cooled slowly, perhaps under great pressure. A minority of meteorites are *stony-irons,* composed of about equal parts of stone and iron.

The existence of the asteroids (many of which are jagged chunks of material, which differ greatly in reflecting ability), the small, rough, dispersed particles of the zodiacal light, and the stony and iron meteorites, can be plausibly linked together. We have only to derive them all from a planet, not very different from the earth in structure, that was fragmented and dispersed throughout the inner part of the solar system. The stones and irons would then be particles of the rocky crust and the metallic core, respectively. How this idea fits in with our picture of the history of the solar system will be discussed in the next chapter.

Meteorites that weigh many tons have been found in various parts of the world. The largest, which must originally have weighed about 80 tons, lies at Grootfontein, South-West Africa. Three large irons, found in Greenland, are now in the Hayden Planetarium in New York city; the largest weighs 33 tons. A number of others now lie in museums, and many more must be still undiscovered. Probably a few thousand meteoritic bodies fall in some part of the earth every year, but most of them are not as large as the ones we have mentioned.

There is evidence that a few masses, even bigger than the largest known meteorites, have struck the earth's surface and produced *meteor craters*. The best-known and most accessible is the Canyon Diablo crater, between Winslow and Flagstaff, Arizona (easily accessible by car, and a most impressive and significant sight). It is about three-quarters of a mile across and 600 feet deep. Its configuration and the structure of its walls suggests that it was thrown up by the oblique impact of a massive body. Although many attempts have been made to excavate the meteorite, no one large mass has been found, and very likely it exploded on striking the ground. The whole neighborhood of the crater was found to be sprinkled with small meteoritic irons, which put the origin of the crater beyond doubt.

A number of other meteor craters, none as large as the one in Arizona, have been identified by the presence of meteorites in their neighborhood. The enormous Chubb crater, in northern Canada, is probably also of meteoritic origin, but conclusive evidence still remains to be collected.

Two spectacular meteoritic falls have taken place in Siberia in the twentieth century, and care was taken to collect the impressions of eyewitnesses of these rare events. The first took place on June 30, 1908 in a desolate region of central Siberia. It caused widespread damage to live-stock and forests, and left a number of craters in the marshy land. The second fall, which took place in 1947 near Vladivostok, was literally a shower of iron.

Fig. 9.4. Meteor crater in Arizona.

The atmosphere provides an efficient shield from the millions of small meteors that strike the earth every day; without it we should be in continual jeopardy. The large bodies, like those that fell in Arizona and Siberia, could cause disastrous damage if they fell in populous regions; fortunately the latter cover only a small percentage of the earth's surface, and the actual danger is very small.

4.   COMETS

The popular conception of a comet is very different from that of an asteroid. At the time when they are best seen—near perihelion—long-period comets have tails, but the line between comet and asteroid is some-times hard to draw.

Kepler surmised that the comets "are as numerous as the fishes of the sea," and they seem indeed to be the denizens of a sea in which the whole solar system is swimming. Over a thousand have been recorded, doubtless but a small fraction of the whole number.

The comets that we observe pass comparatively near to the sun. Some remain visible for years, others vanish in weeks or even days. A comet seen during a total solar eclipse may never be seen again.

Comets travel under the sun's gravitational control in orbits that are conic sections (subject to planetary perturbations). The observed orbits may be elliptical, parabolic, or hyperbolic. A comet whose orbit is an ellipse will be periodic and will return to perihelion at regular intervals (except for perturbations by the planets). Parabolic or hyperbolic orbits are not periodic; a comet that leaves our neighborhood with such an orbit will (provided its motion is not altered) leave the solar system forever. In practice it is difficult to distinguish between an orbit that is a parabola, or even an hyperbola, and one that is an ellipse. About three-fourths of all known comets have apparently parabolic orbits, and less than twenty are known to have apparently hyperbolic orbits. About a hundred comets with elliptical orbits are known, and many of them have been observed to return around the sun several times.

Fig. 9.5. Comet Whipple, 1942 c, photographed on February 25, 1943. (Harvard Observatory)

There are about seventy known comets with periods between three and nine years. They form a distinct group: all have direct motions, moderate orbital eccentricities, and small orbital inclinations. All are rather faint and have small tails or none at all. All have aphelia near the orbit of

Jupiter, and also near to the longitude where the plane of their orbit cuts the orbit of Jupiter. There is no doubt that Jupiter has a directive influence on the motions of these comets, which are known as "Jupiter's Comet Family."

Fig. 9.6. Comet Peltier, photographed August 7, 1946. Notice the trails of the star images, which show the extent of the comet's motion during the exposure. (Photograph by J. S. Paraskevopoulos, Boyden Station, Harvard Observatory.)

The great influence of the largest planet on a comet that comes near it is a result of the large mass of Jupiter and the extremely small mass of a comet, perhaps one ten-thousandth of the mass of the earth. The influence of Jupiter has so altered the motion of Comet Pons-Winnecke that between 1819 and 1933 the comet's period has varied between 5.56 and 6.16 years, and the perihelion distance between 0.755 and 1.10 astronomical units; changes have also taken place in the inclination, eccentricity, and orientation of the orbit. Comet Wolf I experienced similar disturbances. Even more drastic was the effect of Jupiter on Brooks' Comet, which changed its period from 29 to 7 years. Comets, it may be mentioned, are often named after their first discoverer, or serially, with years and Roman numeral, in order of time of perihelion passage.

Jupiter is not the only planet that can affect a comet's motion. Any planet will produce similar effects in proportion to its mass. The earth itself changed the period of Lexell's Comet of 1770, but only by 2½ days, and the fact that the motion of our planet was utterly unaffected shows that the comet's mass could not have been more than 1/13,000 of that of the earth.

About forty comets are known with periods between ten and a thousand years. Because their orbits are large, they rarely come near the major planets, even if their orbital inclinations are rather small, and their motions are not so habitually disturbed as those of the short-period comets. Some of them, including Halley's Comet, have retrograde motions.

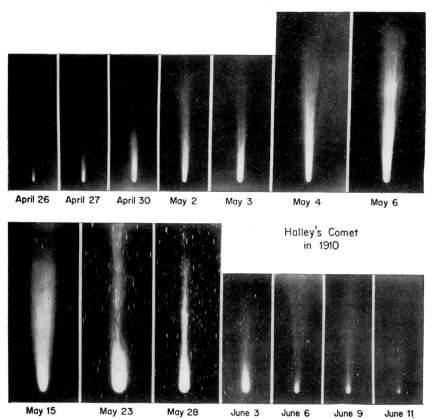

Halley's Comet
in 1910

Fig. 9.7. Fourteen views of Halley's comet at an apparition of 1910. (Mount Wilson Observatory)

Halley's Comet was noted by Halley to be periodic. He recognized that the orbits of the comets of 1531, 1607, and probably 1456, were identical with that of the comet of 1682. He correctly predicted a return in 1758, but did not live to see it. Records have been found of every apparition of the comet since 240 B.C., except that of 163 B.C. It has been credited, in less enlightened ages, with being the precursor of grave events. Its appearance in 1066 is recorded in the Bayeux Tapestry. The apparition of 1453 was associated in the popular mind with the fall of Constantinople, but the story that the pope issued a bull excommunicating it on that occasion is a fabrication. The most recent return in 1910 was an event of scientific, rather than political, moment. Halley's Comet will be seen again in about 1984.

The parabolic and hyperbolic comets are not necessarily visitors from interstellar space that pass through the solar system but once. The orbit of any comet only tells how it is moving *at the moment;* there is no record of any comet's having *approached* the sun in a hyperbolic orbit, and those that have departed in that way have been turned into hyperbolic orbits by perturbations. They may leave the solar system forever, but there is no good evidence that they are newcomers to it. They have probably been diverted inwards from a great reservoir of comets that encircles the solar system.

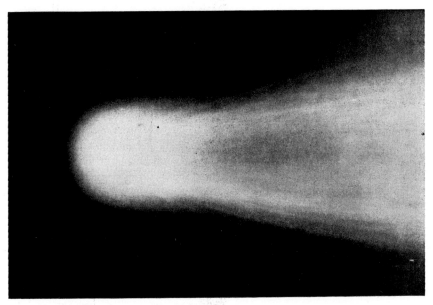

Fig. 9.8. Head of Halley's comet. (Mount Wilson Observatory)

A bright comet when near to the sun is observed to have a complicated structure. The head consists of a large *coma* with a bright *nucleus.* The *tail* always points away from the sun. When a comet is far from the sun, no tail is visible; as it approaches perihelion, the material of the head is volatilized by the sun's radiation, and driven away from the sun by radiation pressure, which affects *small* particles more strongly than gravitation does. At every approach to the sun a periodic comet loses some of the substance of its head and is gradually depleted. Even Halley's Comet, brilliant as it was at the last return, is not as bright as it once was.

The ejection from the comet's head is often spasmodic, and this points to the evaporation of material from irregular particles of various sizes that are bunched together in the nucleus. The loss of material from the head, and the manner of it, may have an appreciable effect on the comet's motion. The speeding-up of Encke's Comet was long an enigma. Recently it has been shown by Whipple that changes in the speed of a comet may be produced by the transfer of momentum from a rotating head to the tail that forms from the vaporization of its icy constituents. The bodies that make up the head are a sort of latticework of nonvolatile material (like the solid part of a meteor) in which the frozen gases are contained. The solid parts of the nucleus act as a shield and slow up the rate at which the gases boil away. If the head of a comet has *direct* rotation, the consequent belated expulsion of the gases will tend to slow its motion down; if the rotation is *retrograde,* the motion will be speeded up. The observed change of speed of Encke's Comet could be produced by a loss of only 0.002 of its mass in each revolution. Other comets, with retarded motion, may be losing comparable amounts of matter from nuclei in direct rotation.

Comets have enormous dimensions. The head may be from 50,000 to 250,000 kilometers across, the nucleus a cluster of icy chunks extending several hundred kilometers. The tail may be from 8 million to 80 million kilometers in length. And yet the small mass (at most a few thousand million tons) shows that a comet's density is about that of a good vacuum! The tails, and even the heads, are quite transparent, and do not perceptibly dim the light of stars seen through them. When the head of Halley's Comet passed between us and the solar disc in 1910, it was quite invisible.

The light of a comet is partly reflected sunlight and partly comes from luminous gases ejected from the head. The spectroscope shows a variety of simple chemical compounds and radicals: C_2, the carbon molecule; CH and CH_2, simple hydrocarbons; CN, cyanogen; NH and NH_2, hydrides of nitrogen, OH, the hydroxyl radical, CO+, ionized carbon monoxide; N_2+, the ionized nitrogen molecule, and OH+, the ionized hydroxyl radical. These chemically unstable molecules are probably products of dissociation (by solar radiation) of icy lumps of ammonia, hydrocarbons, carbon oxides, and water that make up the nucleus, and are volatilized by the sun's rays as the comet comes near perihelion. These compounds form the tail of the comet. The density of the tail is so small that 2000 cubic miles of it contain no more material than a cubic inch of air at the earth's surface. Only an imperceptible fraction of the atomic and molecular contents of the tail would encounter the earth, even if we should pass right through it. Such an event actually took place in 1910 with Halley's Comet, and there were no noticeable terrestrial effects. We have nothing to fear from cometary cyanogen or carbon monoxide and nothing significant to anticipate from any constituents of a comet's head or tail.

Repeated trips about the sun cause considerable wear and tear on a comet, particularly a small one. More than one comet near perihelion has dissipated and vanished forever. Biela's Comet, with a period of about six years, was seen first in 1772; in 1846 it was observed to have split in two, probably the result of a close approach to Jupiter. The twin comets were seen again in 1852. When looked for in 1865, they were not to be found, nor have they been seen since. The debris, however, still circulated for a time in almost the same orbit in the form of a shower of meteors, but even this relic of Biela's Comet is a thing of the past. The heads of all comets are, in fact, regarded as compact groups of meteor-like particles, which can be, and often are, dispersed by an encounter with a massive body. The sun is the most effective disintegrator of comets. Even the massive Jupiter has limited disruptive power. In 1886, Brooks' Comet nearly grazed that planet's surface and survived the encounter. That it did not disturb the motion of Jupiter's moons is further testimony to the smallness of a comet's mass.

A comet that passes near the sun can probably survive several hundred perihelion passages, so the lifetime of a short-period comet may be only a few thousands, or tens of thousands, of years. However, the long-period comets, with greater perihelion distances, may survive many thousand perihelion passages, and their lifetimes are comparable with that of the whole solar system. The interesting question of what keeps up the supply of short-period comets will be mentioned in Chapter X.

5. METEORS*

A "shooting star" can be the brightest object in the night sky, yet these are the smallest bodies in the solar system than can be observed as individuals. If it were not for the resistance of the atmosphere, which produces a glowing train as the tiny particle volatilizes, we should not see them at all. A brilliant fireball may weigh many pounds, a visually observed meteor less than an ounce, and a telescopic meteor, a small fraction of an ounce. Some of these visitors from space are large enough, as we have seen, to survive (at least partially) their hot trip through the atmosphere, and fall to the ground as *meteorites*.

On almost any night a few meteors an hour will be seen from any one place. One observer's range is very limited, and half a dozen an hour seen by him represents over ten million a day falling somewhere on earth. If allowance is made for the fact that the visual observer sees only the brightest, it is calculated that four thousand million meteors brighter than the tenth magnitude strike the earth daily. Yet so small are their masses that they aggregate only about a ton, a negligible contribution to the earth's mass of 6,600,000,000,000,000,000,000 tons.

---

* The accepted nomenclature is as follows: the object itself is a *meteor;* the body in flight is sometimes called a *meteoroid;* the fallen body is a *meteorite.* A *fireball* is bright enough to cast a shadow; a *bolide* explodes with an audible noise.

More meteors are observed after midnight than before it, and they tend to be brighter and to move faster. In the evening the rotation of the earth is working against its orbital motion; in the morning, supplementing it. Therefore before midnight we see meteors that overtake us; in the morning hours we meet them head-on. As the average speed of a meteor is about 65 km/sec, they catch up with us even when we are moving away from them with our orbital speed of about 30 km/sec.

Fig. 9.9. Photograph of the trail of a bright meteor. The picture was made with a stationary camera, and the stars around the pole made segments of circular trails during the exposure. (Harvard Observatory)

The visible path of a meteor through the earth's atmosphere is determined by triangulation. Meteor paths are accurately charted by photography from two stations about twenty miles apart—far enough to provide a good baseline, but not so far that meteors observed from one station will be invisible from the other. Such observations show that individual meteors appear near 100 kilometers up; bright ones remain visible to a height of 55 kilometers, or even less, faint ones fade out at about 80 kilometers. Above 100 kilometers the atmosphere has not raised the particles to incandescence; and at the point of disappearance they are consumed. The bright meteors are the more massive particles, which take longer to "boil away."

An ingenious instrument, which breaks up the photographed meteor trail into segments by means of a periodically occulting shutter, permits the determination of the speed as well as the path. The most rapid meteors traverse the atmosphere at about 70 km/sec; slow ones may have speeds down to 16 km/sec. Photography is not the only means of measuring the trails and speeds of meteors; the relatively new microwave techniques can determine both, and they have the advantage of being usable when the sky is cloudy and even during the day.

The speed of a meteor permits us to determine its orbit, essentially as

Fig. 9.10. A bright fireball, photographed as it passed across the constellation Andromeda. The Andromeda galaxy is visible in the field. (Klepesta)

a comet's orbit is determined. The orbit of an average single meteor, before it strikes the atmosphere, recalls that of an asteroid. The atmosphere, however, has the effect of altering the speed of the meteor, and the observed decelerations are the source of our information about the temperatures and pressures in the high atmospheric layers (p. 21). It was surmised at one time that the meteors, especially the very bright "fireballs," might include some visitors from interstellar space that were not original members of the solar system. But these conclusions were drawn from reports of visual meteors which caught the observer unexpectedly and perhaps biased his judgment of speed. The data from photographic paths and velocities do not point definitely to any visitors from outer space.

If a prism is placed over the objective of a meteor camera it is pos-

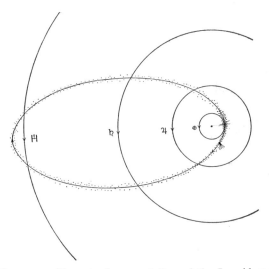

Fig. 9.11. Diagram to illustrate the association of the Leonid meteors with the comet 1866I. The meteors are spread all around the comet's path, most thickly near the position of the comet. The earth meets the orbit once a year, in November, and a shower of meteors is seen at that time. The period of the comet is 33 years, and at this interval the showers are richer.

sible to photograph the spectrum of the trail left by a meteor. Those that have hitherto been so photographed have been predominantly bright ones, comparable with the fireballs that sometimes reach the earth as meteorites. The elements seen in the spectra are very similar to those actually found in meteorites: iron, calcium, manganese, and magnesium predominate in both.

On some nights the number of meteors seen is very great, and moreover the meteors seem to radiate from one point (the *radiant*). The apparent divergence of the trails from the radiant is an effect of perspective and shows that the trails are actually parallel. One of the most spectacular meteor showers of modern times radiated from the constellation Leo on November 13, 1833; it recurs at nearly annual intervals (e.g., November 17, 1950), always from the same area, and is known as the *Leonids*. But its richness varies greatly from year to year. In 1799 it was conspicuous, and also in 1866, but in 1899 and 1932 (which continue the series of 33-year intervals) the shower, though appreciable, was disappointing. It seems as though the Leonids are scattered all the way around an orbit, and more densely distributed in one part of it. The period in the orbit is 33 years and the motion is retrograde. The reason for the poverty of the shower in 1899 and 1932 was that the dense part of the swarm had been perturbed by passing near Saturn and Jupiter.

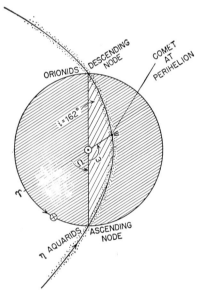

Fig. 9.12. Schematic diagram to illustrate the production of two meteor showers, about six months apart, by the bodies scattered along the orbit of Halley's Comet. The comet's orbit actually passes both nodes at a distance of several million kilometers from the earth; the fact that we see showers of meteors on both occasions shows that (if the meteors are in fact associated with the comet) they are scattered in very broad streams. Notice that the comet's motion is retrograde; its orbit is an elongated ellipse (eccentricity 0.967); its period about 76 years, which varies somewhat because of perturbations.

The most interesting thing about the Leonids is that a small comet, observed in 1866, was shown to have the same orbit as the Leonid shower. There is no doubt that the meteors have originated from the comet, which moves slightly ahead of the dense meteor region in the orbit; they have been gradually spread around it as the comet slowly breaks up.

The association of the Leonids with a known comet was the first of several such connections to be discovered. Many of the well-known showers are now known to be associated with comets. Biela's Comet, which divided in two before 1846 and disappeared before 1865, survived as a meteor shower (the "Bielids"), which has now disappeared, probably because of perturbations. Halley's Comet is associated with two showers three months apart, at the two nodes: the Eta Aquarids and the Orionids; Encke's Comet goes with the Taurid meteors.

EVOLUTION
OF THE SOLAR SYSTEM

This world was once a fluid haze of light,
Till toward the center set the starry tides,
And eddied into suns, that wheeling cast
The planets.

TENNYSON, *The Princess*

The organization of the solar system—its flatness, direct rotation, and apparently orderly arrangement—are convincing evidence that it did not arise in a haphazard manner; ever since its rotation about the central sun was understood, attempts have been made to picture how it may have come into being.

For the study of this question, a knowledge of the present configuration of the system is not enough. It is also necessary to know how long a time interval can be allowed for the operation of the supposed processes.

1. THE AGE OF THE SOLAR SYSTEM

The time is long past when a "date of creation" of a few thousand years ago was seriously considered. There are several ways of estimating the ages of atoms, planets, stars, and galaxies, and (with a few exceptions) their unanimity is surprising. The evidence is summarized in Chapter VIII in connection with the much wider problems of the development of stars and galaxies. It points to an interval of significant change (the only sense in which we can discuss cosmic ages) of something between a thousand million and ten thousand million years. The age of the solar system does not differ materially from the probable ages of the atoms, the building-blocks of the cosmos. Older theories pictured the sun as having developed from a pre-existing nebula and the planets having, in their turn, developed from the sun; perhaps their moons following later by a like process. Today, however, we are compelled to think rather in terms of a fundamental process that gave rise to sun, planets, and satellites at about the same time, and as parts of the same development.

There are a few cosmic objects that seem definitely to be younger than the rest. The supergiant stars can scarcely have been shining for as long an interval as has elapsed since the Carboniferous period. We have seen that the life of a short-period comet is not more than a few thousand years. But most of the stars and galaxies, as well as the main part of the solar system, seem to be between 10^9 and 10^{10} years old.

2. THEORIES OF ORIGIN OF THE SOLAR SYSTEM

The *Nebular Hypothesis* of Laplace held the field for many decades. It pictured the system as having originated from a rotating nebula, which formed gaseous rings that, in their turn, condensed into planets and their satellites. The sun was thought to be finally formed by the concentration of the central part of the nebula.

The Nebular Hypothesis accounted for the axial symmetry of the system and for the direct rotation. But it failed on several counts. The most important one concerns the *angular momentum*. If a rotating nebula had produced the planets in the manner imagined by Laplace, the major part of the angular momentum should have been retained by the sun. Actually the planets all together have fifty times as much energy of rotation as the sun has—a difficulty that is not peculiar to Laplace's Nebular Hypothesis. Further, it was shown by Clerk Maxwell (p. 221) that a fluid ring would not condense into a single planet, but into a zone of small bodies like the asteroids. The hypothesis could, perhaps, account for the asteroids, but not for single planets.

At the beginning of the present century, theories of the origin of the solar system rested on the idea that the planets had been ejected from the sun as a result of a close approach to another star. Huge tides, raised by the passing body, were thought to have risen from the solar surface and become detached from the sun. As the intruder receded, they began to move round the sun and to form the planets. Much of the great tide was lost forever, much fell back into the sun, but some had the right position and speed to move in elliptical orbits. The *Planetesimal Hypothesis* of Chamberlain and Moulton supposed the circulating material to have condensed into small solid bodies (planetesimals), which in course of time coalesced, rounded off into nearly circular orbits, and formed the planets. Another view was that the tidal mass formed a long gaseous filament, which broke up into gaseous masses, and later condensed to form the planets. These tidal theories, as developed by Jeans and Jeffreys, partially met the difficulty with angular momentum by the supposition that the energy of revolution of the planetary system was supplied by the passing star; but mathematical theory shows that they have not succeeded in evading the difficulty altogether. Furthermore, it has been shown that hot gaseous material ripped from the interior of a star cannot possibly condense into solid bodies; on the contrary, it would promptly dissipate explosively into space.

Other variants of the "approach" hypothesis encounter the same objections; the condensation of the filament is not facilitated by supposing that the sun was originally a double star, of which the second component was driven off by a collision, leaving behind some debris that formed the planets. Neither are matters helped by the bizarre suggestion that the disruption was from within: one of the original pair having become a supernova, the recoil of the explosion sending it off into space, and leaving potential planetary stuff behind!

The collision and double-star hypotheses all require a pre-existing sun that produced the planets as a result of a unique experience. We may remark in passing that such a close approach of two stars as would be required can be shown to be so unique as to be practically out of the question. But the more important point is that they require the planets to be younger than the sun, whereas present-day ideas suggest that they are not. The original idea of Laplace does not suffer from this objection, and the current theory of planetary origin has many features in common with it.

The development of a rotating mass of gas is more thoroughly understood today than it was in Laplace's time, and we know now that internal friction tends to produce vortices in a rapidly spinning fluid mass. On this basis von Weizsäcker has suggested that the planets were produced by vortices that must have been set up as the mass rotated; the theory is still approximate, but it seems that the size of the vortices might be about the same as the distances that separate the planets. This is the first theory of the origin of the solar system that predicts something like Bode's Law. However, it is still faced with the problem of getting the rotating gas to condense into planets. According to this theory the origin of the solar system may be pictured as part of the process that also produced the stars and galaxies themselves; von Weizsäcker suggests that the biggest possible eddy has the size of a large galaxy. This theory has hardly passed the qualitative stage as yet (p. 468).

The difficulty of condensing the planets out of gas is evaded by the theory, developed by Whipple, that they may never have been gaseous. He pictures the solar system as having formed out of a cloud of *dust and gas,* such as we observe within many stellar systems. The original cloud may have been 30,000 astronomical units in diameter, and had about the mass of the sun, half in gas and half in dust. The total angular momentum of the cloud must have been very small. Local irregularities, inevitable in a dispersed cloud, would have gradually produced rotation and led to the cloud's collapsing on itself under its own gravitation. At first the collapse would be merely a very slow condensation, taking perhaps ten million years, but as time went on it would speed up, until finally it would attain supersonic velocity. The solid particles would collide and stick together, ultimately balling up to form planets; the variety of their motions would tend to smooth out the original elliptical orbits and lead to nearly circular orbits with direct rotation—its sense determined by the original angular momentum of the cloud. The rapidly collapsing gases in the center would form the sun, which would have rather small angular momentum.

On this theory the earth was never gaseous, and was never part of the sun; in fact the compositions of earth and sun seem to differ slightly (in the ratio of the isotopes of carbon of mass 12 and 13, for example). Moreover, the sun, as we know it today, may actually, according to this conception, be *younger* than the planets.

The theory of the formation of stars from interstellar globules of mixed dust and gas is closely related to this picture of the origin of the solar system. In fact, the formation of a solar system not unlike ours would follow as a possible, even a probable, process in the development of a star by this route. On the other hand, the various types of encounter hypothesis represented the formation of a solar system as an almost vanishingly rare event. Opinion today points to the possibility that planets may be very numerous throughout the stellar system. But their detection, let alone their observation, would tax our observing techniques to the limit.

The theory gives a plausible account of the origin of the sun and planets, but several points remain uncertain. It does not immediately lead to Bode's Law. The retrograde satellites of Jupiter, and of Saturn, are not difficult to understand as captured asteroids; the retrograde motion of Triton is more difficult to fit into the picture. And it does not account for the asteroids and the comets.

The fact that the asteroids move in a zone near to which Bode's Law predicts a planet suggests that they represent such a planet. Their eccentric and inclined orbits set them apart from the larger planets. So does the variety of their albedos, which suggests that their surfaces differ, and the variation of their brightness, which points to irregularly shaped bodies. All these things would be compatible with the possibility that the asteroids are fragments of a planet or planets that broke up at some time in the remote past. Meteorites may well be parts of the same object; and, as we have seen, their physical structure is rather like what would have been produced by the fragmentation of a planet that resembled the earth, possibly as the result of a collision. If there was once such a planet, it was very likely smaller than Mars. The aggregate mass of all the asteroids is not one-hundredth that of the earth. But a considerable part of the exploded planet may have had velocities great enough to carry it way from the sun's gravitational field altogether, and the surviving asteroid zone represents the fragments that had the smallest speeds.

The comets represent another problem; the short lives of those with small perihelion distances require that there be some source from which comets are continually drawn. A theory suggested first by Öpik and recently developed by the Dutch astronomer Oort makes them a plausible part of the same picture.

The solar system, on Oort's view, lies at the center of a great cloud of comets, between 50,000 and 150,000 astronomical units from the sun. There are a hundred billion comets in this cloud, and their aggregate mass is between 1/10 and 1/100 of the earth's. Surprisingly enough, perturbations of the orbits of these remote comets are affected not only by Jupiter, but also by the nearer stars. These perturbations are probably adequate to divert new comets into the central parts of the solar system in numbers that are consistent with observation. It is natural that perturbations should produce a certain proportion of retrograde orbits for long-period comets, which are indeed the only bodies in the solar system whose motion is sometimes not direct.

The cometary population of the outer zone of the solar system must differ physically from the asteroids, which seem to be rocky fragments without the icy constituents that produce the comets' tails. The relatively inconspicuous tails of the short-period comets are probably results of the exhaustion of this material after repeated trips about the sun. Once a comet has entered the inner solar system it is probably only a matter of time before Jupiter perturbs the newcomer into a small orbit and thus produces a short-period comet, which cannot survive more than a few hundred revolutions.

The foregoing account of the origin of the solar system may well seem disappointing to the student. With so rich a fund of data we might have expected greater certainty. But we must recognize that theories of cosmic development, whether of planets, stars, or galaxies, are deployed within an enormous span of time (still uncertainly defined) and involve a sequence of events and conditions that are inaccessible to observation. The great advances in understanding the processes of cosmic evolution in recent years have resulted from the synthesis of many problems into one, and especially from the realization that the development of planet, star, and galaxy are parts of the same process and have probably been spread over comparable intervals of time.

CHAPTER
XI

INTRODUCTION
TO STELLAR ASTRONOMY

Through knowledge we behold the world's creation,
How in his cradle first he fostered was;
And judge of Nature's cunning operation,
How things she formed of a formless mass . . . .

From thence we mount aloft into the sky,
And look unto the crystal firmament,
There we behold the heaven's great hierarchy,
The stars' pure light, the spheres' swift movement . . . .
EDMUND SPENSER
The Tears of the Muses

Fifty years ago astronomy was largely concerned with the geometry and dynamics of the solar system. The planets and the sun were only beginning to emerge as individuals with observable physical and chemical properties. Until the beginning of the present century, little was known about stars except their brightness and color, even less about their distances and apparent motions. As we shall see, speed, at right angles to our line of vision, and distance, can be measured only for the very nearest stars, and the knowledge of the geometry and dynamics of the stellar system was therefore practically confined to the small district near the sun. The extent of our own system of stars was grossly underestimated, and though the existence of other stellar systems was suspected, the study of their properties had scarcely been begun. Astronomers realized that stars differ from one another, but the information was largely empirical and the picture static.

The modern approach to the study of the stars supplements this geometrical and dynamical knowledge with physical information about the stars as individuals. We are now able to measure their "chemical composition," and we can actually probe their internal structure and study the sources of their light and energy.

Today we envisage the stars as developing entities and can even trace the course of their probable development. We estimate the ages and future lifetimes of stars and stellar systems. Emphasis is shifting to the nonstellar

255

parts of the universe: astronomers today study the dust and gas that pervade much of interstellar space, and profoundly affect the birth and the break-up of stars. One might describe the modern picture of the stars as a dynamic one, in contrast with the static conception of the beginning of the century. The difference is a result of the development of *astrophysics*—the application of the laws of physics on the widest possible scale. While the astronomy of yesterday relied upon the applications of geometry and the laws of motion, the astrophysics of today draws upon all our modern knowledge of the properties and behavior of matter and radiation.

The laws of geometry and of motion under gravitation suffice for the study of the motions of stars. But when the motions within stellar systems are to be considered, it is necessary to make use of the far more complicated theories of hydrodynamics and turbulence, which are still incompletely worked out; hence the dynamics of stellar systems, and the development of stars within the nonstellar substratum, are still infant branches of astrophysics and are full of problems and possibilities.

The stars are wholly gaseous, and physical studies of their surfaces and interiors are greatly simplified by this fact. We can interpret the intensity, color, and energy distribution of starlight and relate these to the conditions at the stellar surface by means of the laws of continuous radiation, deduced from the quantum theory and verified in the laboratory.

The laws of the production of line and band spectra (by atoms and molecules) are even more significant for the interpretation of astrophysical data. From the spectra of stars and interstellar material we can tell, not only what chemical elements and compounds are present, but also their relative proportions. The stage (number of electrons) and state (distribution of energy among the electrons present) of the stellar atoms reveal the conditions at the surface of the stars. They can be used not only to find the temperature and density of stellar atmospheres, but also to detect the presence of magnetic and electric fields. Moreover, the comparison of precise measurement of the color (wave length) to our laboratory standards of the light given out or absorbed by known atoms will reveal the total motion of a star toward or away from the observer, rotation or pulsation of the star itself, and even superficial changes, ejections, or explosions on its surface.

The laws of gases, deduced from physical theory and verified in the laboratory, are even more powerful tools for examining the stars. By their aid it is possible to arrive at a very good idea of the internal structure of stars, conditions in their interiors, and even the propagation of deep-seated disturbances through the outer layers.

Finally, the recent burgeoning of our knowledge of nuclear structure opens the door to understanding of the nuclear processes that probably go on within the stars and permits us to state, with some confidence, what is the source of their energy and what the course of their development.

The methods of astrophysics, when applied on the broad scale permitted by the conditions in the stellar and interstellar universe, reveal a surprising uniformity of chemical composition for all the bodies accessible to observation. But a surprising variety of stars is constructed of these uniform materials. Their brightnesses differ by a factor of a million millions; their diameters, by three hundred thousand; their masses, only by a thousand; yet their densities have a range of a factor of four thousand million, and their surface temperatures range from 1700° to about half a million degrees. Nor are the stars all built alike; some are strongly concentrated toward their centers, others far less so. The following chapters will give an outline of the facts on which these conclusions are based.

Stars and interstellar material are the subject matter of astrophysics. Neither stars nor nonstellar matter are scattered at random. They are aggregated into larger systems that have definite structure and definite properties. The latter part of the book will deal with stellar systems and with the greater systems in which they themselves are organized. In the interests of clarity and coherence, we shall anticipate some of the material of the final chapters and begin our study of astrophysics with a description of the broader features of the stellar universe.

1. STELLAR SYSTEMS

The observable universe contains a large number of aggregations of matter, each organized from aggregations of smaller scale. The largest known aggregations are *clusters of galaxies,* or organizations of *galaxies.* Galaxies are organizations of *stars* and interstellar matter (*dust* and *gas*); stars, dust, and gas are organizations of *atoms* and *molecules,* themselves organized of elementary particles (*protons, neutrons, electrons,* etc.). The whole is pervaded by *radiation,* which plays a vital part in the behavior of atoms and stars, and governs the development of galaxies and groups of galaxies.

Galaxies are complex organizations, made up of stars, gas, and dust, all of which are themselves irregularly distributed. Dense groups of stars are *star clusters;* conspicuous aggregations of gas may form *bright nebulae,* whereas dense regions of dust form *dark clouds.* Not all galaxies contain all of these constituents.

There are perhaps ten thousand million observable galaxies. They differ greatly in size, structure, and population, but an average galaxy may contain about a thousand million stars. Our own galaxy is much larger and more populous than the average and contains perhaps a hundred thousand million stars, as well as large quantities of dust and gas.

A comparatively small and simple galaxy is NGC 147, No. 147 in a tabulation of galaxies known as the "New General Catalogue" (Fig. 11.1). It lies near the "great spiral in Andromeda," distant from us rather more than a million light-years. It is not visible to the naked eye, and even with the most powerful photographic telescope, its brightest individual stars are barely resolved. We do not know how many fainter stars it con-

Fig. 11.1. The *elliptical galaxy* NGC 147, photographed with the 200-inch telescope. All the bright stars in the picture are foreground objects, members of our own stellar system. The stars in the distant galaxy are just resolved in the photograph. (Photograph by Baade, Palomar Observatory.)

tains, but the number is certainly large. NGC 147 is a typical *elliptical galaxy* (an obvious description of its shape). It contains only stars and has no gas or dust. It is of neatly symmetrical shape and is probably rotating about its smallest axis.

Another galaxy of simple, though very different, form is the *Small Magellanic Cloud*. It is one of the galaxies nearest to us—about a hundred and fifty thousand light-years distant. It is visible to the naked eye in the

Southern Hemisphere as a bright hazy blur, and even a moderate telescope resolves it into a large number of stars. Besides stars, it contains some gas and perhaps some dust. It is not of particularly symmetrical shape, and is known as an *irregular galaxy*. Another irregular galaxy is the *Large Magellanic Cloud* (a larger, brighter neighbor of the Small Magellanic Cloud), which is in fact a galaxy above the average in size and brightness. It contains stars, gas, and considerable dust. Although classed as irregular, it shows a suspicion of symmetry, and is a step in the direction of the next group of galaxies to be mentioned.

Fig. 11.2. The nearest galaxies, the *irregular systems* known as the Large and Small Magellanic Clouds. On the left is part of the Milky Way, the stars that are crowded towards the central plane of our own system. The Large Magellanic Cloud is near the center; the Small Magellanic Cloud, near the lower right. (Boyden Station, Harvard Observatory)

The great class of spiral galaxies * is illustrated by the "Whirlpool" in Canes Venatici, Messier 51 (i.e., No. 51 in the catalogue of "odd objects" compiled 150 years ago by the French astronomer Messier for the use of observers of comets. In listing the things to be disregarded in looking for some of the slightest objects in the solar system, he enumerated some of the most enormous stellar systems!) We do not know the exact distance of Messier 51, but it is farther away than anything yet mentioned. It consists of a bright center or nucleus and two symmetrically situated spiral arms that coil around the nucleus. Variations on this beautiful structure are extremely common among galaxies.

We see Messier 51 practically face-on. If viewed on edge it would be seen to be very flat, probably like NGC 4565 (Fig. 16.19, p. 450). These spirals, and most others, contain stars, gas, and dust. The stars are too faint to be seen individually in these two systems. The gas is visible as bright blurs, the dust as dark lanes or streaks; gas and dust are confined to the

---

* In the days before the nature of these systems was understood, they were called "spiral nebulae." Today this name is obsolete because it tends to confuse stellar systems with gas clouds, and we shall therefore not use it.

spiral arms and avoid the nucleus. They seem to be distributed in a central plane, like the filling of a sandwich. This spiral type of galaxy is sometimes described as a "pinwheel," and is known to be spinning rapidly, though not like a rigid wheel. Probably the spinning motion has flattened it and produced the spiral structure.

Fig. 11.3. The spiral *galaxy* Messier 51, seen behind the stars of the constellation Canes Venatici. (Mount Wilson Observatory)

Our own galaxy (the "Milky Way system") is one of the largest known. Most of the stars that we observe individually are in it, and it contains a great many faint and distant ones that we cannot see. Our galaxy is much more difficult to study than a distant one, because we are inside it. We happen to be within the central dust-sandwich, which obscures our view. However, by piecing together various scraps of evidence about the distribution and motion of the stars within it, we can conclude that our own galaxy is indeed a spiral. The distribution of the stars shows great flattening and gives a rough impression of spiral arms; their motions point to pinwheel rotation. There is much dust and gas in our neighborhood, but it seems to be confined to the thin layer that constitutes the dust-sandwich.

When we look at the hazy bright band (the "Milky Way") that circles the starlit sky, we are looking at the system along the central plane. Photographs of the Milky Way show the bright band, peppered with innumerable faint stars and mottled by the irregularly distributed clouds of interstellar dust. Large dust clouds, such as the "Coal Sack" near the Southern Cross, are readily visible to the naked eye. Unfortunately the bright nucleus of the galaxy happens to be behind some particularly dense dust clouds, and can be observed directly only with special techniques. We are far from the center of the Galactic system, nearer to the edge than to the nucleus.

The zone of the Milky Way is streaked with dark clouds (as in Serpens and Ophiuchus) and is also full of clouds of glowing gas (as in Orion). The "great nebula in Orion" is a cloud of gas that is caused to shine by a group of hot stars within it. The bright nebulae are often streaked with dark clouds: the dark matter always seems to be accompanied by gas, but the gas will only glow in the neighborhood of an intensely radiating hot star. Such stars are the central bodies of the planetary nebulae, and they cause the gaseous envelopes to glow. Gas is not, however, invariably accompanied by dark matter, as we shall see.

A few bright nebulae are not caused to shine by stars, but are probably the result of the energy produced by the collision of two clouds of gas—the interface of the two clouds is glowing. Probably the Crab Nebula, the debris of a colossal stellar explosion, is at least partly caused to glow by the energy of collision of clouds.

Fig. 11.4. The *galactic cluster* ϰ Crucis (left), and the *globular cluster* 47 Tucanae (right). Both are in the Southern Hemisphere. The former may be seen in Figure 13.11, near the Southern Cross; the latter, in Figure 16.2, near the Small Magellanic Cloud. (Boyden Station, Harvard Observatory)

Clusters of Stars. The galaxies contain many aggregations of stars, from groups of two, three, or more, up to the gigantic groups known as star clusters. Two well-marked types are the galactic star cluster and the globular star cluster. *Galactic clusters* are so called because they tend to lie in the plane of the Milky Way; they are sometimes called "open clusters." Well known naked-eye examples are the Pleiades, the Hyades, and Coma

Berenices. The Pleiades, which contains hot stars, illuminates a nebula, and its stars are embedded in a cloud of dust and gas. Probably most galactic clusters share this property.

Globular clusters, on the other hand, contain little dust or gas. Galactic clusters contain a few hundred stars; globular clusters contain hundreds of thousands or millions and are beautifully symmetrical and compact, though occasionally rather flattened. The contrast between the two types of cluster is of great significance in our understanding of the history and relationship of the different parts of our own and other galaxies.

2. THE STARS

The stars are all completely gaseous. They are held together by gravitation and are distended by gas pressure and perhaps also by the energy of the radiation that flows outward from their hot interiors. Some stars are distorted by their own rapid rotation and by the gravitational influence of other stars near them. Thus, although all stars are roughly spherical, some are flattened by rotation into spheroids or distorted into ellipsoids by gravitational effects.

Probably all stars increase in density and in temperature from the outside inward. All are surrounded by more or less tenuous envelopes—the only part of them that can be studied directly. These envelopes are made of the same substances that are familiar to us on the earth, and no material has been observed on the surfaces of the stars that cannot be found on our planet. The envelopes are often in more or less violent motion. Some stars are throwing off material, either steadily or in bursts. Some stars may be picking up material from their surroundings. The surfaces of some stars are the seat of intense magnetic fields. Chapter XIII will present some of the evidence on which these statements are based.

As mentioned earlier, the stars represent an enormous variety. Diameters range from three thousand times to about one-hundredth of the sun's. Surface temperatures are from about 1700°C to over half a million degrees. All the very large stars are of low surface temperature. Few hot stars are more than ten times the sun's size.

By far the greater number of the stars can be arranged in a sequence of decreasing temperature, size, and brightness, from high-temperature stars about ten times the sun's size to cool stars about a tenth of its size. This sequence is called the *main sequence,* and the sun is a member of it, about intermediate in size and well below the average in surface temperature and in mass. At the hot, bright end of the sequence, the stars are very uncommon, and most main sequence stars are both smaller and cooler than the sun. The very smallest and coolest stars are apparently rare.

Another group of stars, far less common than those of the main sequence, are the large, luminous stars of moderate-to-low surface temperature, roughly called the *"giant stars."* Even less common are the enormously luminous and distended stars—the *"supergiants."* Finer classification of stellar brightness and dimensions has led to the recognition of a

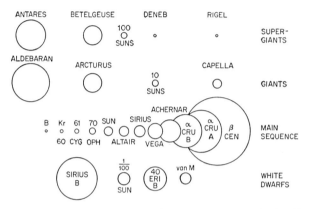

Fig. 11.5. Comparative sizes of some typical stars. B stands for Barnard's star; Kr 60, for Kruger 60; 61 Cyg, for 61 Cygni A; 70 Oph, for 70 Ophiuchi A (brighter components of several well-known stellar pairs); van M is van Maanen's star. In each row, the scale relative to the sun is indicated.

group intermediate between main sequence and giants, the *"subgiants,"* and another sequence, almost parallel to the main sequence but probably differing radically from it—the *"subdwarfs."* The most remarkable group of all, second only in commonness to the main sequence, contains the *"white dwarfs,"* stars of moderate mass and very small luminosity but of almost incredibly high density and small size.

The desirability of grouping the stars in these classes is not indicated by their differences of size, brightness, and surface temperature only. Recent advances in the study of stellar development have shown that stars that are found physically associated—in clusters, for instance—tend to be grouped according to the classes just outlined. Moreover, stars that move in similar fashion (for instance, stars that move in similar orbits about the nucleus of the galaxy) tend to outline certain of the main stellar groups to the exclusion of others. And theoretical studies of the inner structure of stars show that the outer properties of giants, main sequence stars, white dwarfs, and so on, correspond to profound internal differences, which involve not only structure but also the fundamental processes by which the light of the stars is generated and their equilibrium maintained.

The sun may be considered a typical main-sequence star. Because it is so near to us, it gives information on stellar behavior that is inaccessible for any other star. It may prove instructive to re-read the section of Chapter IV that deals with the sun, bearing in mind that *all* stars may well show the same phenomena and that many of them may do so to a far greater degree.

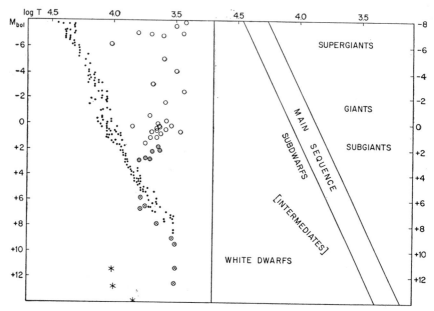

Fig. 11.6. The principal types of stars. On the left are shown a number of typical stars; total brightness (M_{bol}) is plotted against the logarithm of the surface temperature. Main-sequence stars are shown by dots, giants and supergiants by circles, subgiants by shaded circles, subdwarfs by crosses in circles, white dwarfs by stars. On the right the domains of the various types of stars are roughly outlined.

We may, for example, recall the variety of conditions represented by sunspots, photosphere, granulations, spicules and flares, chromosphere, and prominences and corona. The range of temperature for the solar surface runs from perhaps 3500°K for the sunspots to about a million degrees for the corona. The other stars undoubtedly show similar stratifications, perhaps with even greater temperature disparities.

Flares, for example, are shown by many cool main-sequence stars. A solar flare brightens the whole light of the sun only slightly, but on a cool star a flare may more than double the star's total brightness. Prominences are shown by the detailed spectroscopic study of some luminous double stars, and they must be far more massive than the solar prominences. Some of the highly luminous, hot stars have both chromospheres and coronas that compare with, or exceed, their photospheres in brilliance or conspicuousness. Evidences of the million-degree corona are shown by some exploding stars. And although the spectroscopic effects of the general magnetic field of the sun as a whole are so small as to be barely measurable, some stars show magnetic fields (as a whole) that are greater than those displayed by the localized tornadoes that we know as sunspots. The sun probably loses comparatively little of its substance by photospheric and spicular activity, but some stars are spilling and spurting matter into the surrounding space in quantities that may affect their internal economy within calculable times.

The sun, again, is rotating comparatively slowly, but even this slow rotation is far from uniform. Many stars can be shown to be spinning on their axes very rapidly; are these, too, in nonuniform rotation, and if so, what effect will it have on their inner structure and their history? Probably its speed of rotation is a crucial factor in a star's career, especially because this property seems to be intimately linked with other features of behavior.

What is a star's relation to its environment? Here again the sun gives instructive information, for the motion of the gases that surround it is evidently governed by its magnetic properties. Some stars are in really "empty space," others in dense regions of dust and gas, and the difference may well have a crucial effect on their development.

From the physical and astrophysical properties of the stars we shall pass to the account of stellar associations and stellar systems, and we shall attempt to show how the problem of the development of stars is linked with the problem of the history of the systems in which they lie.

STARS:
GEOMETRICAL PROPERTIES

Look at the stars! look, look up at the skies!
O look at all the fire-folk sitting in the air!
GERARD MANLEY HOPKINS

1. STELLAR MAGNITUDES

The first known catalogue of 1080 naked-eye stars was made by the Greek astronomer Hipparchus in about 120 B.C., and it was later edited by Ptolemy in the form of the famous catalogue known as the "Almagest" (1022 stars). Hipparchus enumerated the stars that could be seen in each constellation, described their positions in relation to the mythical constellation-figures, and indicated their brightness by six rough grades (1, 2, 3, . . . , etc.), 1 denoting the brightest. This crude way of assigning the apparent brightness of the stars survives in our present system of describing a star as of the "first magnitude," "sixth magnitude," etc. Note that smaller numbers apply to brighter stars. Hipparchus, of course, had no telescope; when fainter stars were telescopically observed, the same system was extended to them. The faintest magnitudes attainable today go to perhaps the twenty-second.

It was found, when the brightness of stars was measured accurately, that each magnitude is about 2.5 times brighter than the next greater magnitude. More precisely, a first magnitude star is defined as a hundred times as bright as a sixth magnitude star. Everywhere along the scale, a difference of five magnitudes corresponds to a ratio in brightness of one hundred. In other words, *the magnitude scale is not linear but logarithmic.*

This feature of the magnitude scale is similar to those of many scales that express the intensity of a sensation. A general law, known as *Fechner's Law,* states that equal increases of sensation are produced by equal *ratios* of stimulus. This law applies, for instance, to the perception of sounds.

As 5 magnitudes correspond to a ratio of 100 in brightness, the brightness-ratio for 1 magnitude is $\sqrt[5]{100}$. The logarithm of 100 to base 10 is 2.00; 1/5 log 100 is 0.400, and 0.400 is log 2.512. Thus the light ratio for 1 magnitude is 2.512; a first-magnitude star is 2.512 times as bright as a second-magnitude star; a twentieth-magnitude star is 2.512 times as bright as a twenty-first magnitude star.

The classes of Hipparchus were very rough; with more accurate measures it was found necessary to subdivide them, so that one spoke of a star of magnitude 4.2, etc. Some stars were definitely too bright for the first magnitude, and both zero and negative magnitudes were found necessary. The brightest object in the sky, the sun, has a magnitude of -26.72; the full moon, -12.5, the brightest star, Sirius, -1.5. Stars of about the sixth magnitude can be seen with the unaided eye. The world's largest telescope, the 200-inch, permits the photography of stars a little fainter than the twenty-second magnitude.

Measurement of magnitudes is a problem of measuring light-intensity accurately. For the stars it is complicated by the earth's atmosphere, whose absorption must be allowed for. The early measures of magnitude were made with the eye, in conjunction with instruments (photometers) designed for accurate comparison of brightness. During the past century the photographic plate has largely replaced visual work; it has advantages but also introduces its own problems. During the past decade the photoelectric cell, which depends on the photoelectric effect to measure incident energy, has become the most valuable instrument of precision in astronomical photometry. Other devices, such as the thermocouple, are used to measure the *heat* emitted by the stars. Magnitudes measured with the eye are called *visual magnitudes;* the eye responds most readily to yellow-green light, so it measures the yellow-green light of the stars. Magnitudes measured with the ordinary photographic plate are called *photographic magnitudes;* these are measures of the blue light of the stars. Magnitudes obtained with a yellow-sensitive plate and a yellow filter which restricts the range of color as nearly as possible to that of the eye are called *photovisual.*

Special photographic plates may be used to measure light of any color that is photographically accessible; such magnitudes are called *red* and *infrared magnitudes.* Most photoelectric photometers use light near that used for ordinary photographic magnitudes; special filters are used to measure photoelectric magnitudes in various colors. Visual and photographic magnitudes are usually expressed to hundredths of a magnitude, e.g., 8.76; however, the photoelectric cell is more precise, and photoelectric magnitudes are often given to thousandths, e.g., 6.584.

The numbers of stars brighter than a given magnitude go up rapidly for fainter magnitudes. If the stars were uniformly distributed, and if their light were not dimmed at a distance, there would be very nearly four times as many stars for each fainter magnitude. The numbers do not go up as fast as this, because neither condition is fulfilled. Away from the galactic plane, the numbers actually fall off with distance, and in the galactic plane, dimming by interstellar material is mainly responsible for the drop in the observed numbers below expectation. Numbers of stars brighter than several limits of magnitude are given in the next table:

	NUMBER OF STARS	
Limiting Magnitude	Visual	Photographic
5	1620	1030
10	324,000	166,000
15	32,000,000	15,000,000
20	1,000,000,000	505,000,000

The total number of stars in our galaxy has been estimated at 100,000,-000,000. Note that the number of stars down to a given visual magnitude is always greater than the number of stars down to the same photographic magnitude. This is because most of the stars are somewhat red. The *relative* number of stars down to a given visual magnitude increases for faint magnitudes; this shows that the proportion of red stars is even greater among faint stars than among brighter ones.

The starlight in the night sky (no moon) equals the light of 1092 first-magnitude stars. Stars brighter than magnitude 11 contribute 1/2 of this light, stars brighter than magnitude 15, about 3/4. But the starlight contributes only about 1/6 of the total sky brightness; the rest comes from the zodiacal light and the "permanent aurora." This residual sky light puts a limit to the faintness of stars that can be photographed with a given instrument.

Absolute Magnitude. The stars are all at different distances, so their brightness, as seen by us, does not tell us how bright they actually are. In order to compare the real brightnesses of stars, we must line them up (in imagination) all at the same distance, and find how bright they would look at this distance. The distance that has been arbitrarily chosen for this purpose is *ten parsecs.* (See page 274 for definition of a parsec; see page 273 for description of how the distances are measured.) The *apparent magnitude* (as seen by us) and the *absolute magnitude* (as seen from a distance of ten parsecs) are connected with the distance (in parsecs) by the simple formula:

$$M = m + 5 - 5 \log D$$

where M is the absolute magnitude, m the apparent magnitude, and D the distance in parsecs. The data for Sirius are: $m = -1.52$; $D = 2.66$ parsecs, or $\log D = 0.4249$. Thus $M = -1.52 + 5 - 2.12 = +1.36$. The absolute magnitude of the sun is $+4.86$.

A list of the twenty (apparently) brightest stars is given in Table 12.1. They are arranged in order of decreasing apparent brightness. If a star has a close companion, the latter is given in the table too. After the common name of the star, its constellation designation, apparent magnitude, absolute magnitude, surface temperature, and its brightness, size, and mass in terms of the sun's are given.

TABLE 12.1. THE BRIGHTEST STARS

Name	Apparent Visual Magnitude	Absolute Visual Magnitude	Surface Temp. °K	Brightness (suns)	Diameter (suns)	Mass (suns)	Remarks*
Sun / ⋯	−26.72	+4.86	6,000	1.00	1.00	1.00	MS
Sirius / α Canis Majoris	−1.52	+1.36	11,200	26.	1.8	2.4	MS
	+8.44	+11.3	7,500	0.003	0.034	0.96	WD
Canopus / α Carinae	−0.86	−7.4:	8,000	80,000	210.		SG
⋯ / α Centauri	−0.33	+4.7:	6,000	1.12	1.0	1.1	MS
		+6.1		0.32	0.74	0.94	MS
Vega / α Lyrae	+0.14	+0.6	11,200	50.	2.4	(3)	MS
Capella † / α Aurigae	+0.21	−0.6	5,500	150.	12.	4.2	G
Arcturus / α Boötis	+0.24	−0.2	4,100	100.	30.	(8)	G
Rigel / β Orionis	+0.34	−5.8	12,500	18,000	42.		SG
	+6.66	+0.2	12,500	80.	2.6		MS?
Procyon / α Canis Minoris	−0.53	+2.8	6,500	5.4	2.0	1.1	MS
	+10.8	+13.1	?	0.00005	0.04?	0.4	WD
Achernar / α Eridani	+0.60	−0.9	15,000	200.	3.5		MS
⋯ / β Centauri	+0.86	−3.9	21,000	3,100	11.	25?	MS
Altair / α Aquilae	+0.92	+2.4	8,600	9.	1.3		MS
Betelgeuse / α Orionis	+1.58	−2.9	3,100	1,200	290.	15?	SG
⋯ / α Crucis	+2.09	−2.7	21,000	1,000	6.6		MS
		−2.2	21,000	650.	5.2		MS
Aldebaran / α Tauri	+1.06	−0.1	3,300	90.	60.	4?	G
	+13.	+12.	?	0.001	?		?
Pollux / β Geminorum	+1.21	+1.2	4,200	28.	16.	?	G
Spica † / α Virginis	+1.21	−2.6, −2.2	18,000	1,500	8.	?	MS
Antares / α Scorpii	+1.23	−4.0	3,100	3,400	480.	30?	SG
	+5.5	−0.3	16,500	100.	3.		?
Formalhaut / α Piscis Austrinae	+1.29	+2.0	9,000	13.5	1.6		MS
Deneb / α Cygni	+1.33	−5.2	10,000	10,000	42.		SG

* MS = main sequence; WD = white dwarf; G = giant; SG = supergiant.
† Double star seen as one.

The list of the apparently brightest stars is very interesting, especially when it is compared with the lists of the nearest stars (p. 277) and of the stars of largest proper motion (p. 287). It contains a far larger variety

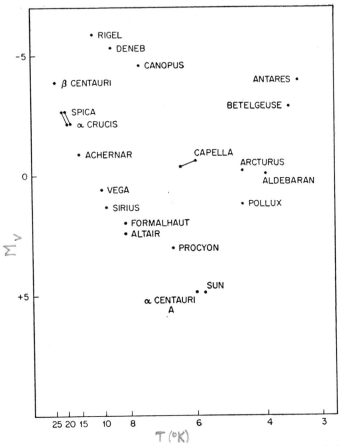

Fig. 12.1. The twenty brightest stars. Temperature (in thousands of degrees) is plotted against absolute visual magnitude. Faint companions are omitted from the diagram. Notice that it contains no stars much fainter than the sun.

of stars than the others: temperatures from 3100° to 21,000°K; brightnesses from eighty thousand times the sun's to one twenty-thousandth of the sun's; sizes from almost five hundred times the sun to about three-hundredths of the sun; masses from four-tenths to thirty times the sun's. Seven of the stars are double (the fainter components are included in the table). Three of these bright naked-eye stars (Canopus, Rigel, and Deneb) are actually so distant that their distances cannot be accurately measured by geometrical methods. The bright naked-eye stars include supergiants, giants, and main-sequence stars; their companions include main-sequence stars and white dwarfs. As we shall see, they are far from being a typical sample of the stellar population—the very bright ones are grossly over-represented.

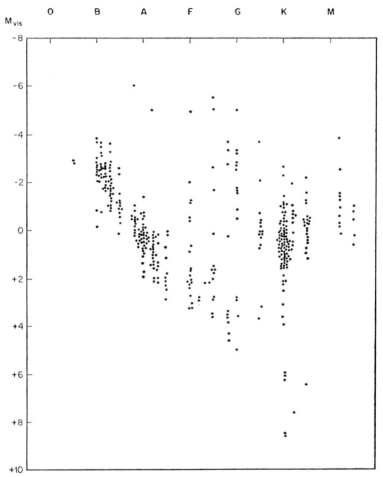

Fig. 12.2. Spectrum and absolute visual magnitude for all stars of known parallax that are visually brighter than the fourth apparent magnitude. Compare this diagram with Figure 12.4, which shows the nearest stars, and notice the great difference. The stars of bright apparent magnitude include the stars of bright absolute magnitude in exaggerated numbers and do not include any of the faint M stars, which preponderate in space. The bright end of the main sequence and the K giant stars are well represented, though these stars are actually quite rare. For relation of spectral class and temperature see page 180.

The main sequence has probably been spread out vertically to some extent by errors in the parallaxes. The horizontal concentration of stars at certain spectral classes is not real, but is a result of lack of detail in the spectral criteria; actually the spectral classes should show a continuous distribution.

Most of the stars of unknown parallax brighter than the fourth magnitude (about fifty in all) are B stars.

2. STELLAR DISTANCES

The measurement of the distances of stars is basically a trigonometric problem, and the methods are the same as those used in surveying. On the earth, the distance of an inaccessible object may be measured by a simple determination of angles and the measurement of an accessible distance, the baseline.

For distances within the solar system, as we have seen, the diameter of the earth is used as a baseline. The earth is not quite spherical, since its polar and equatorial diameters differ. The baseline used to express distances within the solar system is the equatorial diameter of the earth. The displacement of a nearby object (for example, an asteroid such as Eros, or the planet Mars) as seen from two widely separated points on the earth's surface against a very distant background (the stars) is measured. Sometimes the earth is simply allowed to turn, and the displacement is measured by comparing positions as observed twelve hours apart. Sometimes simultaneous observations are made from points far apart on the earth's surface. The equatorial horizontal parallax of Eros was determined from measurements made at the same time from twenty-four different stations in 1930–1931 (p. 173).

The stars are too distant to be appreciably displaced when seen from different points on the earth's *surface*. A larger baseline is found in the diameter of the earth's *orbit*. The earth is 93 million miles from the sun, so in six months it moves 186 million miles from its former position. Even this large difference of position produces only a very small displacement of the nearer stars: the *parallax*.

When Copernicus suggested that the earth describes an orbit around the sun, those who criticized his theory pointed out that the positions of the stars ought to shift as the result of the earth's motion. The fact that no such shifts had been observed was considered a serious drawback to the theory. But those early critics did not realize that the stars are so far away that the shifts that are produced are much smaller than anything that could be measured with the instruments of that day.

Sir William Herschel attempted to measure the parallaxes of stars by comparing the positions of stars that are very close in the sky. He argued that the nearer of the two should show the parallactic shift, relative to the more distant one, every six months. He did indeed find relative motions, but not with the six-month period he expected. He found that many of his pairs of stars were in orbital motion *around one another* with various periods (all much larger than six months). His original mistake lay in supposing that the components of a close pair of stars were at very different distances. He was not slow to realize the significance of his measures: pairs of stars are usually *at the same distance* (i.e., they are really *double stars*), and they are subject to the law of gravitation, revolving under the same rules that govern the motion of the planets. This discovery was of the first importance, for it showed for the first time that gravitation governs not only the solar system but the stars.

It was not until 1838 that the first parallax was successfully measured by Bessel. He determined the distance of the double star 61 Cygni relative to more distant stars. The same year saw the measurement of the parallax of the nearest double star, α Centauri. Since that time, measurement of parallax has become an important and active branch of astronomy. About 5000 stars are near enough for their parallaxes to have been measured by trigonometric means; but this is only a very small proportion of all observable stars.

When the parallax of a star is measured, it must be remembered that the star may have some motion relative to the sun, and a displacement, as seen from opposite sides of the earth's orbit, will include the component of this motion across the line of sight in six months. Thus at least three

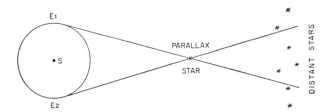

Fig. 12.3. Principle of measurement of stellar parallax. Two positions, E_1 and E_2, represent the earth in its orbit at times six months apart. The nearby (parallax) star is seen displaced against the background of distant stars. The nearer the parallax star to the earth, the larger will be the displacement due to the earth's orbital motion. The size of the displacement may therefore be used to measure the distance of the star. If the parallax star is nearly at the same distance as the comparison stars, the displacement will be inappreciable, and the method is not applicable. The size of the earth's orbit is much exaggerated in the diagram.

measures of position are necessary: the first; the second, six months later; and the third, a year later. At the times of the first and the third measures, the earth is in the same position in its orbit, and any displacement observed will be the true motion of the star. A comparison between the first and second measure will give a combination of the true motion and the parallactic displacement; and as the motion is to be considered uniform, it can be deduced from the first and third measure and used to correct the second measure, which will then give the parallactic displacement, free of the real (or "proper") motion. The *parallax* is half the angular displacement from opposite sides of the earth's orbit.

The parallax of a star must be measured by the displacement referred to distant objects whose parallax can be considered negligible. One selects a number of faint stars in the same field of view and measures the displacement relative to them. All these stars, of course, have *some* parallactic displacement; but for most of the distant stars, it is less than 0".001 and can be neglected. A parallax of less than 0".02 is difficult to measure, and in comparison with this quantity, 0".001 can be disregarded. Even better than faint stars (which are within the galactic system) would be faint galaxies, which are so far away that their parallactic displacements are practically zero. Future work will very likely be based upon these; however, they present difficulties too, for they are not sharply defined like stars, and it is not so easy to measure their positions.

In practice, in modern times, relative parallaxes are measured on photographs. From twelve to twenty photographs of the parallax star are made, and its position is measured on all of them relative to about twenty stars around it, which are supposed to be more distant. Statistical methods derive the best determination from the average of all the measures. The probable error of a determination of parallax will usually be about 0.01". Occasionally it happens that the parallax star is shifted in the opposite direction to that expected; in other words the parallax is "negative." In this case the parallax star is really more distant than the average of the comparison stars—too far away for a direct measure of distance.

The unit of stellar distance is related to the parallax. The distance that corresponds to a parallax of 1" (one second of arc) is known as one *parsec.* The distance of a star, in parsecs, is given by $1/\pi$, where π is the parallax in seconds of arc. The parsec is used in technical astronomical work for the expression of distances; for very large distances, the *kiloparsec* (a thousand parsecs) and the *megaparsec* (a million parsecs) are employed as units.

A more easily visualized unit is the *light-year,* the distance traveled by light in a vacuum in one year. One parsec is equal to 3.258 light-years, or one light-year is equal to 0.3069 parsecs. For very large distances we may use as units the kilolight-year and the megalight-year. In the present text we shall employ the parsec (ps), the kiloparsec (kps), and the megaparsec (mps).

If D is the distance of a star, π its parallax in seconds of arc, and $D = 1/\pi$ parsecs $= 3.258/\pi$ light-years;

$$1 \text{ parsec} = 206,265 \text{ astronomical units} = 3.083 \times 10^{13} \text{ km.}$$

We defined absolute magnitude in the previous section as the magnitude that a star would have if placed at a distance of 10 parsecs. Since

$$M = m + 5 - 5 \log D,$$

where M and m are the absolute and apparent magnitudes and D is the distance in parsecs, and since $D = 1/\pi$, where π is the parallax in seconds of arc,

$$M = m + 5 + 5 \log \pi.$$

Only about 5000 stars are near enough for direct measure of their parallax from opposite sides of the earth's orbit. A longer baseline is needed for stars at greater distances and is provided by the motion of the sun in space. The sun moves in an essentially straight line at about 20 km/sec among the nearer stars. Like a moving train, it produces an apparent backward shift of the landscape (or starscape), the shift being greater for nearer objects. One can measure the apparent backward drift of groups of stars, and thus determine their distance from us. Notice that the method applies only to *groups of stars,* not to individuals, whose random motions would confuse the result. It has the advantage that the baseline increases continually in length, as the sun moves forward steadily, while the method of annual parallax depends on a baseline of fixed length, as the earth goes round and round in its orbit. By means of the "parallactic drift" it is possible to determine the average parallax of large groups of stars, e.g., of stars of different apparent magnitude.

There are indirect methods of determining the distances of many other stars, but all are ultimately based upon the trigonometric method. If we can pick out stars of *known absolute magnitude* by some individual properties (e.g., from some features of their spectra), we can derive their parallaxes from the formula $M = m + 5 + 5 \log \pi$. This time, we know M and can measure m, and therefore we can determine π. About two million stars can have their distances estimated in this manner.

Some groups of stars can have their parallaxes measured if we can deduce their absolute magnitude from their behavior. Such are several of the classes of variable stars: the Cepheid variables (p. 363); RR Lyrae stars (p. 370); and novae (p. 384) and supernovae (p. 390). All these classes of stars (whose absolute magnitudes are determined, fundamentally, by measuring the distance of nearby specimens trigonometrically) are important in permitting us to measure the distances of remote stellar systems in which they are found—globular clusters and galaxies. On them depends essentially our knowledge of the scale of the stellar universe.

Table 12.3 gives a list of all the nearest stars, whose parallaxes are greater than 0″.250; that is, their distances are less than four parsecs. It is based on a compilation by Kuiper. Compare this list with the list of the brightest stars, given in the previous section. Notice that it contains *no giant stars,* and that most of the entrants are fainter than the sun.

There are several points of interest in the list of the nearest stars. There are seven double or triple systems in it, and Barnard's star also is probably an unresolved double. One may infer that double stars are very common. Most of the stars belong to the main sequence (p. 264), but there are two white dwarfs, Sirius B (p. 297) and Procyon B, and one subdwarf (Kapteyn's star). The spectra and radial velocities will be discussed in later chapters. The fainter components of L 726–8 and Krüger 60 are "flare stars" (p. 399).

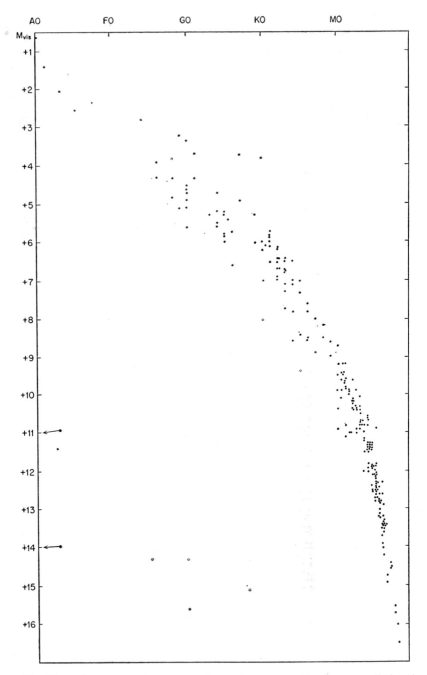

Fig. 12.4. The known stars nearer than 10.5 parsecs (from a compilation by Kuiper). Absolute visual magnitude is plotted against spectral class. Notice that the collection contains no star brighter than absolute magnitude zero, or of earlier spectral class than A0 (except for some white dwarfs). The faint red stars preponderate enormously.

TABLE 12.2. STARS NEARER THAN FOUR PARSECES

Star	Apparent Visual Magnitude (m)	Absolute Visual Magnitude (m)	Parallax "	Annual Proper Motion "	Spectrum	Radial Velocity (km/sec)
Proxima Centauri	+11.3	+15.7	0.762	3.85	Me	. . .
α Centauri A	+0.33	+4.7	0.756	3.67	G4	−22
α Centauri B	+1.70	+6.1	0.756	3.67	K1	−22
Luyten 726–8	+11.9	+16.2:	0.56:	3.37	M	+30
	+12.4	+16.7:	0.56:	3.37	M:	
Barnard's star	+9.46	+13.1	0.543	10.25	M5	−110
Wolf 359	+13.5	+16.5	0.403	4.76	M8	−90
−36°2147	+7.47	+10.4	0.388	4.78	M2+	−87
Sirius A	−1.52	+1.4	0.376	1.32	A0	−8
Sirius B	+8.54	+11.4	0.376	1.32	wA5	(+12)
Ross 154	+10.5	+13.2	0.350	0.67	M5+	. . .
Luyten 789–6	+12.3	+14.9	0.328	3.27	M6+	. . .
Ross 248	+12.2	+14.7	0.314	1.82	M6+	. . .
Eridani	+3.75	+6.2	0.303	0.98	K2	+15
Ceti	+3.66	+6.0	0.298	1.92	G5	−16
1 Cygni A	+5.35	+7.7	0.296	5.20	K3	−65
1 Cygni B	+6.05	+8.4	0.296	5.20	K5	−63
Procyon A	+0.53	+2.8	0.291	1.25	F4	−3
Procyon B	+10.8	+13.1	0.291	1.25		. . .
Indi	+4.70	+7.0	0.288	4.69	K5	−40
−43°44 A	+8.1	+10.4	0.284	2.90	M2	+8
−43°44 B	+10.9	+13.2	0.284	2.90	M5+	. . .
−59°1915 A	+8.92	+11.2	0.282	2.28	M3+	0
−59°1915 B	+9.73	+12.0	0.282	2.28	M4	
−36°15693	+7.23	+9.4	0.278	6.90	M0+	+10
Kapteyn's star	+8.8	+10.9	0.262	8.72	M0+	+242
−5°1668	+10.0	+12.1	0.262	3.75	M4+	. . .
Ross 614	+11.3	+13.4	0.260	1.01	M6+	. . .
−39°14192	+6.65	+8.7	0.257	3.47	M6+	+22
Krüger 60 A	+9.87	+11.9	0.256	0.90	M4	−24
Krüger 60 B	+11.3	+13.3	0.256	0.90	M6	
−12°4523	+9.8	+11.8	0.255	1.17	M4	. . .
Ross 42	+11.4	+13.4	0.250	0.30	M6	. . .

3. COLORS AND TEMPERATURES OF STARS

Temperatures of stars and nebulae are determined in various circumstances by all the methods mentioned in Chapter III. The most direct methods, on which the others are based, depend upon the application of the laws of radiation.

Three relations are available for the application: the Planck Law and its two derivative laws, the Wien Law and the Stefan-Boltzmann Law. The latter can be applied only if distance and dimensions are known and is therefore generally applied to double star systems (p. 293). The Wien Law is theoretically easy to apply, but the practical difficulties are great. It states that the temperature is inversely proportional to the wave length at which the energy curve is at its maximum. The difficulties arise from the very narrow range of wave lengths in which we can measure energies. The range is limited at the ultraviolet end by the absorption in our atmosphere where the ozone cuts off all wave lengths shorter than about 2900 Angstroms. At the red end it is limited by the scope of the measuring facilities, though this limitation is today being removed by the development of new methods of measuring energy in the infrared.

The maximum energy is at about 5000 Angstroms (in the visible region) for temperatures of about 6000°K. Thus, from the inverse proportionality of temperature and wave length, we see that:

Temperature (°)	Wave Length of Maximum (Angstroms)
1,500	20,000 (outside accessible region)
3,000	10,000 (accessible by special methods)
6,000	5,000 (visible light, easily accessible)
12,000	2,500 (cut off by atmospheric ozone)
24,000	1,250 (cut off by atmospheric ozone)
48,000	625 (cut off by atmospheric ozone)

This table sufficiently illustrates the limitation of the Wien Law in practice.

The most useful method depends on a modification of the Planck Law. Even though the whole course of the energy curve cannot be determined, if a limited part of it is measured, we deduce the temperature on the assumption that the radiation is like that of a black body. In many cases the measurement is made only at two points, and these points are most commonly furnished by the photographic and visual magnitudes (p. 267). Although neither photographic nor visual magnitudes are actually confined to a single color (or wave length), each covers only a limited range of color and wave length, and may be regarded as measuring the energy in the middle of that range. Other types of magnitude, such as ultraviolet and red magnitudes, represent the energy received in other regions of the spectrum, which can be isolated by suitable light filters and measured by specially sensitized plates or other light-measuring devices. The average wave lengths of several types of magnitude in common use are:

Photographic	4250A
Visual	5280
Ultraviolet	3700A
Red	6300

The Planck formula, as we saw in Chapter III, expresses the relative amounts of energy in the continuous spectrum at different wave lengths as a function of temperature. Magnitudes measured in light of different colors actually express these relative amounts of energy, and these can therefore be used, in conjunction with the Planck formula, to compare the temperatures of stars. For example, the difference between photographic and visual magnitudes ($m_{pg} - m_{vis}$), known as the *color index,* expresses the difference of intensity between wave lengths 4250 and 5280 Angstroms.

The Planck formula may be transformed, for these two wave lengths, into the more convenient form:

$$\text{color index} = m_{pg} - m_{vis} = (7200/T) - 0.64$$

where T is the temperature in degrees Kelvin.

Notice that from the formula the color index for hot stars approaches −0.64, but even for infinite temperature it cannot exceed this value arithmetically. A few color indexes, calculated from the formula, are given below:

Temperature	Color Index	Temperature	Color Index
infinite	−0.64	3,600	+1.36
14,400	−0.14	2,700	+2.36
7,200	+0.36	1,800	+3.36

The color index is zero for a temperature a little over $11,000°K$.

The preceding relations are approximate. They would be correct only if photographic and visual magnitudes were determined in monochromatic light (i.e., if they referred to a single wave length, not a range of wave lengths). The precise color index depends on the actual range of wave lengths that is integrated to form the magnitudes by any given instrument and can be calculated when this is known. Different eyes and different photographic instruments may have appreciably different color systems, of which account must be taken in accurate work.

Many of the cooler stars do not have color indexes as large as would be expected from their temperatures (as determined in other ways). In the atmospheres of these stars, the molecular bands, which arise from chemical compounds, distort the energy curve; consequently, some of these stars are redder than would be expected, but most of them are bluer. As a result, color index is not a satisfactory method of determining the temperatures of stars whose spectra show chemical compounds. Also, for the very hot stars there are other sources of distortion of the energy curve (some produced by our atmosphere or by interstellar matter, which acts in much the same way, and some inherent in the atmospheres of the stars themselves). But for stars with temperatures between 3500° and 15,000°K, color indexes give a good idea of the temperature.

If the temperature of a star is known, the "magnitude" that would correspond to its *total emission of energy* can be calculated from the Stefan-Boltzmann Law. This magnitude is called the *bolometric magnitude*. The difference between the visual magnitude and the bolometric magnitude is called the *bolometric correction*. It is arbitrarily adjusted to be zero for stars near the temperature of the sun; it has a minimum at the corresponding temperature because it so happens that the visual magnitude for this temperature corresponds to the maximum of the energy curve. For stars hotter or cooler than the sun it is always negative, i.e., *the total brightness is always greater than the visual brightness.* For stars hotter than the sun, the excess comes from the ultraviolet part of the energy curve; for cooler stars it comes from the red and infrared parts of the energy curve.

The following table shows "spectral types" (defined hereafter), temperatures, color indexes, and bolometric corrections for an average array of stars.

The color indexes and temperatures are from Russell, Dugan, and Stewart; the bolometric corrections, from an evaluation by Kuiper.

TABLE 12.3. COLOR AND TEMPERATURE

SPECTRAL CLASS	COLOR INDEX		TEMPERATURE		BOLOMETRIC CORRECTION		
	Main Sequence	*Giant*	*Main Sequence*	*Giant*	*Main Sequence*	*Giant*	*Supergiant*
O5	−0.6:	...	79,000:	...	−5.6	...	...
B0	−0.33	...	25,200	...	−2.70	...	...
B5	−0.18	...	15,500	...	−1.58	...	...
A0	0.00	...	10,700	...	−0.72	...	...
A5	+0.20	...	8,530	...	−0.31	...	...
F0	+0.33	...	7,500	...	(0.00)	...	...
F5	+0.47	...	6,470	...	−0.04	−0.08	−0.12
G0	+0.57	+0.67	6,000	5,200	−0.06	−0.25	−0.42
G5	+0.65	+0.92	5,360	4,620	−0.10	−0.39	−0.65:
K0	+0.78	+1.12	4,910	4,230	−0.11	−0.54	−0.93:
K5	+0.98	+1.57	3,900	3,580	−0.85	−1.35	−1.86:
M0	+1.45	+1.73	3,500	3,400	−1.43	−1.55	−2.2
M5	(+1.45)	(+1.73)	...	2,850	−3.1:	−3.4:	...

Spectral Types of Stars. When the stars are arranged in order of color, i.e., in order of temperature, their spectra show a definite progression. Actually, the arrangement of the stars by colors and by spectral types represent different approaches, which were made independently and later found to correspond to one another.

When the spectra of the stars were examined, first visually, and then photographically, they were seen to represent a great variety, but there were found to be great numbers of each kind that were sensibly similar. During the past century, many astronomers applied themselves to the problem of arranging the spectra of the stars in classes, which could themselves be placed in a series. The names of the Italian Father Secchi, the Englishman Sir Norman Lockyer, and the German Vogel, are especially to be remembered for this work. The system of classification of stellar spectra that is now adopted by everyone is that devised at Harvard in the first years of the present century. It is associated with the names of

E. C. Pickering, Mrs. Fleming, and Miss Cannon. (Miss Maury, who also worked at Harvard at this time, produced a system of classification that was superior in many ways, but it was too complicated to come into general use. It will be mentioned later, page 322).

The Harvard system of spectral classes received the arbitrary letters: B–A–F–G–K–M. Later the letters W and O were added at the beginning, and the letters N, R, and S were found to be necessary at the end. The final form of the system is as follows:

$$M\text{–}S$$
$$\diagup$$
$$(W)\text{–}O\text{–}B\text{–}A\text{–}F\text{–}G\text{–}K$$
$$\diagdown$$
$$R\text{–}N$$

This sequence is found (when compared with the colors of the stars) to be a sequence of descending temperature. All the temperatures can be shown to be consistent with the atomic and chemical equilibrium of the envelopes of stars of essentially similar composition. As the temperatures of the stars of the various spectral classes can therefore be regarded as known, the spectra of stars can be used (on the same assumption of statistical equilibrium) to deduce their temperatures. The spectra of stars will be described, and their physical significance discussed, in Chapter XIII.

4. THE SIZES OF STARS

The sun is the only star that is near enough for us to measure its size directly. No other star shows an observable disc, even with the most powerful telescope. Nonetheless, it is possible to determine the true sizes of many stars, which exhibit a very large variety.

There are three main ways of determining the sizes of stars: from the laws of radiation, from observations of eclipsing stars, and from measures with the interferometer.

Stellar Sizes from the Laws of Radiation. A relationship can be derived from the Planck formula between absolute visual magnitude, temperature, and radius (in solar radii):

$$M_{\text{vis}} = (29{,}500/T) - 5 \log R - 0.08,$$

which is equivalent to

$$\log R = (5900/T) - 0.20\ M_{\text{vis}} - 0.02;$$

and by using the expression for color index $(7200/T - 0.64)$, this becomes

$$\log R = 0.82\ I - 0.20\ M_{\text{vis}} + 0.51,$$

where I is the color index. Thus if the color index and the absolute visual magnitude of a star are known, the radius can be calculated at once. Notice that the determination requires three data: apparent visual magnitude, apparent photographic magnitude, and distance. It is a very useful method of obtaining the sizes of stars, valid only if they radiate like black bodies, but usually a good approximation.

Absolute visual magnitudes of stars range from -8 to $+19$, color indexes from -0.6 to $+1.6$ (color indexes that are falsified by large distortions of the energy curve excluded). The range of $0.20\,M_{\text{vis}}$ is thus from -1.8 to $+3.8$, and that of $0.82\,I$, from -0.4 to $+1.3$.

The largest stars will evidently be those of largest color index and greatest luminosity. If $M_{\text{vis}} = -8$ and $I = +1.6$, then $\log R = 3.1$ and the radius is 1020 times that of the sun. By similar reasoning, the smallest stars will be those of smallest color index and lowest luminosity. The faintest main sequence stars are of low luminosity and large color index; one with a color index $+1.6$ and absolute magnitude 19.0 would have a value of -2.0 for $\log R$, and a radius $1/100$ that of the sun. The smallest stars of all are found among the white dwarfs, which have fairly low luminosities and are somewhat blue (i.e., they have small color indexes). The highly luminous blue stars have only moderately large radii; for example, a star of absolute magnitude -8 and color index -0.6 would have a radius 63 times that of the sun. It should be noticed that the color index contributes more to the radius than the luminosity does; the former is multiplied by $8/10$, whereas the absolute magnitude is multiplied only by $2/10$, in the formula for the radius.

A table of the probable dimensions and other physical properties of a number of typical stars is given in Table 12.4 (adapted from Russell, Dugan, and Stewart's *Astronomy*).

Stellar Sizes from Eclipsing-Stars. It has already been mentioned that a large proportion of the stars are members of physical pairs, which move in orbits around one another (pp. 293 and 304). The orbits of some of these pairs are so oriented in space that the stars pass periodically in front of one another, as seen from the earth. When either of the stars passes in front of the other, it obscures the star in the rear, and we observe a drop in the light received from the pair (which are so far away that they are never separately visible). From the duration of the drop in brightness (the *eclipse*), it is possible to determine the geometry of the pair of stars, and to find their sizes *in terms of the size of their relative orbit* (p. 306). If, in addition, we know the true speeds with which the stars are traveling (information that is obtained from their spectra), it is possible to determine the true scale of the whole system and therefore the actual sizes of the two stars.

Such information is available for a few hundred eclipsing-stars, and the sizes so obtained fit well into the picture presented by the results from

TABLE 12.4. TYPICAL STARS

Name	Magnitude Apparent	Absolute	Temp. (°K)	Radius (suns)	Mass (suns)	Density (suns)	Spectrum
MAIN SEQUENCE:							
Centauri	0.9	−3.8	21,000	11.	25.	0.018	B1
Scorpii	4.3	−0.8	17,000	3.2	5.2	0.16	B3
Aurigae	2.8	0.6	11,200	2.4	2.2	0.13	A0
Vega	0.1	0.6	11,200	2.4	3.0	0.11	A0
Sirius A	−1.6	1.3	11,200	1.8	2.4	0.42	A0
Altair	0.9	2.5	8,600	1.4	1.7	0.6	A5
Procyon	0.5	3.0	6,500	1.9	1.1	0.16	dF5
Centauri A	0.3	4.7	6,000	1.0	1.1	1.1	dG0
O Ophiuchi	4.3	5.7	5,100	1.0	0.9	0.9	dK0
1 Cygni A	5.0	8.4	3,800	0.7	0.45	1.3	dK7
Krüger 60 A	9.2	11.2	3,300	0.34	0.26	9.0	dM3
Barnard's star	9.7	13.4	3,100	0.16	0.18	45.	dM4
Van Biesbroeck's star		19.0	3,100:	0.01	0.1:	100,000	dM:
GIANTS:							
Capella	0.9	−0.1	5,500	12.	4.2	0.0024	gG0
Arcturus	0.2	−0.3	4,100	30.	8.	0.0003	gK0
Aldebaran	1.1	−0.1	3,300	60.	4.	0.00002	gK5
Pegasi	2.6	−1.4	2,900	170.	9.	0.000002	gM5
SUPERGIANTS:							
Betelgeuse	0.9	−2.9	3,100	290.	15.	0.0000006	cM0
Antares	1.2	−4.0	3,100	480.	30.	0.0000003	cM0
WHITE DWARFS:							
Sirius B	8.4	11.2	7,500	0.034	0.96	27,000	A5
O Eridani B	9.7	11.2	11,000	0.019	0.44	64,000	A0
Van Maanen's star	12.6	14.5	7,500	0.007	0.14	400,000	F

the laws of radiation. Eclipsing-stars give information about stars of many kinds—giants, main-sequence stars, supergiants (which we have encountered already), and also a remarkable group known as the *subgiants*, little known until recently except as members of eclipsing pairs. No white dwarfs are known among the eclipsing-stars, but one interesting pair (UX Ursae Majoris) lies well "below" the main sequence.

Stellar Diameters with the Interferometer. Although all stars (except the sun) are too far away to show visible discs, it is possible, by using the *interference* between the waves of light coming from opposite sides of the disc of some of the largest and nearest stars, to measure their apparent diameters in seconds of arc. Essentially the instrument brings together two beams of light coming from the star and caught by two mirrors placed a large distance apart. (The largest interferometer placed the mirrors at the two ends of a 50-foot girder placed across the end of a telescope.) By the reinforcement or cancellation of the waves of light coming from the two edges of the star (according to whether the crests of the waves in one beam coincide with, or fall in the troughs of, the waves from the other beam), the difference of light path to the two mirrors is determined, and this depends on the angular diameter of the star.

Measures with the interferometer determine the *angular diameter* of the star's disc in seconds of arc; if the distance of the star is known, this measure can be converted into actual dimensions. Only a few stars are large enough and also near enough for their angular diameters to be measurable.

Star	Angular Diameter	Parallax	Radius (suns)	Spectrum
Arcturus	0.020″	0.080″	27	gK0
Aldebaran	.020	.057	38	gK5
Betelgeuse	.047–.034	.017	300–210	cM0
Antares	.040	.0095	450	cM0
β Pegasi	.021	.016	40	gM5
α Herculis	.030	.008:	400:	gM8
"Mira" Ceti	.056	.02:	300:	gM7e

Note that all the stars in this list are red giants, or supergiants, of late spectrum and low temperature.

A method similar in principle has recently been applied to measuring stellar diameters by observing them as they pass behind the dark edge of the moon (*occultation*) with the photoelectric apparatus. It is still in its early stages but gives results very similar to those just described. Its application is, of course, confined to objects that pass behind the moon, so these must lie close to the ecliptic.

5.   TRANSVERSE MOTIONS OF STARS

The stars are conventionally distinguished from the planets by saying that the stars are "fixed," while the planets "wander." [The name planet is derived from the Greek: *planētēs,* a wanderer.] All the stars are nevertheless in motion. Yet their great distances make this motion inappreciable to the eye, and it can be measured only with precise instruments. The motions of the stars were first definitely recognized by Halley (of comet fame) in 1718. He noticed that Arcturus was 1° south of the position ascribed to it by Ptolemy in the "Almagest," and that Sirius was 0.5° south of its Ptolemaic position. Stellar motions have been measured in large numbers only since photographic methods were applied to determining the positions of stars. Modern lists contain the measured motions of tens of thousands of stars.

Newton's first law of motion applies to the stars, and indeed to all known matter. It states that "Any body persists in its state of rest, or of uniform motion, except in so far as it is compelled by impressed force to change that state." The school of thought represented by Newton recognized that uniform motion needs no actuation, a contrast to the Aristotelian idea that a body moves only if something is impelling it. The stars may be regarded as moving uniformly with constant speed, essentially in straight lines. If they are changing their speed, or not moving in straight lines, some "impressed force" is indicated. Many stars are in fact moving in orbits around other stars, and their mutual gravitational force is the "impressed force" concerned. This was the discovery that William Herschel made, when, on looking for evidence of stellar parallaxes, he discovered double stars and recognized that they furnish evidence that the law of gravitation is in operation throughout the stellar system, as well as in the solar system. All the stars in a stellar system are also moving under the gravitational control of the whole stellar system. Thus they are actually moving in large orbits (which may be circles, ellipses, or even parabolas or hyperbolas, and will be even more complicated if nearby stars are disturbing the motion too), but the orbits of all stars whose motion we observe are so large that the stars move effectively in straight lines. We can infer the shapes of these galactic orbits, not from the *curvature* of the stars' paths, but from the *direction* of the motions, as will be mentioned in the next section.

The position of a star is expressed in *right ascension* and *declination* (p. 50). The former is usually expressed in hours, minutes, and seconds (24 hours to the circle); the latter, in degrees, minutes, and seconds. Occasionally right ascensions may also be expressed in circular measure, since

$$360° = 24^h, \qquad 1° = 4^m, \qquad 60' = 4^m, \qquad 15' = 1^m.$$

If the right ascension and declination of a star today are compared with the right ascension and declination of the same star fifty years ago, they will usually be found to differ. There are several reasons for the change. *Precession* and *nutation,* which result from the swinging and wobbling of the earth's axis, cause steady and sinuous changes in the point of intersection of the equator and the ecliptic and, therefore, in the apparent position of all celestial objects (p. 39). *Aberration,* which is caused by the changing direction of the earth as it goes around the sun and the consequent changing rate of speed of the earth relative to the star (p. 30), displaces the stars by an amount that depends on their positions in the sky.

Precession and aberration are the same at any one time for all stars in the same region of the sky. Once these small displacements have been detected and measured, they may be allowed for. After this has been done, it is found that most stars have small residual motions, at right angles to the line of sight and relative to one another. These motions are known as the *proper motions* of the stars.

Proper motion is the apparent rate of motion across the sky, expressed in seconds of arc per year (sometimes per century). It must, of course, be corrected for the annual parallax, and therefore it expresses the annual transverse motion of a star *as it would appear if seen from the sun.* The largest known proper motion is that of "Barnard's star," 10.25″ a year, which would amount to 1° in 352 years.

We are apt to think of motions only in two dimensions, because most of our everyday motions are along the earth's surface (which we cus-

OBSERVER

Fig. 12.5. The size and direction of the true speed of a star are determined from its proper motion and radial velocity. The proper motion must be converted into tangential speed in kilometers a second by a knowledge of the star's distance. The radial speed is measured directly in kilometers a second from the spectroscopic Doppler effect. Proper motion and radial velocity are at right angles to one another. The true speed is given by the length of the diagonal of an imaginary rectangle, of which the base and height are numerically equal to the radial and tangential speeds. The angle between the radial direction (from observer to star) and the true direction of motion is determined by the proportion between tangential and radial speeds; the ratio of transverse to radial speed is the trigonometric *tangent* of this angle.

tomarily treat as being flat). However, in thinking of stellar motions we must think in three dimensions. The proper motions of stars are entirely across our line of sight, but the stars are also moving toward and away from us. These latter motions (discussed in the next section) are radial velocities of the stars. The total motion of the star is compounded of the radial velocity and the cross motion, which are put together by the well-known "parallelogram law."

The measured proper motion of a star is its angular speed. It is common experience that the objects viewed from a moving train seem to move faster when they are nearer to us. In the same way, the angular speeds of the stars will seem to be greater if the stars are nearby. In order to translate angular speed (proper motion) into true speed (kilometers a second), the distance of the star must be known. Proper motion × distance in parsecs, or proper motion ÷ parallax = tangential velocity in astronomical units per year.

To convert this to km/sec, we multiply by the number of kilometers in an astronomical unit (149,674,000 km) and divide by the number of seconds in a year, and obtain

$$\text{tangential velocity} = 4.74\ \mu/\pi\ \text{(km/sec)},$$

where μ is proper motion (per year), π is parallax. Some stars of large proper motion are shown in the next table. The data are chiefly from a compilation by Kuiper.

It is of interest to compare this list with the list of stars of large parallax. Both contain a large proportion of faint main-sequence stars, but each also contains a number that are not in the other. The list of stars of large proper motion contains only one star (α Centauri A) that is absolutely brighter than the sun. It contains two white dwarfs, but not the same two as were in the list of stars of large parallax. The radial velocities of the stars of largest proper motion are, on the whole, larger than those of the stars of large parallax.

To have large proper motion, a star must of course be near to us; but it must also be really moving fast. Sirius, for example, one of the nearest stars (parallax 0.571″) has an annual proper motion of only 1.38″, and its tangential velocity is only about 7 km/sec. An average speed for a star in our neighborhood is 20 km/sec. Evidently these stars of large proper motion are traveling faster. Many of them are in fact "high-velocity stars."

TABLE 12.5. STARS OF LARGE PROPER MOTION (over 3″ per year)

Star	Apparent Visual Magnitude m	Absolute Visual Magnitude m	Annual Proper Motion ″	Parallax ″	Spectrum	Radial Velocity (km/sec)
Barnard's	+9.67	+13.1	10.25	0.543	M5+	+110
Kapteyn's	+9.2	+10.9	8.72	0.262	M0	+242
Groombridge 1830	+6.46	+6.6	7.04	0.108	G6	−98
−36°15693	+7.23	+9.4	6.90	0.278	M0	+10
−37°15492	+8.57	+10.3	6.11	0.222	M3	+26
Ross 619	+12.5	+13.4	5.40	0 154	M6	. . .
61 Cygni A	+5.35	+7.7	5.20	0.296	K3	−65
61 Cygni B	+6.05	+8.4	5.20	0.296	K5	−63
+36°2147	+7.47	+10.4	4.78	0.388	M2+	−87
Wolf 359	+13.5	+16.5	4.76	0.403	M8	−90
ε Indi	+4.70	+7.0	4.69	0.288	K5	−40
+44°2051 A	+8.7	+9.9	4.53	0.174	M0+	+64
+44°2051 B	+14.8	+16.0	4.53	0.174	M8	. . .
o²Eridani A	+4.52	+6.0	4.08	0.202	K1	−42
o²Eridani B	+9.62	+11.1	4.08	0.202	WA *	−42
o²Eridani C	+11.1	+12.6	4.08	0.202	M6	. . .
Wolf 489	+14.5	+15.1	3.94	0.130	g8 †	. . .
Proxima Centauri	+11.3	+15.7	3.85	0.762	m	. . .
μ Cassiopeiae	+5.19	+5.8	3.77	0.133	G5	−97
+5°1668	+10.0	+12.1	3.73	0.262	M4+	+22
Washington 5583	+9.1	+6.8	3.68	0.034	G5	+307
Washington 5584	+8.9	+6.6	3.68	0.034	G0	+295
α Centauri A	+0.33	+4.7	3.67	0.756	G4	−22
α Centauri B	+1.70	+6.1	3.67	0.756	K1	−22
−39°14192	+6.65	+8.7	3.47	0.257	M0	+22
Luyten 726–8	+11.9	+16.2:	3.37	0.56:	m	+30
	+12.4	+16.7:	3.37	0.56:	m:	. . .
82 Eridani	+4.29	+5.3	3.25	0.159	G5	+87
van Maanen 2	+12.4	+14.3	3.00	0.243	WG *	+238

* The prefix "W" denotes a white dwarf.
† The small letter "g" indicates a color appropriate to spectrum G.

The true motions of these stars are compounded of the tangential velocity and the radial velocity. The *speed* and *direction* of a star's motion tell us what sort of orbit it is describing in the galaxy. A star traveling with a high enough velocity relative to the sun would escape from the gravitational pull of the galaxy altogether. In our list, only the pair Washington 5583, 5584 is traveling at such a speed; other stars are known to be traveling equally fast (though they are uncommon).

The sun is describing a nearly circular orbit about the center of the galaxy. A star in our neighborhood that has a large velocity *relative to the sun* cannot be in a circular orbit. The shape and orientation of its orbit can be calculated immediately (if we assume that it moves according to Newtonian laws, like a planet, under the gravitational control of the galactic nucleus) from the speed and direction of its motion.

The measurement of proper motions calls for high precision. It is similar in principle to the measurement of parallaxes, already described. The position of the star to be measured is compared with those of a number of nearby stars on several carefully taken photographs, and the actual motion is determined by careful statistical treatment of the measured positions. The resulting motion is relative to the average motion of the comparison stars. If the absolute value is desired, then allowance must be made for this average motion—a laborious and difficult process.

The sun moves among the stars and thus provides the baseline for the determination of group parallaxes. The motion of the sun produces an apparent backward drift of the stars as a whole, and the direction and speed of the sun's motion can be derived from the displacements of the stars. The sun's path appears to be directed toward about right ascension 18^h and declination $+30°$. The speed of the sun relative to the group of stars used in the determination is about 19 kilometers a second. The sun thus has an average speed among the nearby stars; it is not a rapidly moving star. The speed and direction of its motion enable us to say that it is traveling a nearly circular orbit in the galaxy.

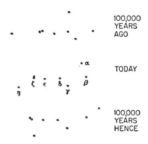

Fig. 12.6. The bright stars of the Big Dipper, as they looked 100,000 years ago, as they look today, and as they will look 100,000 years hence. The motions of α and η differ from the rest, and produce the greater part of the change. The common motion of the other stars show that they are part of a *moving cluster.*

Table 12.5 shows several pairs of stars that have the same motion in space. There are a number of more extended groups of stars that are also traveling through space in parallel paths, and these are of great interest, for they evidently represent groups that originated together and have had similar histories. One of the best known of these groups comprises several of the stars of the Big Dipper, though some other stars, separated from these in the sky, also belong to the family, Sirius perhaps being one. The motions are shown in the diagram; only two of the bright stars of the Big Dipper do not share them. Clearly the shape of the constellation was different in the far past, and will differ again in the far future.

The Hyades are an even more striking example of a moving cluster of stars. Here we see an apparent convergence of the motions; they are acutally parallel but seem to converge to a point like railroad tracks which converge to a point in the distance.

6.   RADIAL MOTIONS OF THE STARS

Of the three-dimensional motions of the stars, the proper motion (p. 285) gives only the component perpendicular to the line of sight. Moreover, before this component can be expressed in kilometers a second, the distance of the star must be known. The component of the motion in the line of sight—the motion of the star toward us or away from us—is known as the *radial velocity*. The measures give it directly in kilometers a second, and it can therefore be directly determined for any star whose spectrum can be recorded with a spectrograph, irrespective of distance.

Measurement of radial velocity depends on two observed facts. First, the spectra of most of the stars when examined with the aid of prisms or gratings (which disperse their light into its component colors) are shown as dark (absorption) lines superimposed on a bright continuum. These lines can be identified with those of the atomic species known on earth; the great majority of the lines shown in the spectra of the stars can be produced in the terrestrial laboratory. In the laboratory, also, the wave lengths (colors) that correspond to these lines can be accurately measured. The wave lengths in the spectra of stars are measured by photographing the spectrum of some terrestrial substance on the same plate at the same time (iron, excited by the electric arc or spark, is commonly used, and since its spectrum contains thousands of lines of accurately known wave length, it furnishes a convenient standard for light of nearly all colors).

Secondly, the *color* of the light of a given atom coming from a moving source is altered. In 1843, Doppler pointed out that colors are thus affected, and the phenomenon usually bears his name, being called the "Doppler effect." In 1848, Fizeau showed that the *wave lengths* of the spectral lines are similarly altered, the change being proportional to the wave length and to the speed with which the source is moving; hence the name "Doppler-Fizeau effect" is sometimes used. Light from a receding source is reddened, and the wave length increased; light from an approaching source is rendered bluer, and the wave length diminished.

The effect may be pictured by an analogy from the world of sound (although sound waves differ from those of light in being compressional waves; light waves are transverse waves). It is readily noticed that the pitch of a whistle or bell on a moving train drops abruptly as the train passes the observer. As the train approaches, the speed with which the source of sound is approaching crowds together the sound waves, so that their frequency for a stationary observer is increased, and the pitch of the sound raised. As it recedes, the speed with which the source of sound is receding spreads out the sound waves, so that the frequency for the stationary observer is reduced, and the pitch of the sound seems to fall. When the train is moving very fast, the change in frequency will be great, and the change of pitch, large; for smaller speeds, both will be smaller, and if the train is stationary, the pitch will be constant. One could use the amount of the change in pitch as a rough measure of the speed of the train.

If the analogy is extended to light waves, it is seen that an advancing source, which crowds the waves together, increases the frequency and diminishes the wave length of the light, which thus becomes bluer. A receding source spreads out the waves, decreases the frequency, reddens the light, and increases the wave length. Evidently the change in the wave length will provide a measure of the speed of the source.

Measures of the wave lengths of the lines in the spectra of stars call for very accurately designed instruments, made with great optical refinements and carefully controlled for rigidity and temperature. The changes in wave length recorded for most stars are very small and must be measured with precise instruments, and the calculations by which the final velocities are derived must be made with attention to all possible sources of error. Allowance must be made for the motion of the earth around the sun and its effect on the relative motion of the sun and star (which is the final product, just as the proper motion is the motion as it would appear from the sun after the earth's motion has been eliminated).

The velocities just described are obtained with a *spectrograph,* which can measure one star at a time; attempts have also been made, by interposing absorbing screens that produce lines of known wave length, to determine the velocities of a large number of stars at once. These methods are very much less accurate than the spectrograph and have been used only for crude work. With good lines, an accuracy of 0.5 km/sec is attainable with a spectrograph of moderate dispersion; with poor lines (and the lines of many stars are naturally fuzzy), only 5 or 10 km/sec; and the mass methods with absorbing screens hardly give an accuracy of 30 km/sec. Special spectrographs of very large dispersion have been designed for the study of the brightest stars, and the accuracy obtained with these is very much higher; unfortunately there are still few of these instruments, and the number of stars that can be reached with them is limited.

Several thousand radial velocities of stars have been successfully measured. They range from 0 to about 400 km/sec; velocities over 100 km/sec are rare, and most stars give values between 10 and 40 km/sec. These speeds are all relative to the sun.

When the radial velocity of a star is known, and when the proper motion and parallax have been determined, the total velocity of the star in space can be calculated. If T is the tangential velocity in km/sec (p. 286) and R the radial velocity, the total velocity is given by the expression:

$$V^2 = T^2 + R^2, \text{ or } V = \sqrt{T^2 + R^2}.$$

The *direction* in which the star is moving is determined geometrically from the direction of the proper motion and the proportion between the tangential and radial velocities.

The observed speed and direction of the motion of a star is a combination of its motion with respect to the average of the motions of all the stars and the motion of the sun relative to the average of all the stars (the "solar motion"). We have already described, in principle, how the solar motion is determined from the proper motions of a large number of stars; it can similarly be derived from the radial velocities of a large number of stars, but the derivation is simpler because the "drift" produced by the sun's motion in this case is independent of distance. The solar apex (the point to which the motion is directed) and the solar speed are found to be about the same from radial velocities as from proper motions. (It differs somewhat when the sun is referred to different groups of stars, but this refinement need not obscure the principle.) When the solar motion is known, one can readily calculate how much it affects the observed total motion of any star and make an appropriate correction. The resulting motion (that of the star considered, relative to the average of all the stars) is known as the star's *peculiar motion* (which is to be distinguished from its proper motion).

The measured radial velocity of a celestial object may convey information other than, or supplementary to, true motion toward or away from the observer. It may vary periodically, thus giving information about the orbital motions of double stars. It may, again, vary periodically because the surface of a single star is rising and falling rhythmically (as with certain "pulsating" variable stars). It may undergo short-lived fluctuations, which tell of the motions of prominences near the star's surface; or it may give evidence of the violent explosive ejection of material from a star.

It is shown in the theory of relativity (by methods beyond our present scope) that in a very intense gravitational field, light sources emit light that is somewhat reddened; the wave lengths are increased in proportion to the gravitational potential. High gravitational fields are to be expected at the surfaces of very small stars, especially if their masses are not very low. The stars that best display this combination of conditions are the white dwarfs, and it was the "red-shift" of their spectrum lines that clinched the case for the extremely high density of these objects. For the companion of Sirius, the red-shift is observed to amount to $+19$ km/sec, which fully confirms the small dimensions of the star.

A few of the very massive blue stars show (as a group) a small red-shift of this kind, which is quite compatible with their large masses and moderate dimensions. Another way of expressing the same thing is to say that the average of the radial velocities of the massive blue stars is slightly positive. The average positive velocity is known as the "K term," and for a long time presented a puzzle. It could mean that, on the average, these stars are receding from us. But since the observed K term is of about the size of the expected relativity shift, it is generally thought to arise from this cause instead.

A red-shift of quite another order, that increases steadily with distance from us, is shown by the spiral galaxies. To what extent it represents a true speed is still a matter of controversy; it will be discussed in Chapter XVII.

When all the information on the transverse and radial motions of the stars is put together, it appears that the stars do not move in a completely chaotic fashion. If our stellar system rotated about its center of gravity like a rigid body, all its parts would preserve their relative positions. However, each star may be imagined to describe its own orbit about the center of the galaxy, just as each planet describes its own orbit about the sun. The further a planet is from the sun, the longer its period of revolution and the smaller its velocity, hence the outer planets steadily fall behind the inner ones. In the same way, the outer stars of a galaxy tend to fall behind those nearer the center, the system does not preserve its radial configuration, and the consequent shearing motion appears in the motions of distant stars.

The actual speed of the sun in its orbit around the center of the galaxy is difficult to determine from a comparison with the slightly different motions of other stars, also going round it in larger or smaller orbits. A standard of reference must be sought in groups that do not share this motion.

The existence of a number of stars that do not share the rotational motion of the sun about the galactic center (the *high-velocity stars*) has long been recognized, but only recently has their significance been appreciated. The planetary nebulae, the cluster-type stars, and the globular clusters themselves are found to have a peculiarity in common. Their motions (some of them very large) are apparently all directed toward one hemisphere of the sky; none is observed to come *from* this direction. The information is complete for many cluster-type stars and planetary nebulae, for which both proper motion and radial velocity are known; for the globular clusters only the radial velocities are measured, since they are too distant to show appreciable proper motions. However, the radial velocities of the globular clusters place them quite evidently in the same category.

These groups of stars have long been known as the "high-velocity stars," and *relative to the sun*, their velocity, on the average, is indeed high. But it is now recognized that the high velocity is in fact that of the sun and its neighbors as they move around the galactic center and that the "high-velocity stars" *seem* to have exceptionally high velocities because the sun and the other stars are sweeping past *them* with high velocities. The *directions* of motion of the "high-velocity stars" are exceptional in comparison with the sun's neighbors; the *sizes* of their motions are not.

It might be thought that the "high-velocity stars" would provide a sort of zero point from which the rotational velocity of the sun around the galactic center might be measured. This would be true if we were sure that, *as a group*, they have no rotational velocity about the galactic center. But even the "high-velocity stars" have, as a group, a rotational velocity about the galactic center. Opinion tends to place the rotational speed of the sun near 200 km/sec. Rotational velocities at the same distance from the centers of comparable galaxies (where they have been measured; see pp. 430, 435) are of the same order.

The sun is probably between 8 and 9 kiloparsecs from the center of the galaxy, and the speed just mentioned would carry it around one complete circuit in about 200 million years. This time interval (uncertainly known, as the preceding remarks indicate) is sometimes called a "cosmic year." Our present guesses as to the age of the stellar system make it between 10 and 20 cosmic years old. But these guesses are rough and unreliable.

7. VISUAL DOUBLE STARS

A pair of stars that appear close together in the sky may have nothing in common save the fact that they lie in the same direction. If their motions are such that they cannot be describing orbits around one another, they are called an *optical pair*. If their motions are evidently connected, they are a *physical pair*. And if, in addition, they show evidences of orbital motion, they are known as a *visual binary*. Comparatively few close double stars are optical pairs. Very likely the majority of stars are members of systems that are at least double, and many multiple systems also exist.

(This statement, while true of the stars that inhabit our neighborhood, is not necessarily true of *all* neighborhoods: for instance, double stars may well be less frequent elsewhere than in our part of the galaxy.)

There are many pairs of stars that are too far away to be seen separately and thus detected as a physical pair or a visual binary. Some of these can be detected spectroscopically by their orbital motions (p. 301), and others, if their orbits are oriented in a certain way, by their mutual periodic eclipses. These classes of double stars are known as *spectroscopic binaries* and *eclipsing binaries*.

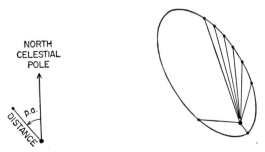

Fig. 12.7. The motion of a visual double star is measured by recording the distance and position angle (*p. a.*), as shown in the diagram on the left, over a number of years. The relative orbit of the fainter star about the brighter is shown schematically on the right for a double star with a period of 200 years. Positions are recorded at intervals of 10 years. Notice that the motion is fastest at *periastron:* it obeys the law of areas.

The measures that must be made, if orbital motion of a visual binary is to be detected and its nature determined, are the distance of the pair apart and their orientation. The *distance* is measured in seconds of arc, and the orientation is fixed by the *position angle,* which is the angle between the line drawn from the principal (brighter) star to the north celestial pole and the line drawn from the brighter to the fainter star. It is measured counterclockwise and expressed in degrees of arc. (If the two stars are equally bright, careful and continuous measures may be needed, to be certain which is which.) Stars about 0.11″ apart can be detected with the 36-inch telescope, if they are equally bright. If they are not equally bright, the limit is larger. Smaller telescopes will have a higher limit that depends on their resolving power, which falls off nearly in proportion to the aperture. The actual measures of distance and position angle are made with an instrument called a *micrometer.*

The motion of a double star is deduced from a number of successive measures of distance and position angle, which are combined into one diagram. Often the observations must be continued for decades, or even centuries, before a complete orbit is obtained. Occasionally the motions of the two stars are evidently unconnected. But more often, evidence appears of orbital motion, and the orbit is drawn by setting off the distances at the correct position angles, from the position of the principal star.

Although the motion of a double star is measured on the "plane of the sky," we must think of it, like the space motions of stars, in three dimensions. Only rarely will the orbit be oriented so that we see it "full face." For most double stars the apparent orbit is a projection of the true orbit.

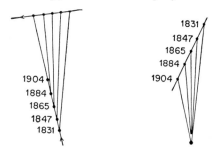

Fig. 12.8. A close pair of stars, forming an *optical double*. The actual motions are shown on the left; the relative motion on the right. The two stars are traveling in straight lines and show no evidence of orbital motion.

The orbits of periodic double stars are ellipses (a circle being the special case). The stars follow the laws of motion: each describes an ellipse about the other as focus, and both describe ellipses around the common center of gravity; and these three ellipses are *similar* (i.e., of the same shape, having the same eccentricity, see Figure 12.9). The law of areas is obeyed: the radius vector sweeps out equal areas in equal times; and the squares of the periodic times are proportional to the cubes of the mean distances divided by the sum of the masses.

When the orbits are seen in projection, they are still ellipses (though the stars no longer lie at the foci of the projected ellipses); the most important point is that *the law of areas holds good for the projected ellipse.* Thus, when the apparent orbit has been plotted from the observations, the first problem is that of drawing through the observed points the best possible ellipse that satisfies the law of areas. Then it is possible to determine how much the orbit is tipped, and to determine the shape and orientation of the true orbit in space.

Many visual binaries have very long periods. The shortest known is about a year, there are about a dozen less than 25 years, about 50 between 25 and 100 years. Other pairs have longer periods, many of them still undetermined because too short an arc of the orbit has been measured.

When he studied the motions of the planets about the sun, Kepler formulated the law that the squares of their periodic times are proportional to the cubes of their mean distances. Newton showed that a gravitational attraction that varied inversely as the square of the distance and directly as the attracting mass was not only adequate, but necessary, to

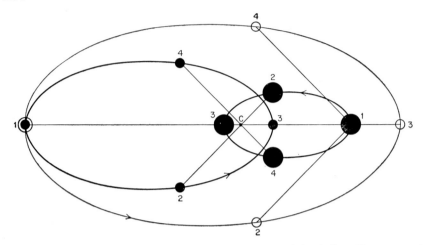

Fig. 12.9. The motion of two bodies under gravitation is in similar ellipses around their common center of gravity and around one another. The ellipses are the same shape, but not the same size.

The center of gravity of the system is marked C. In accordance with Newton's laws, the sizes of the ellipses described by the two stars around C are inversely proportional to the masses of the stars. The larger star represented has a mass twice that of the smaller; its orbit around C is therefore half as large.

Four positions of the two stars are marked 1, 2, 3, 4; the positions of the stars as they move around C are shown by black dots.

The orbit of the smaller, around the larger, star is also shown; the larger star is now supposed to remain stationary at position 1. Positions 1, 2, 3, 4 of the smaller star are indicated by circles, and its relative orbit by a lightly-drawn ellipse; the radius vector is drawn for each of the four positions. Notice that it is always parallel to, and equal in length to, the line joining the corresponding positions of the stars in the actual orbits. Notice also that the major and minor axes of the relative orbit are equal to the sums of the major and minor axes of the orbits of the stars about C.

The orbit of the larger star around the smaller star could be drawn in the same manner, but has been omitted to simplify the diagram. The student may verify that it would be of the same size as the orbit of the smaller, around the larger, star.

describe these motions. Within the solar system the sun's mass is effectively the controlling factor; but to describe the double star motions it is necessary to write the "harmonic law" in the more complete form;

$$(m_1 + m_2)/(m_S + m_E) = A^3/P^2$$

where m_1, m_2 are the masses of the components of the double star, m_S and m_E the masses of sun and earth, A the mean distance of the stars in astronomical units, and P the period in years. But if A is the distance in astronomical units, and a is the measured mean distance in seconds of arc, $A = a/\pi$, where π is the parallax. Thus the previous equation becomes:

$$(m_1 + m_2)/m_S = a^3/\pi^3 P^2.$$

(The earth's mass is so small that it can be neglected.) The mass of the sun is known. If a (the mean distance in seconds of arc) is obtained from the orbit, which is derived in the way just described, and P is the period of the double star in years, and if its parallax is known, the sum of the masses of the two components can be calculated. Note that the parallax enters as a cube, so that a small error in the parallax will cause a large error in the combined masses. For example, a 10% error in the parallax will produce a 30% error in the combined mass.

The *individual* masses of the two stars cannot be determined merely from the relative orbit and the parallax. To determine the mass of one of the two stars, we need to know, not the size of its orbit round the other star, but the size of its orbit *around the center of gravity of the system*. This absolute motion of one of the stars cannot be determined from the relative motions of the pair: it must be referred to a system of reference outside them. When such measures are possible, the masses of the two stars can be determined separately: the sizes of their individual orbits are inversely proportional to their masses.

Sirius consists of a pair of stars for which the individual masses can be determined. The brighter component had been known, since before 1850, to be moving across the sky in a sinuous path, and it was rightly surmised that this sinuous motion was the result of orbital motion around an unseen companion. In 1862 the companion (the celebrated white dwarf) was first seen visually by Alvan Clark. Like its primary, it describes a sinuous path (see Figure 12.10). The motions of the two stars are consistent with the idea that the center of gravity of the whole system is moving in a straight line, just as a single star may be thought to do. If the direction from Sirius to its companion is laid off, with the correct distance, for various times, the relative orbit is obtained, and if the distance of each from the straight line is similarly drawn, the true orbits are found. Evidently the true orbit of the companion is larger than that of the primary, so its mass must be smaller. The *ratio* of the sizes of the orbits gives at once the ratio of the masses, and the sum of the masses is given by the *size of the relative orbit*. So the individual masses are determinate: Sirius has about two and a half times the sun's mass; the companion is almost as massive as the sun.

The presence of unseen companions can be detected by sinuous motions, and Barnard's star, for example, has been shown by this method to be double. Occasionally it is even possible to detect the presence of a third body in a double star system because the elliptical orbit determined from measures of distance and position angle shows a sinuosity.

When the orbits of visual binaries are collected, it is found that most of them are not circular. For 500 stars, of average period about 200 years, the average eccentricity is 0.61, and for nearly 800 more, with average period of about 5000 years, it is 0.76. Longer periods (and larger orbits)

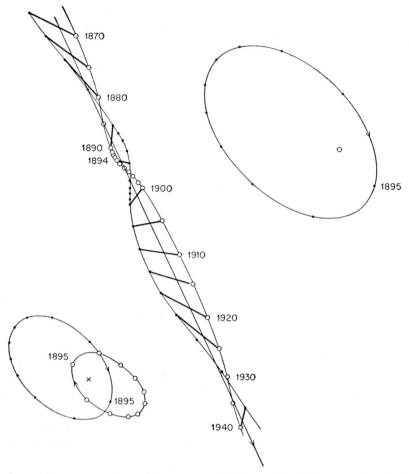

Fig. 12.10. The motions of the system of Sirius and its faint companion. Sirius is shown by a circle, the companion by a dot.

Upper right: the apparent orbit of the companion around Sirius; dots are at five-yearly intervals.

Lower left: apparent orbits of Sirius and its companion about their common center of gravity, shown by a cross. The orbit is projected at an angle of 43°; notice that the center of gravity (the focus of the elliptical orbits) is not at the focus of the projected orbits. The orbits are similar ellipses, both in projection and in actuality.

Center: absolute motions of Sirius and its companion. The center of gravity moves in a straight line in the direction of the arrow. The two components move in wavy paths, projections of a distorted "corkscrew" motion. The center of gravity always cuts the line that joins the positions of the two stars in the same ratio (nearly 2:5). Notice that the pattern repeats itself at intervals of nearly fifty years.

The orbits in the upper right and lower left are shown on a much larger scale than the motions in the central part of the figure.

seem to go with larger eccentricities. Only the most widely separated pairs are detected as visual binaries. The closer pairs, found as spectroscopic binaries and eclipsing-stars, tend to have smaller eccentricities, as we shall see later (p. 304).

The orbit of a double star is specified in much the same way as the orbit of a planet. The important quantities (the "elements") are:

a. *Period* of revolution (given in years).

b. *Time of periastron passage* (the time at which the stars in an elliptical orbit are closest to one another: it corresponds to perihelion for a planet).

c. *Eccentricity* of the orbit (departure from a circle; the eccentricity of a circular orbit is 0.0; a closed orbit must have an eccentricity less than 1.0; if $e = 1.0$, the orbit is a parabola and if it exceeds 1.0, a hyperbola).

d. The *semi-major axis* of the true orbit (corrected for tilt; in seconds of arc).

e. The *position angle of the node.*

f. The *distance from the node to periastron.* (These quantities fix the orientation of the orbit.)

g. The *inclination* of the orbit (its tilt relative to the "plane of the sky").

h. The *mean motion,* in degrees per year, or $360°/P$, where P is the period.

TABLE 12.6. SOME WELL-KNOWN VISUAL BINARIES

Star	Apparent Magnitude	Spectrum	Period (yr)	Eccentricity	Inclination (°)	A (A.U.)	Masses (suns)	Absolute Visual Magnitudes
Capella	0.8, 1.1	G0, F5	0.285	0.01	40	0.85	4.2, 3.3	−0.2, 0.1
Procyon	0.5, 13	F5...	39.	0.32	14	13.	1.1, 0.4	3.0, 15.5
Krüger 60	9.7, 11.3	M3, M3	44.3	0.38	26	9.6	0.27, 0.18	11.8, 13.4
Sirius	−1.6, 8.4	A0, F0	50.0	0.60	43	20.4	2.44, 0.96	1.3, 11.3
α Centauri	0.3, 1.7	G0, K5	78.8	0.51	79	23.3	1.10, 0.94	4.7, 6.1
o²Eridani	9.7, 11.4	A0, M6	248.	0.40	72	34.	0.44, 0.20	11.4, 12.9
α Geminorum	2.0, 2.8	A0, A0	306.	0.56	67	80.	5.5	1.4, 2.2

8.   SPECTROSCOPIC DOUBLE STARS

Out of the large proportion of stars that are double, and therefore traveling in orbits round one another, only a few present their orbits to us "face-on." All those whose elliptical paths are seen in projection will be moving toward us to some extent over half their paths and away from us over the other half. The visual binary 42 Comae presents its orbit to

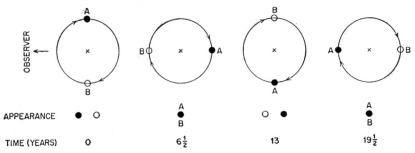

Fig. 12.11. Four positions of the components of a double star that has a period of twenty-six years. The upper figures show the orbit face-on; the observer is supposed to be on the left in the plane of the orbit, so that he sees it exactly on edge. The lower part of the diagram shows what he sees at the four dates indicated below. The components seem to converge, separate, and converge again. Notice that on the third date their positions are reversed as compared with the first date.

us edge-on, so that the two stars (similar in brightness) seem to be moving in a straight line, along which they periodically separate and approach again. The whole period of 42 Comae is 26 years, hence times of greatest separation occur every 13 years (see Figure 12.11). (The diagram is drawn as though the orbits were circular; actually the eccentricity is 0.52, so the spacing of the times of greatest separation and apparent coincidence are actually affected by the way in which the long axis of the orbit is pointed.)

If we fix our attention on star A in the diagram, it is clear that at one extremity the star is moving toward the observer with the whole orbital velocity, and at the other extremity it is moving away from him at the same speed. At the times when the separation is zero, star A is moving entirely across the "line of sight," and is thus neither approaching nor receding from us. When the star is approaching, the Doppler effect (p. 290) must be shifting its spectral lines to the blue and reducing their wave lengths; when it is receding, the light must be reddened and the wave lengths increased. A study of the wave lengths of the light coming from star A, throughout the whole circuit, would clearly show a continuous fluctuation of the wave length (and therefore of the line-of-sight velocity) with a period of 26 years.

The orbital velocity for a star with a period of 26 years is just detectable. It is larger for stars with shorter periods, and also increases with the masses of the components. For stars that are close together, especially if their masses are large, the velocity may reach several hundred kilometers a second, and the Doppler shifts that go with such large velocities are readily detectable. A large number of systems are known

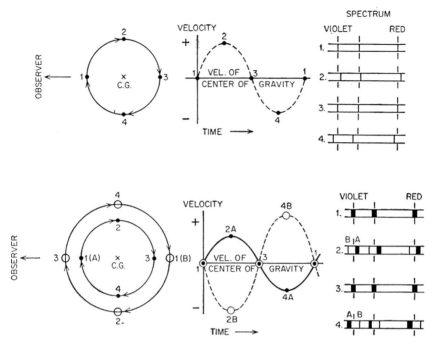

Fig. 12.12. Illustration of the production of the radial velocity curve of a star in orbital motion.

Upper figure: a star in motion in a circular orbit around the center of gravity common to it and an invisible companion. The observer sees the circular orbit on edge (compare Figure 12.11). At positions 1 and 3, the star moves *across* the line of sight, and has no orbital motion in the observer's direction. Therefore its velocity (relative to that of the center of gravity of the system) is zero. On the right is shown a schematic representation of part of the spectrum; three lines are common to the comparison spectrum (above and below) and the star's spectrum (center). For simplicity, the velocity of the center of gravity of the system is taken to be zero. Therefore the lines of comparison spectrum and star coincide exactly.

At position 2, the orbital motion causes the star to *recede* from the observer, the Doppler effect causes the light to be reddened, and the lines in the star's spectrum are displaced to the red.

At position 4, the star is approaching the observer, and the lines in its spectrum are displaced to the violet.

The continuous change of velocity with time, as deduced from the sinuous changes of wave length produced by the Doppler effect, are shown as the *velocity curve* of the star.

Lower figure: two stars, with different masses, in circular orbits that are seen edge-on by the observer. At positions 2 and 4, the lines of stars *A* and *B* are displaced *in different directions* (because *A* approaches as *B* recedes, and vice versa), and *by different amounts* (because the orbits are of different sizes, and star *B* must move faster than star *A* in order to make the circuit in the same time). Therefore the two velocity curves are different in sign and amplitude.

for which the radial velocities fluctuate periodically in this way: the changes of velocity provide evidence that the systems consist of double stars in orbital motion, even though the components are too close together, and too far from us, to be seen separately. Such double stars are known as *spectroscopic binaries*. Capella (p. 299), a "visual binary" detected as such by the interferometer, was first discovered as a spectroscopic binary.

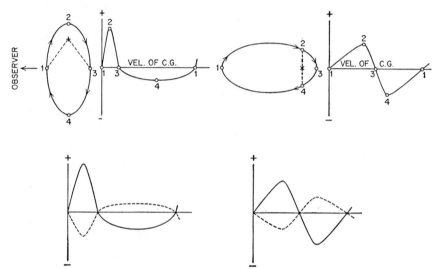

Fig. 12.13. How the velocity curves of a binary are affected by elliptical orbits. Above, on the left, a star travels in an elliptical orbit whose minor axis points toward the observer. The speed in the orbit varies in accordance with the law of areas, and the velocity curve is distorted from the symmetrical sine curve of Figure 12.12. The time taken to travel from 1 to 3 is now much smaller than from 3 to 1.

On the right, the major axis of the orbit points toward the observer. The time from 1 to 3 is now equal to the time from 3 to 1, but the time from 1 to 2 and from 4 to 1 is longer than that from 2 to 3 and from 3 to 4.

Below are shown the velocity curves for two components with ratio of masses 2:1, and orbits pointed as in the two upper diagrams. Notice that for each pair the curves are of the same shape but different amplitude (compare Figure 12.12).

If the orbit of the star is not circular but elliptical, the speed of the star in the orbit is not uniform, as in a circular orbit. It travels fastest at periastron, slowest at apastron, in accordance with the law of areas (p. 165). Therefore the orbital velocity, in kilometers a second, will no longer be constant, and the component of the velocity in the line-of-sight (the radial velocity) will no longer be a symmetrical curve. The amount and type of distortion enable one to determine both the shape and the orientation of the orbit but *cannot reveal its tilt*.

Newton showed, and the observations confirm, that two bodies going around one another in accordance with the law of gravitation will travel in *similar ellipses*. When a "double-lined spectroscopic binary" (two spectra visible) shows evidence of an eccentric orbit for one star, the velocity-curve of the other star is distorted in exactly the same way, and the eccentricity is the same as that for the first component. The ranges of the velocity-curves will be in inverse proportion to the masses of the two stars, as for the case of circular orbits described above.

The details of the determination of an orbit from the velocity-curve will not be discussed here. What has been said will make clear that the shape of an orbit can be determined from the form of the curve; a symmetrical curve means that the orbit is circular, and deviations from this symmetry suffice to determine the eccentricity and the direction in which the long axis of the orbit is pointing. But the velocities that are measured are only the *projection in the line-of-sight:* for an orbit that is "face-on" they will always be zero, and the whole effect will only be observed if the orbit is seen on edge. There is no way to tell from the velocity-curve alone at what angle we see the orbit. Therefore all conclusions that are drawn from the velocity-curves contain an unknown factor—the "inclination."

If the velocity-curves of both the components of a spectroscopic binary are observed, the ratio of their ranges tells the *ratio* of the masses; however, the actual masses can only be determined if we also know the tilt of the orbit. If one considers the results from the measurement of a large number of velocity-curves, one can make the assumption that the tilts of the orbits are distributed at random and can obtain an idea of the *average* mass of star in the group. We cannot, however, obtain information about *individual* masses in this way. Only in rare cases are there independent ways of determining the tilt. For example, if the stars of the pair pass in front of each other periodically and produce eclipses, we know that the orbit is presented nearly edge-on; and the *actual* tilt can often be determined with great precision for an eclipsing-star. In such a case the actual masses can be determined, and double-lined spectroscopic binaries that are also eclipsing-stars are the cornerstone of our knowledge of stellar masses.

A few stars that show periodic change of radial velocity cannot possibly be binary systems; calculation shows that the stars would be one within the other! For such stars (which always show only one spectrum) the fluctuations of velocity are attributed to periodic rising and falling of the star's surface. They include the *intrinsic variables* (pp. 363–383).

Over a thousand spectroscopic binaries have been discovered. One of the first to be detected was β Aurigae, for which the periodic doubling of the spectral lines was noted at Harvard in 1889. Orbits have been determined for a few hundred. The characteristics of the orbits are very different from those of the visual binaries (p. 297). Over half the periods are less than 10 days, over three-quarters, less than 100 days. The shortest periods are those for stars of "earliest" spectral class. Seventy-one per cent of the O and B stars have periods of less than 10 days, 61% of the K and M stars have periods greater than 100 days. The eccentricities are not so large as those found for visual binaries; the general average is 0.17; as for the visual binaries, larger eccentricities tend to go with longer periods. Really large eccentricities are quite rare, though they do occur: β Arietis has an eccentricity of 0.88.

Spectroscopic binaries and visual binaries differ only in method of detection. Large separation, which goes with small orbital velocity, favors the discovery of a visual binary. Large orbital velocity, which goes with small separation, favors a spectroscopic binary. Therefore a given system is more likely to be discovered in one of the two ways, depending upon whether the separation of the components is large or small. But a suitably tilted visual binary may also be detectable as a spectroscopic binary, and this has proved to be possible in a few cases. Moreover, many spectroscopic *pairs* are also known to be components of visual binaries, thus furnishing evidence of multiple stars. The bright star Castor, in Gemini (α Geminorum) is visually triple. The two brighter components of the visual triple are *spectroscopic binaries,* and the faint component is an *eclipsing-star.* Thus Castor has six detectable components. Many other more complex systems are known, and there are undoubtedly many more undetected multiple systems.

9. ECLIPSING BINARY STARS

The two previous sections have dealt with the detection of the orbital motions of pairs of stars, across the line of sight and in the line of sight, respectively. When the orbit of a double star is seen on edge, or very nearly

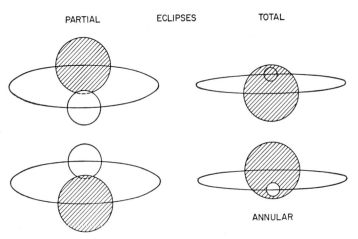

Fig. 12.14. Whether an eclipse is *partial* or *total* depends on the amount by which the orbit is tipped toward the observer. The actual orbits are supposed to be circular and seen in perspective. Notice that *two* eclipses occur in each circuit of the orbit.

on edge, one of the stars will pass periodically between the observer and the other star. It will thus cut off the light from the star in the rear for a short time, producing an *eclipse*. If the star in front conceals the other star completely, the eclipse is *total;* if it crosses the other slightly off center, so that part of the rear star is not concealed, the eclipse is *partial.* If the stars are very close together, a partial eclipse may take place when the orbit is as much as 30° from the edgeways position.

Although the tilt of the orbit greatly restricts the direction from which eclipses can be observed for a given pair of stars, the known eclipsing-stars are very numerous: over fifteen hundred have been catalogued. These eclipsing-stars are very important, for they amplify the knowledge obtained

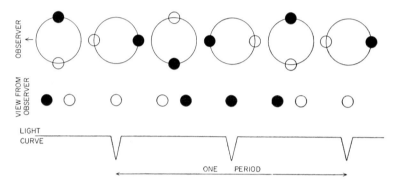

Fig. 12.15. Eclipses for stars traveling in circular orbits. The upper row of pictures shows the orbits face-on. The stars are of the same size and brightness, but one is shaded to distinguish it from the other.

The second row shows the view of the stars from an observer who sees the orbits exactly on edge, from the left. The two stars are never seen separately because of their distance.

The third row shows the changes in brightness that are produced as the stars pass alternately one before the other. The fall in brightness is the same at both eclipses, because the stars are of the same size and brightness, and therefore each can cover the other completely, and cut off all its light. There are two eclipses in each circuit, or *period.*

from the observations of radial velocity and enable us to measure the actual masses of stars. Even when the radial velocities have not been measured, a study of the variations of brightness that take place during eclipses (and even, in some cases, between them), gives valuable information about the physical properties of the two stars. Eclipsing-stars present a great variety: they include the stars of highest known mass, and also stars of very low mass; they are found among supergiants, giants, and main-sequence stars and provide most of our knowledge about the subgiants. No white dwarfs are yet known to be eclipsing-stars, but there is one interesting subdwarf system.

The sizes of the two stars, both in relation to each other and to the sizes and tilt of their orbits, combine to produce a large variety in the manner in which the light of the system varies during the cycle.

If the orbits of the two stars are not circular, then the motion in the orbits is not uniform but is most rapid near periastron (stars closest to-

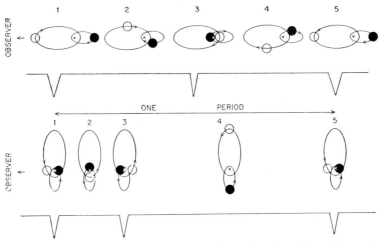

Fig. 12.16. Eclipses of stars that move in elliptical orbits. The orbits are in the ratio 2:1, therefore the masses are in the ratio 1:2.

The upper figure shows the orbits face-on, their long axes pointed toward the observer, who views them exactly on edge. The intervals between eclipses are equal, because the parts of the orbits traversed between positions 1 and 3 and positions 3 and 5 are exactly equivalent. However, at position 1 the stars are at apastron and moving most slowly, therefore the eclipse lasts longer than the one at position 3, when the stars are at periastron, and moving most rapidly. The faster the stars move, the faster will the eclipse be over.

The lower figure shows the orbits aligned with their minor axes in the observer's direction. The time from position 1 to position 3 (stars near periastron) is now much shorter than the time from position 3 to position 5 (stars near apastron), therefore the eclipses are no longer equally spaced. Eclipses take place when the two stars are in line with the observer and the center of gravity (marked with a cross). At both eclipses the stars are equally distant from periastron; they are thus moving with equal speeds at both eclipses, which are therefore of equal duration.

Notice that the periods are the same for the cases represented.

gether, see *perihelion*) and least rapid near apastron (stars farthest apart). Therefore the eclipses may not be equally spaced, as they are in the case of circular orbits. Figure 12.16 shows the motion of two stars in elliptical orbits around their center of gravity.

Elliptical orbits will, in general, be pointed in directions that lie between the two extremes that have been illustrated in Figure 12.16. Thus the *spacing* of the minima and the *durations* of the minima will both differ, and by combining the information from both sources, it is possible to deduce both the eccentricity and the orientation of the orbits.

There are some systems with elliptical orbits that do not keep their major axes ("line of apsides") always pointed in the same direction. The line of apsides may rotate steadily—an effect that has already been noted in the motions of the moon and the planets that have eccentric orbits. Such a rotation is detected by noting that the spacing of the minima, as well as their relative duration, change steadily. For some stars the line of apsides has been observed to go through a complete turn: the orbits of the star GL Carinae turn on themselves in twenty-five years (the most rapid rotation known); Y Cygni takes forty-six years for a complete turn. But for many systems the change is so slow that only a small part of it has been recorded, and all we can say is that the period must be hundreds, if not thousands, of years. The rotation of the line of apsides is an important item of information. It is known to depend on the internal structure of the stars concerned and is the *only direct source* of information on the subject.

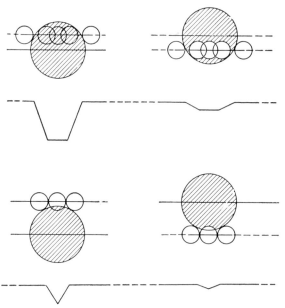

Fig. 12.17. Eclipses of unequal stars. A small portion of the orbit is shown, and the larger star is arbitrarily kept in one spot. Actually they move in opposite directions. For most pairs of this kind, the larger star has a lower surface temperature, and is therefore the fainter *per unit area*. The pair represented is supposed to be of this kind; the star that is fainter per unit area (i.e., has a lower surface brightness) is shaded.

The two upper diagrams show a totally eclipsing system. Notice that the eclipses, and the intervals of totality, are of the same length, *whichever star is in front,* but that the deeper eclipse occurs when the star of greater surface brightness is behind.

The two lower diagrams show a partially eclipsing system. Notice that the obscured area changes continuously in size, so that there is no interval of constant brightness during the eclipse. Notice the difference in depth of the two eclipses.

The shapes of the light-minima have been arbitrarily drawn with straight lines, but actually they are curved, partly because the discs of the stars are round, and partly because they are less bright at the edges than at the centers (*limb-darkening*).

All the systems that we have hitherto used to illustrate the changes of brightness of eclipsing pairs have consisted of two similar stars. But the members of many systems are quite dissimilar in size and temperature, luminosity and mass. The apparent loss of light during an eclipse depends, of course, on the sort of eclipse (total or partial), and for a given sort of eclipse, it is determined by the temperature of the eclipsed star. For the amount of light that is cut off during an eclipse is the amount of the light from the star in the rear that is cut off by the star in front. Therefore it is the amount of light given out by an area of the rear star's surface that is equal to the part of the area of the front star that obscures the rear star. Clearly, then, it depends on the amount of light *per unit area* given out by the star in the rear, and this (by the Stefan-Boltzmann law) depends only on the temperature of the eclipsed star. Each star conceals an equal area of the other at alternate eclipses. But if the two stars differ in surface temperature the *amount* of light lost must differ at the two eclipses; more light is lost when the star of higher temperature is behind, and this eclipse is the "deeper" of the two. Both eclipses, however, are of the same *duration* because (in circular orbits) the time taken for one to pass completely across the other (from "first contact" to "last contact") is the same, whichever star is in front. If the eclipse is central, or nearly central, the two stars will wholly overlap for some time when the larger is in front. Then the total light will remain constant until the smaller star begins to emerge from behind the other.

Without going into technical details, we may say that a study of the form, depth, and timing of the eclipses can be made to give information about the relative temperatures of the stars, their combined sizes (in terms of the sizes of their orbits), the individual sizes, and the shape and orientation of their orbits. The problem, though complicated, is purely geometrical (at least in the terms hitherto stated). One can assume that the stars present uniformly bright discs, or one can make plausible assumptions as to how the brightness of the disc is distributed. We have already noted that the disc of the sun is rather less bright towards the edge, and effects of the same kind must be expected for all stars and allowed for in precise studies of eclipsing-stars.

As the orbits of eclipsing-stars are seen nearly or quite on edge, they would be expected to show radial velocity effects that are repeated once in every circuit; in other words, eclipsing-stars will be observable as spectroscopic binaries. If the two stars are about equally bright, two velocity-curves will be observable, and the ratio of the masses can be determined. The sum of the masses can only be measured (see page 303) if the inclination of the orbit is known, and for eclipsing-stars this is known from the form of the eclipses; for a central eclipse the inclination is $90°$, and for partial eclipses it can be determined geometrically. Thus the eclipsing-stars that are also double-lined spectroscopic binaries give us complete information about stellar masses. This is one of the reasons that they are so energetically studied.

The relation between light curve and velocity curve should be clear from what has been said already. The velocity curves will cross the zero line (stars neither approaching nor receding) at the times of eclipse, and the greatest to-and-fro velocities will appear between eclipses.

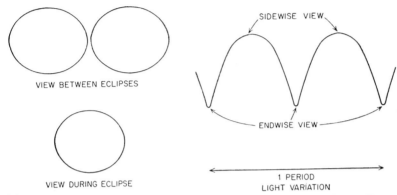

Fig. 12.18. Light-curve of an eclipsing-star whose components are tidally distorted. On the left are shown the sidewise view (between eclipses) and the endwise view (middle of eclipse). On the right is shown the resulting light-curve.

In all that has been said so far, we have supposed the stars to be spherical, so that their discs appear circular. But all stars are not spherical. We noticed that rapidly rotating planets bulge at the equator, and rapidly rotating stars will bulge even more, because they are not solid. Moreover, many double stars are so close together that they distort each other tidally, so they may be spheroids, or even ellipsoids (spheroids have one circular and two elliptical cross sections; ellipsoids have three elliptical cross sections). Clearly, such stars will present discs of different sizes and shapes when viewed from different directions; when seen "sideways," they will present larger surfaces and therefore seem brighter than when seen "endways."

Like the moon, which always presents the same face to the earth because of tidal forces, close double stars always seem to present the same face to each other for the same reason (though for wide pairs this is not always true). Therefore their long axes tend to be lined up. When a close elliptical system is seen from the side, both stars present larger areas to the observer than when they are seen endways, and more light is received from each star in the sideways, than in the endways, position. Clearly the stars are seen from the side *between* eclipses, and *during* eclipses they are seen endways. This "presentation effect" makes us receive most light exactly between eclipses, and the brightness of an eclipsing pair of this kind is not constant between eclipses, but varies continuously. The amount of this variation enables us to determine how much the two stars are distorted. For some stars the distortion is quite large; the ratio of the long axis to the short may be as great as 10:6.

The foregoing chapter has described the study of the geometrical properties of stars: their dimensions, masses, luminosities, and temperatures. We have seen that stars display considerable variety in all these properties. Figure 12.19 illustrates the relations between them, and shows that, despite the variety, physical characteristics of stars are not distributed at random.

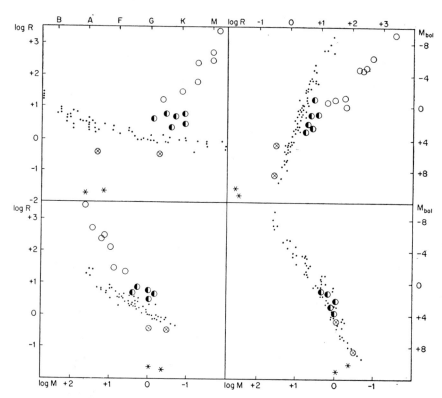

Fig. 12.19. Relations between some of the physical properties of the stars. Upper left: relation between logarithm of radius and spectral class—radii are in terms of the sun, so that if log R is 3, the star has 1000 times the sun's radius; if log R is −1, the star has 1/10 the sun's radius. In this and the other diagrams circles denote giants and supergiants; half-filled circles, subgiants; dots, main-sequence stars; crosses in circles, subdwarfs; stars, white dwarfs. Lower left: relation between logarithm of mass and logarithm of radius, both in solar units. Upper right: relation between logarithm of radius and absolute bolometric magnitude. Lower right: relation between logarithm of mass and absolute bolometric magnitude. Notice that the only one of the four correlations that is at all close for all stars is that between absolute bolometric magnitude and mass, which is nearly linear (except for the white dwarfs). This is the mass-luminosity law. All the other relations differentiate the various groups of stars rather sharply.

Distinctive symbols in Figure 12.19 denote the giants and supergiants, the subgiants, the stars of the main sequence, the subdwarfs, and the white dwarfs. Three of the diagrams illustrate the physical distinctions between these groups. Only one, the diagram that compares absolute bolometric magnitude (p. 279) and mass, shows the same relationship for all stars except the white dwarfs. This correlation is the *mass-luminosity relationship,* and indicates that the mass of a star is its most significant property, since it appears to govern the star's total luminosity. Because the only stars whose masses can be measured directly are double stars, and most stars of known mass lie on the main sequence, we can state with certainty that the mass-luminosity relationship applies to main-sequence stars; some stars of other groups have been shown by recent researches to deviate significantly from it, and these deviations promise to give important evidence that bears on stellar development, as will be mentioned in the next chapter.

PHYSICS OF THE STARS

Roll on, ye stars! exult in youthful prime,
Mark with bright curves the printless steps of time . . . .
Flowers of the sky! ye too to age must yield,
Frail as your silken sisters of the field.
ERASMUS DARWIN *

The preceding chapter has introduced us to the external physical properties of the stars. Distance, motion, brightness, and color reveal a tremendous range in gross physical properties. Straightforward trigonometrical methods of celestial surveying, and a knowledge of the laws of continuous radiation (the same for all substances) have already shown us that stars differ enormously in size, mass, surface temperature, density, and total brightness.

The properties of the stars are not, as we have seen, distributed in a haphazard manner. Certain types of stellar physique are very common. Many thousand known stars, for instance, are indistinguishable from our sun, and cooler, fainter stars, all essentially similar, are found in even greater numbers. We have learned to place the stars in several great groups on the basis of physical properties—the *main sequence,* backbone of the local stellar population, a series of stars from hot and bright to cool and faint; the distended *giant stars;* the very diffuse *supergiants;* the *subgiants* that lie between giant and main sequence; the subdwarfs that are smaller and denser than the main sequence stars; and the *white dwarfs* of almost incredibly high density. These stellar groups are distinguished not only by gross physical properties, but also by finer details of structure and composition and by "the company they keep." The present chapter will describe how the surface conditions of stars can be analyzed in almost incredible detail, and how the methods of modern physics are performing the apparently impossible feat of probing their inner structure.

Modern astrophysics, the science that may be said to bring the stars into the laboratory, is the "child" of spectroscopy. The study of the spectra of atoms and molecules on which it depends is little more than a century old; the theoretical analysis of spectra (briefly described in Chapter III)

* Grandfather of Charles Darwin, the naturalist.

312

has made its great strides only during the last three decades, and is still actively expanding. So it comes about that astrophysics is still a "lusty infant" among the sciences. Not only does it abound in unsolved problems, but its basic facts are still incompletely garnered. Well-known techniques continually add to the facts that require classification and analysis, and the new techniques (such as infrared spectroscopy and microwave astronomy) are bringing in not only new facts but new types of facts to enrich and complicate the picture.

An even newer branch of astrophysics is today being born of nuclear physics. The astronomer of today speaks of the conditions at the centers of stars (forever inaccessible to direct observation) with more confidence than he would have spoken of conditions at their surfaces fifty years ago.

1. THE SPECTRA OF THE STARS

The spectrum of the solar disc, which was described in Chapter IV, is a typical stellar spectrum. It consists of a bright continuum that is overlaid by large numbers of "absorption lines." Most of the atomic absorption lines in the spectrum of the sun have been successfully identified with the characteristic lines given by atoms known on the earth. For some atoms the solar spectrum actually gives a more complete representation of the possible spectrum transitions than has ever been obtained in the terrestrial laboratory, because the "atmosphere" of the sun has such an enormous volume and therefore constitutes a much larger specimen of material than any laboratory could contain. Many more lines of the iron spectrum, for example, have been observed in the solar spectrum than on the earth. We are confident that these lines are produced by the atom of iron, for their frequencies agree exactly with those that are predicted for transitions between *known* energy states of the atom of iron. Later in the chapter we shall encounter other examples of spectrum lines that are observed in astrophysics because of the enormous cosmic specimens that are available for analysis.

Although many lines in the solar spectrum have not yet been matched with lines observed in the laboratory, there is little reason to suppose that the sun's atmosphere contains any atoms unknown on earth. Most of the unidentified lines are weak, and have probably not been observed terrestrially because we can analyze only small samples of matter in the laboratory. Many of them probably originate not from atoms but from molecules, and our analyses of molecular spectra are still very incomplete; this is especially true of the unidentified features in the sunspot spectrum which (from the nature of their Zeeman effect) evidently emanate from molecules.

Most stars show spectra that resemble the solar spectrum in that they consist of a bright continuum and absorption lines. A few show bright-line spectra as well, and some have only bright lines and no absorption lines; both these groups are exceptional. But the absorption spectra of stars differ enormously. Some show practically nothing except lines of hydrogen and helium; others, like that of the sun, are rich in metallic lines; still others are dominated by molecular bands that show the atmospheres to be rich in simple chemical compounds.

When stellar spectra were first observed, the striking differences were thought to show that stars differed widely in chemical composition—that some were made principally of hydrogen, some, of metals, some, of metallic oxides. When it was realized that stellar spectra can be arranged roughly in a continuous sequence, Norman Lockyer suggested that the differences arose from a gradual evolution of the chemical elements, in which hydrogen developed steadily into more complex atoms.

In 1922 the true interpretation of the sequence of stellar spectra was realized by the Indian astrophysicist Meghnad Saha. He saw that the process of ionization of an atom—the loss of one or more electrons by the absorption of energy—is analogous to chemical dissociation, and that the degree of dissociation must depend primarily on the *temperature* of the stellar atmosphere. The higher the temperature, the greater the tendency to ionization, and therefore the greater the number of atoms of a given kind that will be put out of action by the loss of an electron.

Suppose that the atmosphere of a star is at a low temperature, so that all atoms are in the neutral (un-ionized) state, and disregard for the moment the formation of compounds (molecules). Then, if the temperature is raised, atoms begin to lose electrons, and those that have done so will be unable to absorb their characteristic lines. Therefore the said lines will be weakened in the spectrum, since fewer atoms are available to produce them.

The atoms that will lose their electrons most readily will be those of lowest ionization potential (p. 73). An atom of iron will be put out of action more easily than one of hydrogen, and will therefore fade more rapidly as the temperature rises. But as the iron atom loses one of its electrons, it becomes able to absorb the totally different spectrum of ionized iron. Thus, as the temperature rises, lines of neutral iron will weaken, those of ionized iron will become stronger.

Saha was able to show that the whole series of observed stellar spectra could be thus interpreted in terms of differences of temperature. The atmospheres of all stars were seen to be *essentially similar* in atomic composition, and the spectral differences to be primarily the results of difference of temperature. The theory has been greatly refined since it was first formulated; we know today that density, as well as temperature, plays a major role in determining the degree of ionization, and therefore the lines represented in the spectrum. Some differences of atomic composition have been established among stars, and, though minor, they are of enormous importance, as we shall learn. But the amazing thing is not the difference of atomic composition among stellar atmospheres, but the similarity. The uniformity extends not only to stars, but also to the nonstellar material that lies between them and makes up the nebulae.

The sections that follow will summarize the main features of the sequence of stellar spectra and will describe in elementary terms the way in which the spectra yield information about the physical conditions at the surface of stars.

2. THE SEQUENCE OF STELLAR SPECTRA

The spectra of most stars can be arranged in the continuous sequence already mentioned: O–B–A–F–G–K–M. A few form a branch sequence R–N parallel to K–M, and Class S is in some ways another parallel to Class M. Class W is placed before Class O, but we shall discuss it later, because is differs radically from the others. The sequence will be seen to represent a descending order of temperature.

Typical spectra of representative classes are shown in Figure 13.1. At Class B0 the only conspicuous lines are those of hydrogen and neutral helium (Fig. 13.2). Ionized helium appears only in O stars and is strongest in the "earliest" class observed, O5.

As the spectra pass from B0 to A0, the lines of ionized oxygen and ionized carbon are seen (strongest at Class B3); neutral helium is strongest in the middle of this range. A line of ionized magnesium appears at Class B8 and grows stronger at lower temperatures. Hydrogen is progressively stronger from Class B0 to Class A0.

At Class A0, hydrogen is at its strongest and so is ionized magnesium. Helium has disappeared; the lines of ionized oxygen are no longer seen. Lines of ionized metals (iron, titanium, calcium) are present, but they are quite weak. They strengthen progressively as we pass toward Class F0, and hydrogen weakens over the same interval. This progressive weakening of lines requiring high excitation and strengthening of those of low excitation supports the view presented earlier that the spectral sequence represents a descending temperature scale.

The spectrum of Class F0 is rich in the lines of ionized metals, and a line of neutral calcium (wave length 4227) is also visible. Hydrogen is still strong. The strongest lines are the two lines of ionized calcium at wave length 3933 and 3968 (the H and K lines). All these metallic lines strengthen, and the hydrogen lines weaken, between F0 and G0.

At Class G0 the neutral lines of metals are becoming strong; the H and K lines continue to strengthen and the hydrogen lines, to weaken. The molecular bands of cyanogen (CN) and the hydrocarbon "G band" (CH) appear.

At Class K0 the neutral lines and the molecular bands are far stronger;

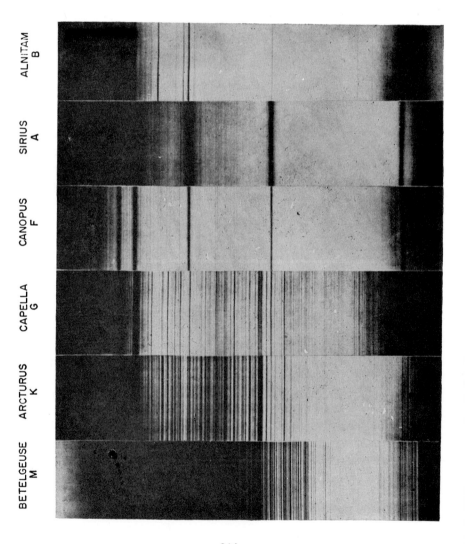

Fig. 13.1. Typical stellar spectra. The stars are arranged in order of descending temperature. (Harvard Observatory)

316

the ionized metallic lines in general weaker, though the H and K lines are at their strongest (Figure 13.5). The lines of hydrogen are weaker than at Class G. At Class K5 the bands from the molecules of the metallic oxide of titanium (TiO) are weakly visible.

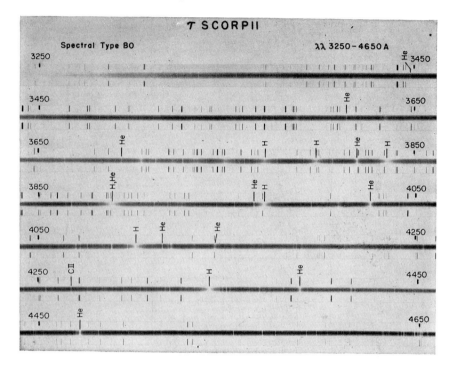

Fig. 13.2. Spectrum of the star τ Scorpii, Class B0. Lines of a number of characteristic atoms are marked. The presence of once and twice ionized atoms points to a high temperature, between 25,000° and 30,000°. (Mount Wilson Observatory)

The characteristic feature of the M stars is the presence of molecular bands; the strongest are those of titanium oxide. The lines of neutral atoms are even stronger than at Class K0, and those of hydrogen, weaker. In the cooler M stars the molecular bands are extremely strong; the lines of hydrogen, very weak.

The spectra of the parallel sequence of Classes R and N resemble those of Class M in showing prominent molecular bands, but these are the bands of carbon compounds—principally cyanogen (CN) and the carbon molecule (C_2). The difference between the M stars and the R–N stars is probably one of composition, the former being richer in oxygen and the latter, in carbon. The S stars are very like M stars, except that they show prominent bands of zirconium oxide (ZrO); very likely the difference in this case is not one of composition, but of temperature and density. All the spectra that show molecular bands are also extremely strong in the lines of neutral atoms, and the hydrogen lines are always weak.

The spectral sequence can best be understood if we trace along it the changes of the lines of some common atoms. We select hydrogen, helium, oxygen, calcium, and iron.

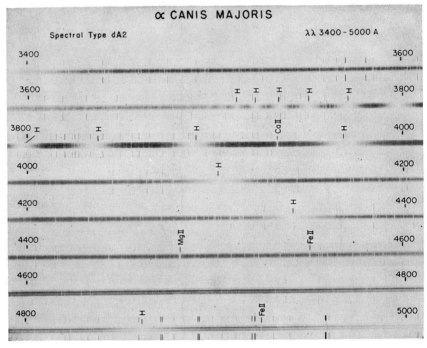

Fig. 13.3. Spectrum of Sirius, Class A0, temperature about 11,000°. A number of typical atoms are marked; notice that the excitation is less than in Figure 13.2. Notice the great breadth of the Balmer lines of hydrogen. (Mount Wilson Observatory)

The only lines of hydrogen that appear in the photographed spectrum of Figure 13.1 are the Balmer lines. Another series of hydrogen, the Paschen series, can be studied by special means in the infrared. Neither of these series arises from the ground state of the atom of hydrogen, and both require considerable energy of excitation before the atoms can be raised to a state from which they can be absorbed. At the temperatures of the coolest stars, the fraction of hydrogen atoms excited to a state appropriate to the Balmer line is very small, hence the weakness of the lines. The hydrogen lines that are associated with the ground state of the atom are the Lyman series in the inaccessible far ultraviolet.

As temperature increases from Class M through A to B, fraction of excited hydrogen atoms able to absorb the Balmer series increases steadily, and the Balmer lines grow stronger, until at Class A0, they are the strongest feature of the spectrum. At still higher temperatures, ionization begins, the electrons are separated from the atoms of hydrogen, and the Balmer lines become weaker again.

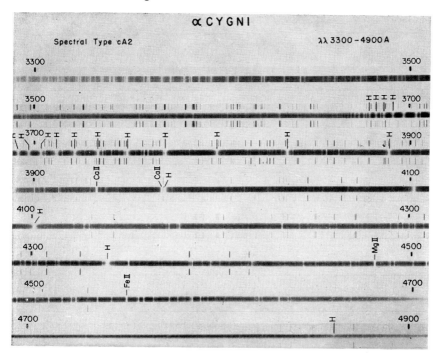

Fig. 13.4. Spectrum of Deneb, a supergiant of Class A2. A number of typical lines are marked. Though nearly of the same temperature of Sirius (a little lower), it has a spectrum of very different appearance. Especially marked is the appearance of the Balmer lines, and the presence of lines of neutral helium, absent from Sirius. Both features are marks of high luminosity. The student will find it an interesting exercise to match the two spectra (the comparison spectrum of iron, on either side, is a helpful guide) and to compare them in detail. Compare also Figure 3.13. (Mount Wilson Observatory)

Even when the lines of hydrogen are at their strongest, only about one hydrogen atom in a hundred thousand is producing the Balmer lines. Therefore one might tend to underestimate the role played by hydrogen in the atmospheres of stars. Actually there are about a thousand times as many hydrogen atoms in stellar atmospheres, and in the stars as a whole, as all other kinds of atom put together, except helium, which is 10 to 20% as common as hydrogen. A few stars are unusually poor in hydrogen— though even these usually show the Balmer lines in their spectra.

The occurrence of helium along the spectral sequence is in some ways parallel to that of hydrogen. For neutral helium, again, the lines associated with the ground state of the atoms are in an inaccessible region of the spectrum. The observable helium lines have even higher excitation potentials than the Balmer lines, and in the spectra of the coolest stars, they do not show at all. They are first seen for stars a little hotter than Class A.

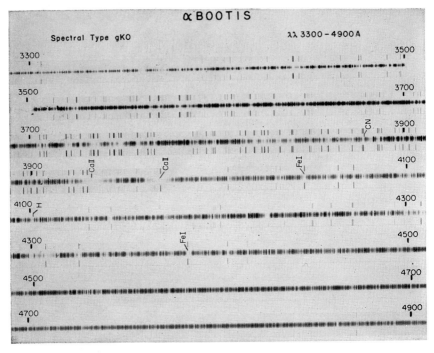

Fig. 13.5. Spectrum of Arcturus, a giant star of Class K0. Notice the vast number of lines of neutral and ionized metals. Arcturus is about 1000° cooler than the sun. Compare the part of the spectrum common to this Figure and to Figure 4.15, and notice that the lines of hydrogen here are weaker, and the metallic lines stronger, than in the sun. (Mount Wilson Observatory)

At Class B5 they begin to decline in strength, because the high temperature is ionizing the helium atoms. As the number of ionized helium atoms grows, at the expense of those of neutral helium, the lines of He II do not at once appear, because the lines associated with the ground state are in the far ultraviolet. The Pickering lines have high excitation potentials, and are not seen below the temperatures of O stars.

The accessible lines of neutral oxygen (O I) are of high excitation potential, and are not seen in the spectra of the coolest stars. Evidence of the presence of oxygen, however, appears in the molecular spectra of titanium, zirconium, and scandium oxides in stars of low temperature. The strongest lines of O I are in the red part of the spectrum, and are strongest at about the same temperature as those of hydrogen. Very many of the lines of O II appear in early B stars, though here again the lines associated with the ground state are not observable. In some of the very hottest stars (classes O and W), lines of OIII, OIV, OV and even OVI can be observed.

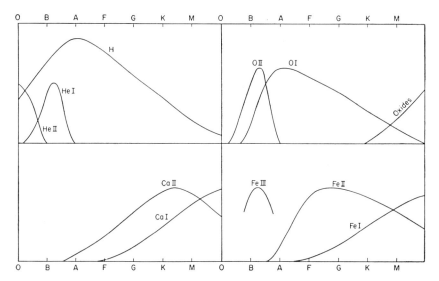

Fig. 13.6. The changes of intensity of lines of hydrogen, helium, oxygen, calcium, and iron throughout the spectral sequence.

A different state of affairs is found with the spectra of calcium. The line of Ca I that is associated with the ground state is in the photographic region, at wave length 4227. It is strongest in the coolest stars. The H and K lines of Ca II are also ground-state lines, and as the ionization potential of Ca I is small (6.09 volts), appreciable numbers of ionized calcium atoms are present in the atmospheres even of the coolest stars. The H and K lines are in fact at their strongest at about Class K0. They grow progressively weaker, as does the line of Ca I, with rising temperature, and are not normally seen for stars of temperature much higher than that of Class A0. They are actually observed in the spectra of many of the hotter stars; but here they are not part of the stellar atmospheres; they are produced by calcium atoms in interstellar space (p. 331).

Lines of iron are one of the most conspicuous features of the spectra of all stars cooler than Class A. In the coolest stars the neutral lines are strong, and the ground-state lines are in the near ultraviolet and readily observed. But the ionization potential of iron is low (7.858 volts), and even in K and G stars some lines of ionized iron are visible; they become progressively stronger, relative to those of neutral iron, with higher temperature. Lines of Fe III are found in the spectra of some exceptional hot stars, and iron spectra up to Fe VII have been observed in the spectra of some novae. We have mentioned in an earlier section the highly ionized iron atoms (up to Fe XIV) that have been identified in the spectrum of the solar corona.

The system of naming spectra that has just been described is called the *Henry Draper system;* it originated at Harvard Observatory under E. C. Pickering and was extensively used by Miss Cannon. At about the same

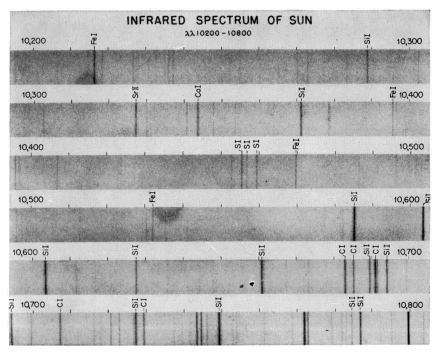

Fig. 13.7. Portion of the infrared color spectrum. Notice the presence of strong lines of iron (Fe), silicon (Si), sulphur (S), calcium (Ca), and carbon (C), which are most easily detected in this part of the solar spectrum. The rest of the strong lines originate in the earth's atmosphere. Compare Figure 4.13. The photograph is a *positive;* the absorption lines are dark.

time, also at Harvard, Miss Maury devised a much more detailed system of spectral classification—so detailed indeed that it did not come into general use. Its main divisions were roughly parallel to those of the system just described, but each of them was subdivided into three groups, not on the basis of the spectral lines present, but of the *quality* of those lines. Stars of division "a" had well-marked lines, but those of division "c" had lines that appeared to have sharp edges; the whole texture of the spectrum looked quite different. It was promptly pointed out by the great observer Hertzsprung that the "c" stars were of exceptional properties, and that the "a" and "c" subdivisions enable one to differentiate the normal from the highly luminous stars, the supergiants. The classification of Miss Maury was, in fact, the first step in picking out spectroscopic criteria for the luminosities of stars.

It was not long before Arnold Kohlschütter and Walter Adams recognized a number of other details in otherwise similar spectra that made it possible to decide, *from the spectrum alone,* whether the star was a giant

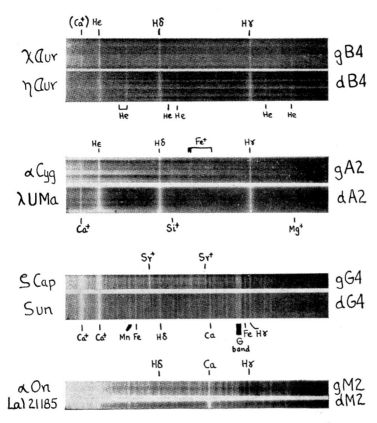

Fig. 13.8. Differences between the spectra of giant and dwarf stars of the same spectral class. These differences are the basis of the measurement of spectroscopic parallaxes. (Photograph by W. W. Morgan, Yerkes Observatory.)

or a dwarf (main-sequence star). Figure 13.8 shows the spectra of highly luminous and main-sequence stars of spectral classes B4, A2, G4, and M2. For the hotter stars the most obvious difference is in the qualities of the hydrogen lines, though other features also differ. In the G spectrum, the most evident difference is in the lines of ionized strontium, strong in the supergiant. This is primarily a result of the much lower density in the atmosphere of the supergiant, which discourages recombination and thus enhances the number of ionized atoms. For the coolest stars in the figure, the two obvious differences are the strength of hydrogen in the supergiant (primarily as a result of its having much more atmosphere than the dwarf) and the strength of neutral calcium in the dwarf, a consequence of the higher density and the resultant greater tendency to recombination of ionized calcium atoms. The differences between stars of similar spectral class and different luminosity are many and intricate. They are embodied in the two-dimensional spectral classification that is standard today. It has refined the Henry Draper classes and put them on a quantitative basis, and is known as the "MKK" system after its originators W. W. Morgan, P. C. Keénan, and E. Kellman.

The difference between the spectra of stars of different luminosities is not merely of qualitative use in telling to which category a star belongs. It can be used quantitatively to estimate the luminosity of a star. The luminosity thus determined is called the *spectroscopic absolute magnitude,* and can be used, in conjunction with the apparent magnitude, to determine the star's distance, or the *spectroscopic parallax.*

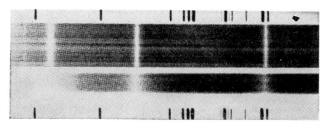

Fig. 13.9. Difference between the spectrum of a normal star and a white dwarf. Notice the greatly broadened hydrogen lines in the spectrum of the latter. (Photograph by W. W. Morgan, Yerkes Observatory.)

An even more striking difference than that between the spectra of giant and main-sequence star is that between main-sequence star and white dwarf. The hydrogen lines in the white-dwarf spectrum are extraordinarily widened, because of the high density in the atmosphere and the consequent mutual electric Stark effect (analogous to the magnetic Zeeman effect) among the hydrogen atoms. A white dwarf can be picked out with certainty by its spectrum (Figure 13.9). The spectrum of a subdwarf is quite different from that of a white dwarf and can also be distinguished without difficulty from that of a main-sequence star.

What has been said of the spectral sequence applies to the usual star that shows absorption lines upon a bright continuum. In every spectral class an occasional abnormal star shows *bright* lines. Bright lines are usually a sign of a distended atmosphere, a sort of chromosphere (though there are exceptions, as will be seen for the long-period variables). Some bright-line stars will be discussed among the variable stars that show shells and flares. Here it must be mentioned that bright-line stars become increasingly common in Class B, especially among the early B stars and the O stars;

and the stars of Class W (the Wolf-Rayet stars) are entirely, or almost entirely, bright-line objects. Their bright lines are usually very broad and hazy. The Wolf-Rayet stars are not well understood. They are certainly of extremely high temperature, for their spectra are rich in ionized helium and the highly ionized spectra of carbon, nitrogen, and oxygen. They seem to differ among themselves in the relative abundance of these three latter elements. For a long time it was regarded as certain that their broad, bright lines were evidence that matter is streaming out from them in all directions with high velocity and broadening the lines by Doppler shifts. But it is difficult to reconcile this idea with the dimensions of the atmospheres, otherwise deduced. A large proportion of the Wolf-Rayet stars have recently been found to be spectroscopic binaries, and some are even eclipsing-stars. Thus it is becoming possible to make direct determinations of the dimensions of both stars and envelopes. The problem of the Wolf-Rayet stars is not yet solved, but the materials for its solution are being rapidly accumulated.

3. COMPOSITION OF STELLAR ATMOSPHERES

Observations of the intensities of the lines in the spectra of stars, in conjunction with knowledge of the ionization and excitation potentials of the corresponding states of the atoms and the inherent probabilities of the transitions that produce the spectrum lines, should suffice to determine the atomic compositions of the parts of the stars that give rise to the spectra. The intensities can be readily measured (at least in the spectral regions that are accessible), but the steps in deriving chemical compositions from these intensities are intricate. The number of atoms that produce a line is not directly proportional to the strength of the line. All spectral lines are naturally slightly fuzzy (i.e., the light is not monochromatic), even when the atoms concerned are at rest; however, when the atoms are in motion, as they must be if the temperature is above absolute zero, the lines will be further blurred by the Doppler shifts produced by the motions of individual atoms. If the lines had merely their natural fuzziness, the intensity of a given line would be simply governed by the number of relevant atoms and the inherent probability of the relevant transition (except for very small numbers of atoms).

If, however, the lines are blurred by the motion of agitation of the atoms (as they must always be at appreciable temperatures, since their velocity is proportional to the square root of the temperature) the intensity does not increase in this simple way with the number of atoms and the probability, but more slowly. The consequence of the inevitable coexistence of both types of blurring is that the relation between line intensity and number of atoms acting is not expressed by a straight line, but by a curve (the *curve of growth*). The latter can be worked out for any particular temperature and used to deduce the atomic composition, provided that the inherent probabilities of the transitions are known, and that no other factors affect the fuzziness of the lines concerned.

Many transition probabilities have been evaluated theoretically, others have been determined empirically in the laboratory; and work on the curve of growth must be confined to these. However, the absence of other disturbing factors on "line profile" cannot be assumed. Lines may be blurred by the Zeeman effect (p. 113), by the Stark effect (p. 324), by mutual collisions, or by large-scale vertical atmospheric motions (*turbulence*). Collisions are similar in effect to natural blurring, and Stark and Zeeman effects are usually negligible (though the former is important for the lines of hydrogen in many stars, particularly the white dwarfs). Turbulence, however, may have an important effect and distort the curve of growth; it is found to be large in the atmospheres of many supergiants. The turbulence can in fact be roughly evaluated from the distortion of the curve of growth, but it makes uncertain the quantitative analysis of the atmospheres concerned.

If the number of atoms that are producing a line (deduced from the curve of growth) is determined, a knowledge of the temperature and density of the star's atmosphere (which govern ionization) makes possible a quantitative analysis of the atmosphere. Occasionally, as with stars that have extended envelopes (*shell stars*, p. 397), the density of the envelope and the distance from the star have spectacular effects in differentiating the intensity of lines that originate from *metastable states* (in which the atoms tend to remain for exceptionally long periods) and normal lines.

Analysis of the sun's spectrum by the methods just described leads to the result that hydrogen preponderates enormously, helium follows (about one-fifth as common by volume), and all other atoms are much rarer. They fall off, roughly, in order of atomic number; but both oxygen and magnesium seem commoner than other atoms (except, of course, hydrogen and helium). Lines of sixty-one of the hundred known chemical elements are identified in the solar spectrum.

Analysis of stellar atmospheres can be made in the same way, but the relative faintness of stars cuts down some detail because the spectra cannot be studied on such a large scale. The surprising fact emerges that most of the stars have atmospheres that are very close in composition to the sun's; *the great variety of stellar spectra is a result of differences of temperature and density, not of composition*. The stars fill in some gaps in the solar data: neon, unobservable at solar temperatures, proves to be about as common as oxygen in the atmospheres of hot stars.

The rule that the abundance of atoms falls off with atomic number is broken by the third, fourth, and fifth members of the periodic table— lithium, beryllium, and boron. All three seem to be very uncommon in the atmospheres of the sun and most stars—probably an indication that they have been consumed by nuclear reactions (p. 347). However, lithium is very common in the atmosphere of at least one cool giant star.

There are other apparent departures from uniform atmospheric composition. Some of them may be tricks of local conditions, but some are undoubtedly real. We have already mentioned that both W stars and M, R, and N stars differ in carbon and oxygen (probably also nitrogen) content. Some stars have atmospheres poor in hydrogen, some (such as R Coronae Borealis, p. 401) are exceptionally rich in carbon. The critical evaluation of differences of atmospheric composition and its interpretation are important and active branches of astrophysics, and their results cannot yet be foreseen.

As we shall see later, the atomic composition of the gaseous and planetary nebulae is very much the same as that of the average stellar atmosphere.

The observed spectra give information of compositions *on* the stars; it does not necessarily follow that the atoms *in* the stars are present in similar proportions. The approach to the internal composition of stars does not repose on spectroscopic information, but on conclusions drawn from the luminosities, masses, and dimensions of stars. How this comes about will be discussed in a later section. As we shall see, hydrogen dominates the composition of most stellar interiors, just as it dominates most stellar atmospheres. The universe, it has been said, consists of hydrogen "with a smell of other elements."

4. SPECTROSCOPIC STUDY OF INTERSTELLAR MATTER

A large amount of material lies in the spaces between the stars. In our neighborhood there is probably almost as much nonstellar as stellar material, partly in the form of solid particles, partly in the form of atoms (gaseous matter).

The solid material is readily recognized in the form of obscuring clouds. They impart a variegated appearance to the "Milky Way"; if they were not present, the sky would be encircled with a brilliant band of light in the directions where we look toward the edges of our flattened stellar system. One side of this band (toward Sagittarius) would be much brighter and wider than the rest, since in that direction lies the center of our galactic system. We are situated toward one edge, and if it were not for the obscuring clouds, our eccentric position would be evident. As it is, only during the past twenty-five years have we realized the extent to which the obscuring clouds distort our view of the galactic system. The clouds are almost entirely confined to the galactic plane, though a few isolated ones are found some distance from it; there is even one that lies near the north celestial pole.

Fig. 13.10. Composite photograph of the Milky Way in the neighborhood of the galactic center. Notice the bright star clouds in the middle of the picture, and the apparent division in them that is caused by layers of nearby obscuring material in the galactic plane. The bright stars of the constellation Scorpio (the Scorpion) occupy the center. The two bright stars on the right are α and β Centauri (the components of the former are too close together to be seen separately on the photograph). (Harvard Observatory)

Some of the obscuring clouds, like the Coal Sack near the Southern Cross, are of regular shape, but many are irregular streaks. We see as individuals only those that are quite nearby; their diameters are of the order of ten parsecs or less. But we also see the cumulative effect of a great many more distant ones of comparable size, and it is this cumulative absorption that shuts off our view of a great part of the Milky Way itself. The effect of the absorption is to make distant stars fainter than they would be from their distance alone; and the more distant a star, the more will its light be dimmed.

The progressive dimming of stars with distance makes it appear at first sight as though they are thinning out in all directions from us. You will remember that if the stars are *uniformly* distributed, each fainter magnitude should include nearly four times as many stars as the next brighter magnitude. But the numbers of stars do not mount up as fast as this, and the reason is that the remoter ones are dimmed by the interstellar material, at least in the plane of the Milky Way (*the Galactic plane*).

It might be thought that matter that is observed only in the form of obscuring clouds would elude analysis; but a surprising amount can be found out about its nature and condition. Although without doubt the dark clouds consist of a mixture of particles from the size of molecules up to dimensions comparable with those of planets, the obscuring is done principally by *fine dust*. The particles not only obscure the light of the stars behind them, but also have the effect of reddening it—in other words, they transmit the red light and intercept (or scatter) the bluer light. The molecules in our atmosphere do the same thing, with the result that the setting sun is red and the sky looks blue. The actual law according to which the light is transformed by passing through the interstellar matter is determined by careful measures of the energy distribution of the light of obscured stars and comparison with the light of unobscured stars. The

Fig. 13.11. Part of the southern Milky Way. The four stars of the Southern Cross are in the center (the two upper ones are red, and therefore do not appear as bright on the photograph as they do to the eye. Below and to the left of the Southern Cross lies a large obscuring cloud, the Coal Sack (compare Figure 13.12). Notice the streaks of obscuration that pervade the photograph. At the extreme left is β Centauri; the bright object at the upper left is the globular cluster ω Centauri (see Figure 15.8). At the extreme right is the bright nebulosity that surrounds η Carinae (see Figure 13.13). (Harvard Observatory)

effect is much the same everywhere. Its size permits us to determine the dimensions of the scattering particles that are the most effective, though this also depends somewhat on what the particles are made of.

Fig. 13.12. The Coal Sack (compare Figure 13.11). Notice the sharp edges of the obscuring cloud, and the absence of faint stars within its boundaries. Near the upper edge of the photograph is the bright galactic cluster ϰ Crucis (see Figure 11.4). The bright star, lower right, is one of the stars of the Southern Cross. (Harvard Observatory)

The first studies of the nature of the scattering material took the view that it consisted of small metallic particles, but more recent work suggests that it consists mostly of nonmetallic "ices," perhaps real ice (frozen water) to a great extent.

In the past few years another striking property of the interstellar particles has been discovered; they polarize the starlight that passes them. This, one of the most remarkable facts of modern astronomy, is still not completely understood. It seems to require that the particles are of elongated (crystalline?) shape, and aligned by some external factor—perhaps magnetic fields in space. Even if the particles are nonconducting, they must have a magnetic (i.e., metallic) core if they are aligned by magnetic fields.

The recognition of the role played by the interstellar particles that dim the light of distant stars has transformed our ideas of the distances of the faint stars in the Milky Way. In some spots the stars are dimmed by three, four, or even six magnitudes, and very much reddened. The ninth magnitude eclipsing-star RY Scuti would be a bright naked-eye object of the third magnitude if it were not for the obscuring material between us and it; this star even seems to be *inside* the cloud.

Red light comes more freely through obscuring clouds than bluer light, and infrared light more freely still. Thus it is possible, by using very long wave lengths, to penetrate clouds that would be difficult to penetrate by photographic methods. The brilliant center of the galaxy, quite hidden by dark obscuration, is most easily observed with microwaves (short radio waves).

In addition to the solid particles, much of the matter between the stars consists of isolated atoms and molecules, which can be detected and identified by the absorption lines and bands that they produce. Such absorption lines were first noted in the spectra of some spectroscopic binaries (p. 299). While most of the lines in the spectra of these binaries showed the typical periodic changes of wave length (and therefore of radial velocity) with the orbital period, some lines always had the same wave length and therefore evidently came from atoms that were not moving with the stars. It was soon recognized that the atoms that produced these lines were not connected with the stars and were scattered in space between us and them. The stars that showed these lines most strongly in their spectra tended to be the stars whose light was most reddened by interstellar matter. The atoms that gave the lines were evidently part of the obscuring clouds.

The first interstellar absorption lines to be observed were the H and K lines, the lines coming from the ground state of the atom of Ca II, and the D lines from the ground state of Na I. Later studies have revealed a number of other interstellar lines that come from atoms of cosmically common substances such as iron and titanium, and from molecules such as those of the hydrocarbon (CH) and cyanogen (CN) radicals. Some interstellar bands are of unknown origin.

All the observed absorption lines accounted for by interstellar atoms are those of very low excitation potential, and the molecular bands are found to have degenerated into the single line that comes from the lowest electronic configuration of the molecule.

Probably the gas in interstellar space is of nearly the same composition as that in the atmospheres of stars; but it produces a very different absorption spectrum. The reason lies in the great difference in physical conditions. The energy-distribution of the radiation that is traversing the interstellar medium may correspond to a temperature of $10,000°$, comparable to that of the envelope of an A star. But this radiation is greatly attenuated: it is about one ten-thousand million millionth (10^{-16}) as intense as in the stellar envelope, and the material is about 10^{-16} as dense. The consequence is that although the interstellar atoms tend to be rather highly ionized, they are so far apart that the extremely dilute radiation does not produce appreciable *excitation,* and only absorption lines associated with their

Fig. 13.13. The bright nebulosity around η Carinae. Notice the small dark areas with sharply defined edges. (Harvard Observatory)

ground states can be observed. For this reason the Balmer lines of hydrogen (commonest of all interstellar atoms, whose presence is attested by its occurrence in molecular form as CH) do not appear as interstellar absorption lines: their excitation potentials are too high. For the same reason the common atoms, carbon, nitrogen, and oxygen do not produce interstellar lines, though the presence of both carbon and nitrogen are revealed by the cyanogen and hydrocarbon bands.

A careful study of the interstellar lines has revealed the important fact that many of them are complex; they consist of several lines, all from atoms of one substance, with slightly different radial velocities. These complex lines give evidence that the light from the distant star has passed through *several different interstellar clouds of atoms* which have slightly different velocities in our line of sight. Here we find an indication that the interstellar matter is not uniformly spread, but tends to "bunch." The single clouds may be the beginnings of condensations that will later aggregate into stars. Their relative speeds are from 10 to 20 kilometers per second.

Interstellar matter does not appear only in the form of absorbing clouds. It is also seen in the form of *bright nebulae,* such as the nebulosity round the Pleiades and the "Great Nebula in Orion."

The Pleiades nebulosity is an example of a "reflection nebula": it shows the same spectrum as the stars that are imbedded in it and is evidently scattering their light (Figure 13.14). All the reflection nebulae lie near bright stars (almost always fairly hot stars) and their brightness falls off with distance from the stars that illuminate them, as would be expected.

Fig. 13.14. The bright nebulosity that surrounds the Pleiades. Notice the concentrations near the brighter stars, and the narrow, nearly parallel streaks that cross the picture. (Lick Observatory)

The bright nebulae near the very hottest stars, however, do not shine simply by reflected light. When their spectra are examined, they are found to consist of a series of *bright lines* (often accompanied by some bright continuum), even though the nearby stars show absorption spectra. The Orion Nebula (Figures 13.15, 13.16) is of this type; the nearby stars (the multiple system of the "Trapezium") are O and early B stars of very high temperature.

Many of the bright lines in the spectra of these diffuse nebulae are easily identified; they are lines of hydrogen, helium, and oxygen for

Fig. 13.15. The Great Nebula in Orion, photographed in blue light. The nebula is faintly visible to the naked eye as the hazy central star of the giant's sword. Notice the bright streaks, which suggest wind-swept clouds, and the numerous dark markings. This nebula is a bright concentration in a glowing cloud that envelops the whole constellation. (Mount Wilson Observatory)

instance. But the lines that are usually the most conspicuous in the spectra of gaseous nebulae are not found as absorption lines in the spectrum of any star, nor (with few exceptions) are they known in the laboratory. They presented a puzzle for many years, and were long supposed to come from a substance otherwise unknown, which was given the name "nebulium."

Other such substances have occurred in the history of astronomy, as we have seen. The spectrum of the chromosphere, observed at the solar eclipse of 1868, was observed to contain a strong bright line in the yellow region, which was at first thought to come from a substance peculiar to the sun and hence called "helium." "Nebulium" suffered a similar fate in

1927, when it was shown by Bowen to consist principally of oxygen and nitrogen (Figure 13.17). The unknown lines in the spectrum of the solar corona were long ascribed to "coronium," but were shown by Edlén to come from the atoms of several of the common metals in a very high state of ionization. To understand the way in which the "nebular lines" are formed, it is necessary to turn again to our pictures of the origin of atomic spectra.

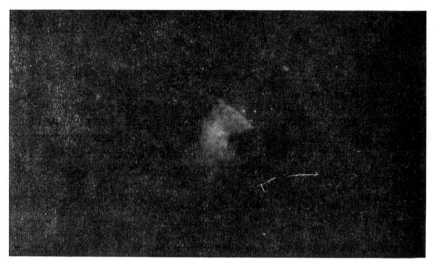

Fig. 13.16. The central parts of the Orion Nebula, photographed in red light. The picture is on nearly the same scale as Figure 13.15, and the two may be matched by referring to the little group of bright stars just above the "fish-mouth."

The photograph shows the source of illumination of the nebula, the group of four high-temperature stars in the center, known as the "Trapezium" (actually six or more stars, for at least two of the four are spectroscopic binaries). The light by which the nebula is photographed is principally that of the red line of hydrogen; notice that the distribution of the nebulosity in light of this color is not the same as in Figure 3.15, which integrates the photographic spectrum from about 3900 to 4900 Angstroms. (Boyden Station, Harvard Observatory)

In the description of the emission of a line spectrum by an atom, it was stated that light is emitted when an electron slips from one energy state to another of lower energy; the properties (i.e., energies) of the states are predetermined by the quantum conditions. The quantum conditions (which may be expressed in terms of very beautiful and intricate arithmetical rules, depending on the properties of the electron configurations) go further; they also enable us to tell which transitions are more likely, which less likely, for a given configuration of electrons.

The more likely a transition is, the stronger will be the corresponding line in the spectrum of a group of atoms of the kind considered. From some states of an atom the transition (with the emission of radiation) is relatively unlikely; in other words, when the atom gets into such a state it will linger there for a relatively long time before emitting energy. Such states are known as *metastable states,* and the spectrum lines that are

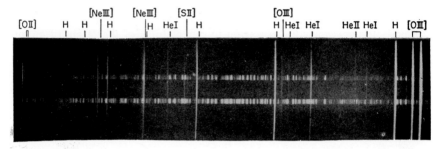

Fig. 13.17. Spectrum of the Orion Nebula. The sources of the more prominent lines are indicated. "Forbidden" lines are denoted by square brackets. (Mount Wilson Observatory)

emitted by transitions from such states are known as *forbidden lines.* The name is unfortunate, for the transitions are not forbidden, only relatively improbable; moreover, the degree of improbability differs for different states.

The five lowest energy levels in the spectrum of twice-ionized oxygen (O III) are shown in Figure 13.18. Their positions have been found from an analysis of the whole spectrum of the atom, which involves transitions

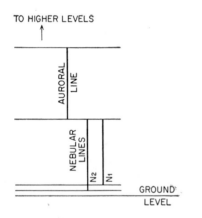

Fig. 13.18. Lowest energy levels of the atom of twice ionized oxygen (O III), to show the origin of the "nebulium" lines—the so-called "forbidden" lines of the atom.

from each of these levels to a number of levels of higher energy (not shown in the diagram). The two uppermost of these five levels are metastable. Downward transitions, with the emission of energy, have not been observed in the laboratory; compared to the transitions between higher levels they are very improbable and therefore relatively infrequent, but they do nevertheless occur in the appropriate small numbers, too small to produce an observable laboratory line.

Bowen identified the lines of "nebulium" when he noticed that the wave lengths of the "nebulium" lines were those that would be given by downward transitions from the two metastable states shown in the diagram for O III, and for several similar transitions for atoms of O II, N II, and others. In a gaseous nebula, such as the Great Nebula in Orion, an enormous volume of O III atoms is scanned, and, even though the "forbidden transitions" are relatively rare, enough atoms are present to make the lines visible, and even intense; whereas the same transitions are not seen in the small sample of matter that can be collected in the laboratory. Actually it is possible to observe some forbidden lines of fairly high transition probability even in the laboratory. Here we see, as in the solar atmosphere, the advantage that astrophysics possesses in the enormous samples of matter that it makes available for analysis.

In the spectrum of a bright-line nebula, the forbidden lines are *relatively* much stronger than the permitted lines of the same atom. Actually the permitted lines are weakened by the physical conditions that prevail in the nebula. Just as the absorption spectra of interstellar atoms are confined to the very lowest states of the atoms because of the diluteness of the radiation and the tenuity of the gas, the observable permitted lines (which arise from states of higher excitation than do the forbidden lines for O II, O III, N II, and other atoms in the nebulae) are relatively weak because of inadequate excitation. The chief source of the excitation of the atoms to the metastable states that give rise to the forbidden lines is probably collision with electrons, and the distribution of energy among the metastable states (as shown by the relative intensities of the forbidden lines that arise from different metastable states of the same atom) gives a measure of the numbers and speed (i.e., temperature) of these electrons.

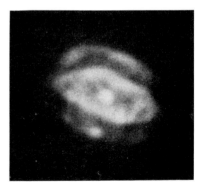

Fig. 13.19. The planetary nebula NGC 7009. The central star, of very high temperature, excites the surrounding nebulosity. Many other planetary nebulae show similar intricate structure. (Mount Wilson Observatory)

Besides the forbidden lines, certain permitted lines (of O III, for instance) appear in the spectra of gaseous nebulae, though the rest of the corresponding spectrum does not. Probably the atoms are excited to the levels from which these lines arise by the absorption of certain other lines in the ultraviolet, such as the Lyman lines of hydrogen, which happen to be exactly attuned to the corresponding frequency.

The Balmer lines of hydrogen are seen in the spectra of all emission nebulae. They are neither forbidden lines nor of low excitation potential, but hydrogen atoms are by far the commonest in interstellar space. Some of them absorb radiation of high frequency and short wave length, are excited or even ionized, and, in reradiating the energy so absorbed, some, at least, of them emit the Balmer lines. Here again the visibility of the lines depends upon the enormous number of atoms that populate a nebula, even at the fantastically low density.

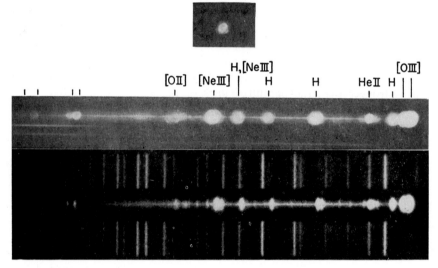

Fig. 13.20. Three photographs of the planetary nebula NGC 7662. The upper picture is a direct photograph. The second row shows a "slitless spectrogram," which forms an image of the nebula in the light of each color. Prominent lines are marked (compare Figure 3.17). Notice that light from different atoms shows different patterns and sizes, an indication that the degree of ionization and excitation varies through the envelope of gas. In general, the lines of highest excitation (such as He II) give the smallest images, those of lowest excitation (H, O II), the largest. The lowest row shows a spectrum made with a slit spectrograph. The central broadening of the bright lines shows that the envelope is expanding. (Lick Observatory)

The spectra of diffuse nebulae show the forbidden lines of the atoms of N II, O II, O III, Ne III, Ne IV, Ne V, S II, and probably argon. This was the first time that the "rare gas" neon was found in astrophysical spectra, but it has since been found in ordinary form in the spectra of stars and is now known to be quite a common substance cosmically.

The forbidden lines of O I are found in the earth's atmosphere in the spectrum of the Aurora. Forbidden lines of the metals are not known in nebular spectra proper; but a great many forbidden lines of various stages of the atom of iron are found in the quasi-nebular atmospheres of novae and some stars, as will be described later.

There are several types of bright nebulae. The Orion nebula consists of a diffuse and ragged cloud near the group of bright, hot stars. The *planetary nebulae,* which consists of symmetrical shells round a bright central star, have essentially similar spectra. There are other bright nebulae which seem to derive their energy not from stars, but from the collision of two interstellar clouds. All these nebulae appear to have similar atomic composition, very like that of stars.

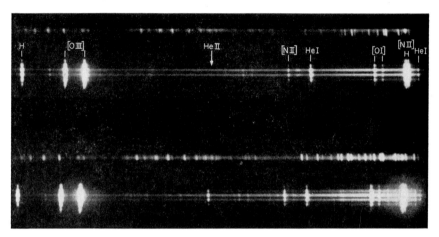

Fig. 13.21. Slit spectra of the two planetary nebulae NGC 6572 and NGC 7027. The upper strip in each case is the comparison spectrum. Prominent lines are marked. Compare Figures 13.17, 13.20. (Lick Observatory, Bowen and Wyse)

There are other diffuse nebulae, even more tenuous than those just described, which pervade the whole Milky Way region and are discovered by the fact that they radiate the hydrogen emission spectrum and the strongest nebular lines. They are essentially similar to the less diffuse nebulae and probably extend, as a thin atomic cloud, throughout the whole central plane of the Milky Way.

5. NUCLEAR REACTIONS AND THE SOURCE OF STELLAR ENERGY

Things which are seen were not made of things which do appear.
Hebrews, XI:3

In ultimate analysis everything is incomprehensible, and the whole object of science is simply to reduce the fundamental incomprehensibilities to the smallest possible number.
T. H. HUXLEY, *Darwiniana* (1893)

The only thing that we can say about the properties of the ultimate particles is that we know nothing whatever about them.
ROBERT OPPENHEIMER

In the last few sections we have been concerned with the spectroscopic behavior of atoms and, to some extent, with their chemical dissociation. These properties depend on the number, arrangement, and energy of the external electrons. None of the chemical or spectroscopic changes that we have considered has any effect upon the nuclei of the atoms. The present section will deal with the reactions that affect the nucleus. These reactions involve greater amounts of energy than have been encountered in our dealings with chemical dissociation or the excitation of spectra.

Nuclear Structure. The atomic nucleus may be regarded as made up of neutrons and protons: the nuclear mass is practically equal to the sum of the masses of neutrons and protons. The number of protons determines the number of external electrons and, therefore, the atomic species (defined by chemical and spectroscopic properties). But it is possible for several nuclei of different mass to contain the same number of protons, and therefore to belong to the same atomic species. Such nuclei are known as *isotopes* [Greek: *isos,* the same, and *topos,* place] because they occupy the *same place* in the periodic table, even though they differ in atomic mass. Notice that isotopes contain the same number of protons, but different numbers of neutrons.

Some Well-known Nuclei. This principle is ilustrated by the known isotopes of the three lightest atoms. The atomic number (number in the periodic table) is equal to the number of protons; the mass number ("atomic weight") to the protons plus neutrons. A shorthand way of condensing this information about a nucleus is to write the atomic number below the chemical symbol to the left, and the mass number above and to the right.

Thus there are three kinds of hydrogen, two kinds of helium, and three kinds of lithium known. All the hydrogens have one orbital electron, all the heliums, two, all the lithiums, three; therefore, they have similar external properties (chemical and spectroscopic) even though their nuclear masses differ. Note that one of the hydrogens (tritium) has actually the same mass number as one of the heliums; they are therefore known as *isobars* [Greek: *isos,* the same; *baros,* weight].

Symbol	Protons (atomic number)	Neutrons	Mass Number	Percentage of Nuclei *	Common Name
$_0n^1$	0	1	1.00893	100.	neutron
$_1H^1$	1	0	1.008130	99.98	hydrogen
$_1H^2$	1	1	2.014722	0.02	deuterium
$_1H^3$	1	2	3.01705	<0.01	tritium
$_2He^3$	2	1	3.01711	10^{-7}	helium
$_2He^4$	2	2	4.00386	100.	helium
$_3Li^6$	3	3	6.01684	7.9	lithium
$_3Li^7$	3	4	7.01818	92.1	lithium
$_3Li^8$	3	5	8.82510	0.1	lithium

* of corresponding atomic number

Evidently there can be several kinds of atom, with different nuclear masses but the same external properties. There are six isotopes of chlorine:

Nuclear Mass:	33	34	35	36	37	38
Percentage:	very low	very low	75.3	very low	24.7	very low

There are thirteen kinds of silver nucleus, eight kinds of gold nucleus (only one of which occurs in nature), etc. The discovery of isotopes was primarily the work of F. W. Aston, who devised the mass spectrograph for their detection.

6. TRANSMUTATION OF ATOMIC NUCLEI

> All preconceived notions he sets at defiance
> By means of some neat and ingenious appliance
> By which he discovers a new law of science
> Which no one had ever suspected before.
> All the chemists went off into fits,
> Some of them thought they were losing their wits,
>> When quite without warning
>> (Their theories scorning)
>> The atom one morning
>> He broke into bits.
>> A.A. ROBB (on J.J. Thomson)

> What's in an atom, the innermost substratum
> That's the problem he is working at today.
> He lately did discover how to shoot them down like plover,
> And the poor little things can't get away.
> He uses as munitions on his hunting expeditions
> Alpha particles that out of Radium spring.
> It's really most surprising, and it needed some devising
> How to shoot down an atom on the wing.
>> A.A. ROBB (on Rutherford)

As all nuclei are to be imagined to consist of neutrons and protons, it might be thought possible to induce nuclear reactions that would actually transform one atomic species into another by altering the makeup of the nucleus. This was the unconscious goal of the medieval alchemists, whose efforts were directed to transforming "baser" metals into gold. Their methods, however, were hopelessly inadequate for the purpose of nuclear transmutation. Modern techniques have succeeded in producing nuclear transmutations by the bombardment of nuclei by particles of suitable properties. The energies involved are expressed in *electron volts* (1 ev $= 1.6 \times 10^{-12}$ ergs). However, the quantities are so large that they are usually expressed in millions of electron volts (1 Mev $= 10^6$ ev $= 1.6 \times 10^{-6}$ ergs).

The subject of nuclear reactions is one of the most active branches of modern physics. We cannot touch upon its theory and shall mention only a few results that bear on astrophysics.

The first reaction to be illustrated is produced by firing an alpha particle (helium nucleus) into an atom of nitrogen. The first product is an isotope of fluorine, $_9F^{18}$. This isotope is inherently unstable; ordinary fluorine $_9F^{19}$

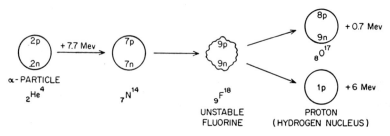

Fig. 13.22. A diagrammatic representation of the nuclear reaction produced when a helium nucleus (α-particle) is fired with energy 7.7 Mev into a nitrogen nucleus (atomic number 7, atomic mass 14). In this and the following diagrams, the nucleus is represented by a circle within which are written the numbers of protons, p, and neutrons, n, that it contains. Arrows show the course of the reaction. The unstable fluorine nucleus is shown in this picture with a wavy outline. It disintegrates into an isotope of oxygen and a proton.

is stable. The unstable isotope of fluorine breaks down spontaneously into an oxygen isotope and a proton. The energies of the original alpha particle and the final products, the oxygen nucleus and the proton, should be noted. The original alpha particle had 7.7 Mev; the total energy of the two final particles is $(6.0 + 0.7)$ Mev $= 6.7$ Mev. Thus the energy has diminished by $(7.7 - 6.7)$ Mev $= 1$ Mev. *This energy has been turned into mass.* The masses of the reacting nuclei, in units of the mass of the proton, are as follows:

$_7N^{14}$	14.0036	$_8O^{17}$	17.0001
$_2He^4$	4.0028	$_1H^1$	1.0076
	18.0064		18.0077

The increase of mass during the reaction is $(18.007 - 18.0064) = +0.0013$.

The transformation of energy into mass is governed by the Einstein equation: $mc^2 = E$, where c is the velocity of light $(3 \times 10^{10}$ cm/sec), m is expressed in the unit of mass number $(1.66 \times 10^{-24}$ grams); E for one mass unit is 1.43×10^{-3} ergs, or 931 Mev.

The transformation of Figure 13.22 *absorbs* energy, which is converted into mass: it is *endothermic*. Figure 13.23 represents an *exothermic* reaction: *mass has been turned into energy.*

Spontaneous Transformations: Radioactivity. Some nuclei are inherently unstable, and disintegrate to produce other nuclei. Most of the very heavy nuclei are unstable, as are many lighter ones, such as $_9F^{18}$. Radium, the first to be thoroughly studied, gave the name *radioactivity* to this type of nuclear reaction. Many unstable nuclei not found in nature are produced artificially in the laboratory; over seven hundred types of nucleus are thus known at the present time, and most of them are unstable.

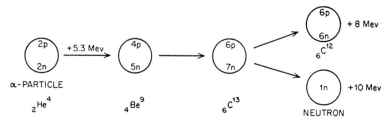

α-PARTICLE

$_2He^4$ $_4Be^9$ $_6C^{13}$

Fig. 13.23. An exothermic nuclear reaction, in which the final products have more energy than the initial reagents. An α-particle with energy 5.3 Mev strikes a beryllium nucleus and forms an unstable carbon nucleus, which emits a neutron and becomes a stable carbon nucleus.

A given radioactive nucleus has a definite probability of disintegration, which governs the rate at which such nuclei are transformed into others. A convenient measure of this probability is the *half life,* the time in which half the nuclei will have disintegrated. The nucleus $_{92}U^{238}$ transforms (Figure 13.24), by the well-known succession of disintegrations of the uranium-radium series, into an isotope of lead, $_{82}Pb^{206}$. The half lives of successive stages differ enormously: that of $_{92}U^{238}$ is 4.56×10^9 years; for radon $_{86}Rn^{222}$ it is 3.85 days, for $_{84}Po^{214}$, 1.44×10^{-4} seconds. If a succession of reactions that involves spontaneous disintegrations is of astrophysical importance, the corresponding half lives determine its progress; such a group of reactions will presently be mentioned.

Nuclear Fission. Sometimes a nucleus may break down into several of comparable size; and each of the products may undergo a succession of breakdowns. Figure 13.25 shows the chain of events that is produced when a suitable neutron impinges on a uranium nucleus. It has been suggested that the actual relative frequencies of the various atomic species in the cosmos may have resulted from the fisson of primitive nuclei of enormous mass; however, other ideas are in greater favor at the present time.

The process of *nuclear synthesis* (Figure 13.26) may be produced by neutron bombardment, with the production of a nucleus heavier than the original one. Such a reaction is involved in the production of plutonium.

Nuclear Bombs. Nuclear fission, as applied to the plutonium bomb, is illustrated in Figure 13.27. The reaction is self-sustaining because it liberates neutrons, which continue to impinge on plutonium nuclei.

A hydrogen bomb would depend on a simpler reaction, illustrated in Figure 13.28. The energy released by this exothermic reaction is measured by the change of the masses of the reacting nuclei:

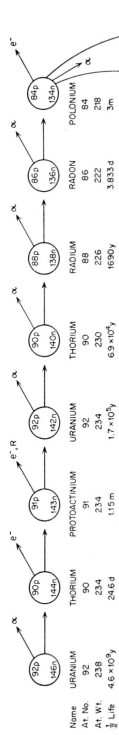

Fig. 13.24. The successive changes by which the uranium nucleus of atomic weight 238 disintegrates into a stable isotope of lead. The ejection of an α-particle (helium nucleus), a negative electron, e^-, and radiation, R, are indicated by arrows. Within each circle are shown the numbers of protons and neutrons in the nucleus. Atomic number, atomic weight (in units of the mass of the proton) and half-life are shown beneath each nucleus. Three of the nuclei ($_{84}Po^{218}$, $_{83}Bi^{214}$, and $_{83}Bi^{210}$) disintegrate in two different ways in fixed proportions with the same half-life.

Notice that the loss of an α-particle diminishes the atomic weight by 4; the atomic number also changes, and the nuclear constituents are redistributed among neutrons and protons. The loss of an electron increases the atomic number by 1, and the atomic weight is unaltered. Notice the great variety of half-lives, from over a thousand million years to about a ten-thousandth of a second for different nuclei.

By similar series of disintegrations, thorium $_{90}Th^{232}$ transforms into stable lead, $_{82}Pb^{208}$; plutonium $_{94}Pu^{241}$ into stable bismuth, $_{83}Bi^{209}$; and uranium $_{92}U^{235}$ into stable lead, $_{82}Pb^{207}$. Notice that the three stable isotopes of lead that terminate the two uranium series and the thorium series have *different atomic weights* (206, 207, 208) but, of course, the same atomic number, 82. Series of disintegrations of this kind permit

344

$_1H^2$	2.014722	$_1H^3$	3.01705
$_1H^2$	2.014722	$_1H^1$	1.008130
	4.029444		4.025180

The energy released is the equivalent of the loss of mass, (4.029444 — 4.025180) = 0.004264 mass units, equivalent to 3.5 Mev per nucleus. An extremely high temperature, difficult to produce on earth, would be required to initiate the reaction just discussed.

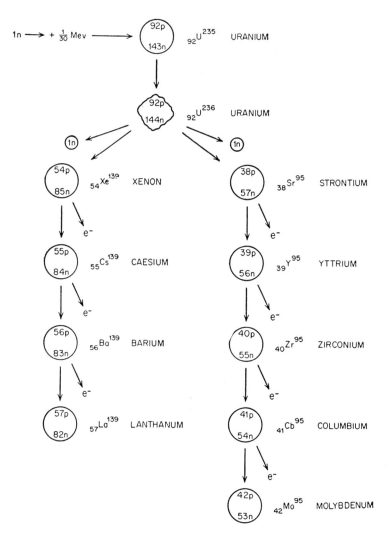

Fig. 13.25. Series of nuclear reactions that follow the fission of a uranium nucleus after collision with a neutron. The final products, lanthanum and molybdenum, are stable.

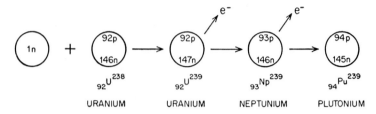

Fig. 13.26. An example of nuclear synthesis; production of neptunium and plutonium from uranium.

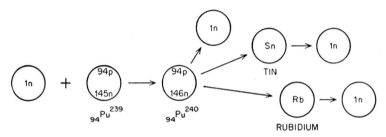

Fig. 13.27. The basic reaction of the uranium bomb. Neutrons are emitted and continue to impinge on other uranium nuclei if there are enough near at hand.

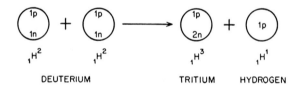

Fig. 13.28. The basic reaction of the "hydrogen bomb."

7.   THE SOURCE OF STELLAR ENERGY

The hopeless inadequacy of any other known source of energy to produce the radiation of the stars and sustain it over long enough periods made it certain decades ago that nuclear reactions must be involved. The problem was to find reactions that were known to be possible, that involved nuclei known to exist in adequate quantities, and that would proceed at the temperatures of stellar interiors with the production of energy.

Several possible reactions proceed at comparatively low temperatures, astrophysically speaking. Figure 13.29 illustrates five such reactions, which involve the lighter elements of the periodic table. All proceed with loss of mass and, therefore, with evolution of energy. All these processes use up the reaction products completely. All the nuclear species involved, except the proton, are uncommon in the universe (perhaps just because they have been used up in this way). This makes it unlikely that such reactions are the main source of stellar energy. They proceed at low temperatures— lower than we ascribe to the centers of main-sequence stars. Even though

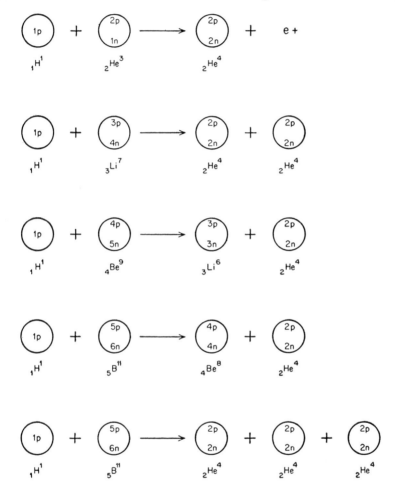

Fig. 13.29. Possible nuclear reactions in stellar interiors.

some giant stars may perhaps have such temperatures at their centers, it is now thought more probable that they draw on sources of energy similar to those of the main-sequence stars, and that their central temperatures are actually higher than might be inferred from their dimensions (p. 355).

A process that probably makes the major contribution to the energy production of most stars is one that regenerates its products. Hydrogen is consumed; helium is ultimately produced. The reactions follow a definite order, as shown in Figure 13.30.

In this process, the so-called "carbon cycle," the nucleus $_6C^{12}$ is regenerated (acts as a "catalyst"). Two spontaneous disintegrations are involved: those of $_7N^{13}$ and $_8O^{15}$; and the successful operation of the process in the stars

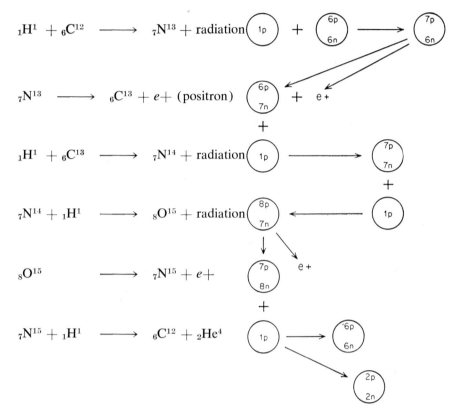

Fig. 13.30. The carbon cycle—probably important in stellar interiors.

depends on their half lives, which appear to be of the right order. The length of time during which the process can continue is limited only by the amount of hydrogen present. Since the universe, including most of the stars, consists largely of hydrogen, one of the criteria for a suitable process is therefore satisfied. The required temperature is of the order of 15 million degrees, probably found within main-sequence stars (p. 353) and perhaps also within giants, thus satisfying the second criterion. The third criterion, possibility of the reaction, is also satisfied, for it has been verified in the laboratory.

Only a small loss of mass is involved in the operation of the carbon cycle: the difference in mass between four hydrogen nuclei and one helium nucleus $(4.02352 - 4.00386) = 0.01966$. Thus, even if a star were to convert all its hydrogen into helium, it could lose less than 1% of its mass in this way. As most of the stars still consist principally of hydrogen, the process has not gone far for them, and the stellar universe as a whole is "young." The white dwarfs, however, must have consumed almost if not quite all their available hydrogen.

Another nuclear process, the "proton-proton reaction," is of potential importance in the interiors of stars. It involves the direct building of helium nuclei from those of hydrogen. Two possible courses for the reaction are as shown in Figure 13.31 ($_{-1}e^0$ is an electron; $_1e^0$ a positron).

Which, if any, of the nuclear reactions just described will operate within a star depends upon the internal temperature of the star. We therefore proceed to consider whether it is possible to reach any conclusions about the conditions in stellar interiors.

8.   THE INTERNAL STRUCTURE OF STARS

Our knowledge of the stars, deduced from their brightness, distance, color, and motions, may be summarized by the statement that we know, for many stars, the size, the total brightness (absolute bolometric magnitude), and the mass. The other known properties (density, surface temperature, etc.) may be expressed by combinations of these three fundamental properties. In what follows we shall refer to them by the symbols R (radius), L (luminosity), and M (mass).

In order to deduce the internal state of the stars from the superficial properties, six fundamental assumptions will provide a starting point.

(1) Stars are in equilibrium; they are held together by gravitation and distended by gas pressure and perhaps also to some extent by the radiation that pours out from their interiors.

(2) For a first approximation, it is assumed that gas pressure bears most of the burden of distention and the pressure of radiation is negligible. This turns out to be probably true for all but the most massive stars.

(3) The star behaves throughout as a perfect gas, so that at any point within it

$$p = CT\varrho/\mu,$$

where p, T, and ϱ are the local values of pressure, temperature, and density, μ is the mean molecular weight of the material, and C, a constant. In other words, the local temperature is proportional to the product of mass and mean molecular weight divided by the radius:

$$T \sim \mu M/R.$$

(4) The opacity (which represents the resistance of the star's substance to the passage of radiation) is primarily *photoelectric* and depends on the interplay of radiation with electrons, either attached to atoms or (temporarily) detached from them.

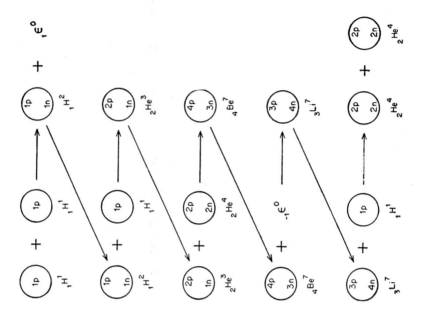

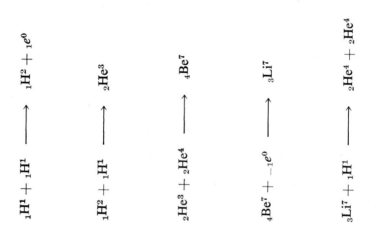

$$_1H^1 + {}_1H^1 \longrightarrow {}_1H^2 + {}_1e^0$$

$$_1H^2 + {}_1H^1 \longrightarrow {}_2He^3$$

$$_2He^3 + {}_2He^4 \longrightarrow {}_4Be^7$$

$$_4Be^7 + {}_{-1}e^0 \longrightarrow {}_3Li^7$$

$$_3Li^7 + {}_1H^1 \longrightarrow {}_2He^4 + {}_2He^4$$

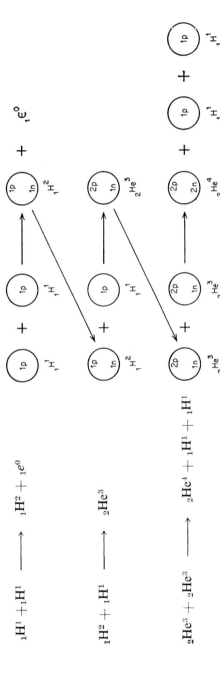

Fig. 13.31. Two possible courses for the proton-proton reaction in stellar interiors.

$_1H^1 + _1H^1 \longrightarrow _1H^2 + _1e^0$

$_1H^2 + _1H^1 \longrightarrow _2He^3$

$_2He^3 + _2He^3 \longrightarrow _2He^4 + _1H^1 + _1H^1$

351

(5) The energy is carried outwards in the star by radiation only.

(6) The star derives its energy from nuclear reactions, such as those described in the preceding section. No other adequate source of energy can be envisaged.

Structure of the Sun. The sun, a main-sequence star of medium size, mass, and luminosity, provides a starting point. From the weight of a column of the solar material, between the edge and the center, the central pressure must be about 10^9 atmospheres, or about a million tons per square centimeter. Assumption (3) enables us to calculate that at this pressure, the temperature is at least 10 million degrees. At such a temperature the atoms must be almost completely ionized—nearly all the electrons must be stripped from the atoms, and the sun's interior must consist of a gas made up of nuclei and free electrons in a state of great agitation. The fact that the atoms are stripped of their electrons makes them effectively smaller, and therefore the gas retains its compressibility and still obeys the gas laws, despite the high pressure, since the particles will pack more tightly.

For the purpose of calculating the "mean molecular weight," μ, at the sun's center, each electron counts as a "molecule"; there are almost as many free electrons as nuclear protons (since almost all are detached from their nuclei). If the star were all hydrogen, the consequent mean molecular weight would be ½ if all atoms were stripped. If it were all helium (2 electrons, atomic weight 4), the mean molecular weight would be 4/3, or 1.33, if all the atoms were stripped. Even a star of stripped uranium atoms (92 electrons, atomic weight 234) would have a mean molecular weight 235/92—less than 3. A star that consists mostly of hydrogen will have a mean molecular weight at the center of less than 1, and the value is scarcely likely to exceed 2 for any star.

At a temperature of 10 million degrees, Wein's Law (p. 64) shows that most of the energy will be in the form of soft X rays. From Planck's Law we recall that *at every wave length* the energy for high temperature is greater than at lower temperatures; hence even the radiation in the visual range is more intense than at the surface, per unit area, but it is much less in proportion to the short wave lengths. From Stefan's Law, which states that the energy per unit area varies as the fourth power of the temperature, the radiation at the center is 8×10^{12} as intense, per unit area, as at the outside of the sun.

We have postulated (assumption 5) that energy travels outward by radiation only. The outward flow of radiation is governed by the intensity at the center and by the *opacity* (assumption 4), which can be calculated in terms of known physical laws. The radiation forces its way out through successive layers which differ in density and temperature. The maximum of intensity gradually moves toward longer wave lengths as the radiation travels outward; by the time the radiation has reached the outside of the sun, maximum energy is at the visual wave lengths. The emerging radiation exerts an appreciable pressure (proportional to the fourth power of the temperature). Within the sun, this is but a small fraction, about 5%, of the gas pressure (assumption 2), but for some stars the pressure of radiation may bear a greater part of the burden of distention. Our assumptions must be restricted to examining sun-like stars.

The Sun-like "Family" of Stars. If our six fundamental assumptions are adopted, and the formulas expressing Nos. (3) and (4) are taken from theoretical physics, it is possible to calculate how the properties of a star *like the sun in structure* will depend upon its mass, radius, and luminosity. An additional requirement is that we shall know *how* the thermonuclear reaction, postulated in (6), depends on temperature. If the sun behaves like a perfect gas, we have already seen that its central temperature must be of the order of 10 million degrees. The exact temperature will depend on the mean molecular weight μ (assumption 3), but, since μ has been seen not to vary widely, the temperature is not likely to differ from ten million by a factor greater than two. We therefore look for nuclear reactions that will proceed at temperatures of the order of 10 million degrees.

Nuclear reactions bear a rough analogy to ordinary chemical reactions; whether or not they will take place will depend on the likelihood of encounters "close" enough to induce nuclear changes. The measure of this "closeness" is the *nuclear cross section,* which determines the tendency of the nucleus to react under given conditions. For some reactions it has been determined in the laboratory; for others it can only be surmised at present. Knowledge of nuclear cross sections makes it probable that the "carbon cycle" (Figure 13.30) will operate at the center of the sun, and conjecture indicates the possibility that the "proton-proton reaction" will also take place there. The reactions involving lithium, beryllium, and boron (Figure 13.29) proceed at much lower temperatures, and, if (as we may suppose) the center of the sun has risen through these lower temperatures long ago in its slow development, these low-temperature reactions must have run their course, using up the three atoms concerned, and incidentally accounting for the virtual absence of such elements from the sun of today.

The carbon cycle and the proton-proton reaction remain as the two likely thermonuclear sources of the sun's energy. Theory shows that they depend very differently on temperature; the former varies as about the eighteenth power of the temperature and the latter, as about the fourth power. If the temperature is 13 million degrees, the carbon cycle is probably unimportant, and the proton-proton reaction may do most of the work. At about 16 million degrees, both would contribute about equally. At 20 million degrees, the carbon cycle is the main source of energy.

The actual temperature at the center of the sun depends upon the mean molecular weight, which, in practice, depends principally on the proportion of hydrogen and helium that the sun contains. The more hydrogen, the lower will the mean molecular weight be, the lower the central temperature, and the greater the relative contribution of the proton-proton reaction.

The six assumptions concerning processes in stellar interiors may be used to compare the sun with other stars of similar structure. With these assumptions, and for the carbon cycle, it can readily be shown that approximately:

(1) Luminosity should be proportional to the fifth power of mass,

$$L \sim \mu^7 M^5 / \varepsilon^{1/40} \varkappa$$

where μ is the mean molecular weight, M the mass, ε a constant that determines the energy generation, and $\varkappa$ a constant that determines the opacity, which is also a function of local density and temperature.

(2) Luminosity should be proportional to the seventh power of radius,

$$L \sim \mu^{7/2} R^7 / \varepsilon^{1/3} \varkappa^{4/3}$$

where R is the radius.

The first relation represents the *mass-luminosity law,* which, as we have already seen, is nearly fulfilled for *all* stars except the white dwarfs (for which our fundamental assumptions are inapplicable). The second represents a *radius-luminosity relation.* Reference to Figure 12.19 shows that a radius-luminosity relation of almost exactly this slope exists *for all main-sequence stars;* another parallel radius-luminosity relation is seen for the subgiants. We may infer with some plausibility (1) that the main-sequence stars are similar to the sun and are probably fed by much the same process; (2) that the subgiants are similar among themselves, but differ in some particular from the main-sequence stars. Very likely they differ in mean molecular weight and perhaps also in opacity.

We may infer that the main-sequence stars are essentially similar to the sun in structure and energy sources. If the law of temperature dependence appropriate to the proton-proton reaction had been used instead of the carbon-cycle law, we should have obtained a similar mass-luminosity relation, but a different luminosity-radius relation. We can also calculate the relation to be expected between surface temperature (and therefore spectrum) and absolute magnitude—at least within the current uncertainties of the nuclear cross sections. This relation is embodied in the Hertzsprung-Russell diagram (Figure 11.6). Here again we see that the main sequence roughly follows the slope that would be expected with the carbon cycle. The observed range of luminosities for a given surface temperature (which opinion tends to narrow for the main sequence proper) is quite within the possible range that would be produced by the likely variations in μ and, therefore, of hydrogen content.

Just as reliance on the carbon cycle and the proton-proton reactions would require different hydrogen and helium contents for the sun, similar conclusions can be drawn for other main-sequence stars. The dependence of temperature on mass and radius (assumption 3) may be used, with plausible values of μ, to estimate the central temperatures for stars of the main sequence other than the sun. It appears that the hotter, brighter stars of the main sequence must have appreciably higher central temperatures than the faint red dwarfs at the lower end. For the O9 stars that comprise the binary Y Cygni, the central temperature may be as much as 35 million degrees, and for Y Cygni we know that the internal structure is nearly the same as that of the sun (from the apsidal motion, p. 307), so that the same formulas may safely be applied. For the coolest dwarf stars, however, the temperature may be nearer to 10 million degrees, at which temperature, as we have seen, the proton-proton reaction should preponderate. We may conclude that for the bright B stars the carbon cycle is the important one, but that the proton-proton reaction may grow in importance for cooler, less luminous stars. It may indeed be the principal source of energy of the fainter dwarfs, though precise calculations are uncertain; for example, the factors that govern the opacity are certainly not the same all the way down the main sequence. We feel at the present time that main-sequence stars are fairly similar in structure to the sun and draw their energy from similar sources, perhaps carbon cycle and proton-proton reaction in different proportions at different parts of the sequence.

9. GIANT AND SUPERGIANT STARS

Such giant and supergiant stars as are of known mass—extremely few —conform to the mass-luminosity law. But they deviate from the predicted spectrum-luminosity law (as is seen from their situation in the Hertzsprung-Russell diagram) and from the predicted radius-luminosity law (Figure 12.19). The conclusion is inescapable that *they are not built like the main-sequence stars.* A simple calculation from the formula for central temperature:

$$T_c = \mu M/R$$

leads to central temperatures too low for either proton-proton or carbon cycle reaction to operate. However, this formula leads to valid comparisons with the sun only if the stars are built on the same model.

It is generally felt that no reaction other than the two just mentioned would be adequate to supply the supergiant or even the giant stars with the vast amount of energy per unit mass that they radiate. The reactions depending on deuterium, lithium, beryllium, and boron would go on at some of these temperatures, but could maintain the stars only for a short time—so short that we should hardly expect to see any of these stars, even in the small numbers in which they occur. An alternative is sought, as D. H. Menzel has indicated, in regarding the luminous stars of low density as like main-sequence stars inside, but surrounded by vast, distended envelopes; in other words, they are much more condensed at the center than the sun and main-sequence stars.

Several "models" for giant stars have been proposed. In one, suggested by Gamow and Keller, the core of the star consists of helium that has been produced by the operations of one of the main reactions; the constituents of the star do not mix in this core, and since the hydrogen is gradually exhausted, the region in which energy is generated moves outward in a zone of appropriate temperature, within which mixing (convection) takes place. Outside this zone is the distended envelope, through which the generated energy is passed out by radiation. Here we have a model that is built up in three layers with different properties; apparently giant stars can be explained only in some such terms.

A still more plausible model has been constructed by Li Hen and Schwarzschild. Energy is produced in a convective core, with $1/200$ of the star's radius, of temperature appropriate to the carbon cycle. Outside this core is a zone, $1/66$ of the star's radius, which transmits the radiation and produces no energy, and about a third of the star's radius is occupied by the tenuous envelope, of different composition from the core. By means of such a "model" it is possible to predict a luminosity-radius relation that is more like that shown by the giant stars, but even this model does not suffice to reproduce the extremest supergiants. It does, however, show the lines on which such stars may ultimately be interpreted. No doubt the picture must involve at least three zones of different properties, perhaps even more.

It is significant that the envelopes of giant stars, and even more, those of supergiants, seem to be distended to a greater extent than would be expected from the conditions, even when the sustaining power of radiation pressure is taken into account. Some unknown factor is involved in keeping the supergiants "blown up." We may mention the rising and falling surfaces of the pulsating variables (p. 367), and the "umbrella" that seems to overlie the surfaces of the long-period variables (p. 382). Here is a fruitful and enigmatic province in the study of stellar atmospheres and stellar envelopes.

Menzel has suggested that the forces of intense electric currents and associated magnetic fields act to oppose gravitation and hold up the stars' atmospheres.

Structure of White Dwarfs. The only stars that deviate greatly, as a group, from the mass-luminosity law are the white dwarfs, which, it may be recalled, are second only in commonness to the stars of the main sequence. Hence, in dealing with their anomalies, we are not considering a small group of abnormal objects, but rather a major stellar phenomenon.

The most striking feature of the white dwarfs is their enormous density, over a hundred thousand times that of the sun. At such densities the electrons and nuclei are tightly jammed together; the state is known as that of a *degenerate gas,* whose properties are such that it is incapable of yielding any energy by any processes of the sort that we have considered for other stars. The central density may be 10 million times that of the sun and the central temperature, of the order of 10 million degrees, but despite this high temperature, nuclear energy is not liberated; the white dwarfs have exhausted all their sources of such energy. The only way in which they can produce energy is apparently by contraction, and probably their very feeble luminosity is produced in this way.

The best known of the white dwarfs is the companion of Sirius, with a mass nearly that of the sun, one-fiftieth of the solar radius, over two thousand times the solar surface gravity, and a spectral class A5, or a surface temperature about 10,000°K.

The low luminosity shows that no nuclear energy is being liberated, and at the existing central temperature, there would be a liberation of energy if the raw materials were available. The inference is that a white dwarf has consumed all its internal hydrogen—the ultimate source in both proton-proton and carbon cycle processes.

The spectra of white dwarfs show immensely broad lines of hydrogen, and this is ascribed to the possession of a superficial layer that contains hydrogen at high density (hence the Stark broadening of the Balmer lines) and overlies the main body of the star, where all the hydrogen has been consumed. The high surface gravity produces a "relativity red-shift" (proportional to surface gravity), which is $+19$ km/sec for Sirius B and confirms beyond doubt the fantastic properties of radius and density.

That the white dwarfs are extremely common stars is a cardinal fact in considering them, and the development of stars in general. They may be thought of as representing almost the end-point of a course of stellar development (perhaps not the only possible course). When they have used up all their potential energy of contraction they must disappear completely, and we have no way of knowing how many such stars exist that are totally invisible and (except by possible gravitational effects) undetectable. A scheme of stellar development that does not provide for a large number of white dwarfs is glaringly unsatisfactory. The giants and supergiants, which are far less common (though they force themselves on our attention by their brilliance), might better be disregarded.

Our survey of the problems of stellar structure has been very superficial. Its main features have been to point out the essential similarity of the stars of the main sequence; the difficulty of understanding the structure of giants, and, *a fortiori,* of supergiants; and the importance of the "exhausted" white dwarfs. The two main themes that run through the problem of stellar constitution are: hydrogen and helium content, and the nature of the nuclear energy-sources; and these themes are closely interwoven. They should provide a *motif* for the story of stellar development. We shall later converge on stellar development from another side—the consideration of families of stars, which we may suppose to be similar in origin and history.

VARIABLE STARS

All that I know of a certain star,
Is, it can throw, like an angled spar,
Now a dart of red, now a dart of blue.
ROBERT BROWNING

Some of the stars have been known for centuries to be variable in brightness. "New stars" have been recorded throughout the history of astronomy. The "new star" that roused the interest of Hipparchus, and resulted in the catalogue of stars that we know as the "Almagest," was very likely a bright comet. But the new star that aroused the interest of Tycho Brahe, and led to the series of observations that marks the beginning of modern astronomy, was a true nova, and one of exceptional brilliance.

Many other variable stars have been discovered during the past two centuries. The variable star Algol was successfully interpreted in the seventeenth century in terms of periodic eclipses, and for some time astronomers attempted to interpret all variable stars in the same way. But we know today that many variable stars are *intrinsic:* they are single stars that fluctuate in brightness. Variable stars are of many luminosities, colors, and sizes. Some show a regular rhythm in their variations; others are quite erratic. Variable stars are of enormous interest to astronomers, not only because of their individual peculiarities, but also because they can be used, as we shall see, in the study of the distances, dimensions, and structure of stellar systems. In fact our whole modern picture of the arrangement of the cosmic scene depends on our knowledge of variable stars.

1. CLASSIFICATION OF VARIABLE STARS

Over eleven thousand variable stars are now known within our own galaxy, and many more are found every year. Several thousand others have been found in the globular clusters (subsystems of our galaxy) and in other galaxies. Their properties allow us to classify them in a number of major groups with common physical characteristics.

All variable stars are comprised in two large groups: *extrinsic variables,* for which the changes of brightness are caused by external conditions such as eclipses, and *intrinsic variables,* single stars that vary in brightness because of physical changes in a single star.

The *extrinsic variables* comprise the eclipsing-stars, which we have already discussed in Chapter XII. Occasionally one or both components of an eclipsing-star are also intrinsically variable. Among the extrinsic variables we may also place the stars within dark nebulosity. This very interesting group may represent stars that are in process of formation, or stars that are undergoing alteration as a consequence of picking up interstellar material. The nebular variables may afford us a glimpse of the early stages of stellar development; they will be described at the end of the present chapter.

The *intrinsic variables* may be quite erratic (irregular variables); they may be regularly rhythmic (*periodic*) or irregularly rhythmic (*cyclic, semiregular*). The distinction between periodic, cyclic, and irregular variables is sometimes hard to draw.

INTRINSIC VARIABLES

NAME	Range of Period	Approximate Cycle	Remarks
A. *Pulsating variables*			
RR Lyrae stars:	<1 day		White stars
Cepheid			Yellow stars,
variables:	1–50 days		high luminosity
Long-period			
variables:	100–1000 days		Red giants
Semiregular			
variables:		40–150 days	Red giants
B. *Exploding variables*			
Dwarf novae:		Several weeks	
Recurrent novae:		Several decades	
Classical novae:		Millions of years?	
Supernovae:		?	
Shell stars:		Decades?	Bright, hot stars
C. *Other variables, perhaps extrinsic*			
R Coronae stars:		Irregular	Luminous yellow stars

The eclipsing-stars, the major part of the known extrinsic variables, have already been treated. Most eclipsing-stars are strictly periodic (as would be expected from the dynamical cause of their variations), but a number are variable in period. Some of the variations of period can be ascribed to apsidal motion: they are periodic in nature and affect the two minima in opposite senses. But many of them are quite erratic and have not yet been adequately explained. It is possible that they are caused by small changes in the masses of the components, which are known in many cases to be shedding material into space and perhaps even interchanging it.

The division of the intrinsic variables into the main classes tabulated above is not arbitrary; although the groups are contiguous (and even over-lap slightly) in period, they stand out clearly from one another. Period, luminosity, color, distribution, and motion differentiate them.

About seven thousand intrinsic variables are known in the galaxy. About two thousand of the rest are eclipsing-stars; the others are as yet incompletely investigated. It is one thing to discover that a star is variable, quite another to determine its period and properties. An enormous amount of work remains to be done in analyzing the known variables of unknown properties, in improving our knowledge of the known variables, and in discovering new ones. We have very likely discovered only a few per cent of the galactic variables that are accessible to us; and an enormous number of the variables in the galaxy are too distant, and therefore too faint, to be studied at all with the existing equipment.

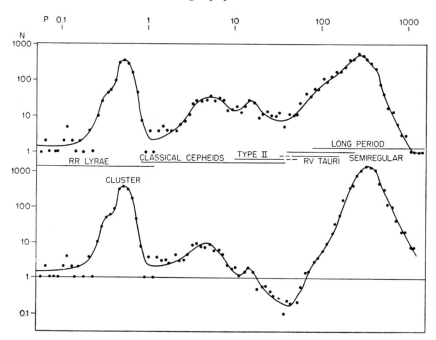

Fig. 14.1. Comparison of numbers of intrinsic variables of different periods. The upper figure shows the relative numbers of known galactic variables of different periods; the scales are logarithmic. The lower figure presents an attempt to derive the actual relative numbers of variable stars of various periods. The number of RR Lyrae stars has been kept the same as in the upper diagram. Notice that the relative numbers of very luminous variables (such as Cepheids) has been much reduced as a consequence of the allowance made for the greater volume of space throughout which they have been discovered. The less luminous long-period variables have become more conspicuous than in the upper figure.

The numbers of regular intrinsic variables of different periods differ greatly. The upper part of Figure 14.1 shows the relative numbers of known intrinsic variables plotted against period. Clearly the variables of longest period are the most numerous, and then those of shortest period; stars of intermediate period are relatively uncommon.

But the observed numbers of intrinsic variables of different kinds cannot be taken at face value. As we shall see later, they differ greatly in luminosity: the Cepheids are the most luminous; the long-period variables of longest period, the least so. Therefore the Cepheids are seen at greater distances, represent a larger volume of space, and thus are overrated in our census. Rare as they seem, they are actually even rarer. The lower part of the diagram shows a rough attempt to estimate the true frequencies. All the numbers are expressed in proportion to the number of RR Lyrae stars. Note that the periods are arranged according to their logarithms in both figures, so that equal ratios of numbers are represented by equal distances in the co-ordinates; in the lower part of the figure the numbers of stars are arranged logarithmically also. The logarithmic scale is a useful device in representing data that cover a large range. The numbers of Cepheid variables fall to a small fraction of those of the RR Lyrae stars, and the semiregular stars are even fewer. The numbers of variables of different kinds are counted for intervals of 0.05 of the logarithm of the period in days.

Accurate figures are impossible because of the incompleteness of the survey of variable stars and the fact that some kinds are much easier to discover than others; but evidently the long-period variables are about four times as common as the RR Lyrae stars and the Cepheids, less than one-tenth as common. This is true for the parts of the galaxy nearest to us, for which the survey is completest; it is not necessarily true even in other regions of our galaxy; and in some other systems of stars it is certainly not true. The relative number of variable stars of given kinds is an important index of the history of the system in which they occur.

When a star is pulsating or vibrating under gravitational control, *its period of vibration will be proportional to the reciprocal of the square root of its mean density.* In other words the product $P\sqrt{\varrho}$ is constant (where P is the period and ϱ the mean density). This relation is found to be approximately fulfilled for the intrinsic variables and lends strong support to the theory that their variations are evidence of something akin to physical pulsations. We have recently learned that this relation is nearly independent of the internal build of the star.

The variable stars, when inserted in the Hertzsprung-Russell (spectrum-luminosity) diagram, show a definite pattern, which is illustrated in Figure 14.2. Note that all the pulsating variables are of rather high luminosity; there is none, in the classes now discussed, that is fainter than absolute magnitude $+2$; *thus they avoid completely the domain where stars in our neighborhood are commonest*—the lower half of the main sequence, near where the sun is situated.

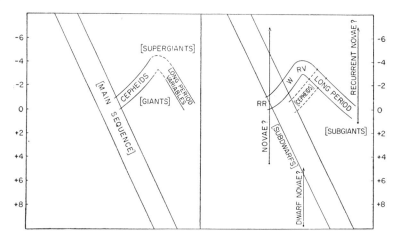

Fig. 14.2. Color (effective temperature) and absolute bolometric magnitude for the intrinsic variables. To the left are the Type I variables; to the right, the Type II variables. The domain of the Cepheids is indicated in the right-hand diagram by broken lines. The course of the main sequence is shown in both figures (compare Figure 11.5).

2. CEPHEID VARIABLES

The Cepheid variables take their name from δ Cephei, one of their number that is a naked eye star, almost the brightest of them in apparent magnitude. Most of the classes of variable stars are similarly named from one well-known and bright member.

The Cepheid variables range in period from just over a day to about forty-five days. In our galaxy the commonest period is about seven days; Cepheids of period less than three days, or more than thirty, are very rare. This is not, however, true everywhere. In the Small Magellanic Cloud (p. 422) the commonest period is rather more than two days—one of several reasons for thinking that the history of the Small Magellanic Cloud differs from that of our galaxy.

The variation of brightness of a star is illustrated by its "light-curve," a picture in which brightness is plotted against time. We cannot usually observe a variable star continuously through a whole cycle of variation because of the alternations of night and day; it would only be possible for a star of very short period, which went through its whole cycle during one night, or by a system of international co-operation, which would keep the star continuously in view as the earth turned. Therefore the light-curve of a variable star has to be pieced together from information obtained on a number of different nights, a process often both difficult and laborious.

The light-curves of Cepheid variables are not all similar. Some show an almost sinuous, smooth variation, but most of them brighten more rapidly than they grow faint, and a star that brightens more slowly than it grows faint is almost unknown. The light-curves tend to be characteristic of the periods, as was first pointed out by Hertzsprung (the so-called "Hertzsprung relation," probably very significant of the condition of the star, but not as yet understood). A few typical light-curves for various periods will illustrate this important effect.

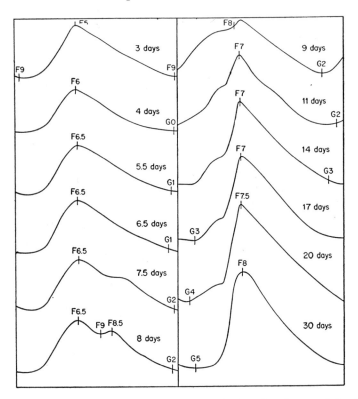

Fig. 14.3. Typical light-curves for Cepheid variables of various periods. Spectra at maximum and minimum brightness are indicated. Notice that the spectral class advances somewhat with period—more at minimum than at maximum. The spectral range is thus larger for long periods than for shorter ones. Notice that the range of brightness, also, is greatest for the longest periods. The light-curves and the spectra are means for about 150 bright Cepheids.

There is an even more important relation between the period of a Cepheid variable and its luminosity. Most of the galactic Cepheids are too far away for their parallaxes (and therefore their absolute magnitudes) to be accurately measured, and a relation must be looked for in a distant stellar system, where all the stars are approximately at the same distance

of Cepheids in our own system and also to measure the distance of another stellar system if Cepheid variables are observed in it.

The period-luminosity curve is so basic to the measurement of cosmic distances that we must remember its limitations. The apparent magnitudes that fix its slope are very difficult to measure, and our values may still be somewhat in error, especially for fainter stars. The values now accepted for the absolute magnitudes may still be in error by some tenths of a magnitude.

A Cepheid variable changes in color (temperature), spectrum, and radial speed with the same rhythm as its changes of brightness. The velocity curve has a superficial resemblance to that of a single-lined spectroscopic binary. Indeed for some time the Cepheids were thought to be true spectroscopic binaries in orbital motion. But the orbits thus indicated proved to be physically improbable, if not impossible, and another interpretation was sought. For several decades astronomers have held the view that the variations of velocity are evidence of a periodic rise and fall of the star's own surface, so that the star is alternately expanding and contracting. This interpretation (the "pulsation theory") is, however, a very schematic way of looking at the stars' behavior, and the theory has never succeeded in reproducing many of the observed complexities, such as the progression in form of light curve with period. Moreover, the exact timing of the radial velocities relative to the changes of brightness still presents a puzzle. In general, maximum brightness comes at the time of minimum velocity (most rapid approach); minimum brightness, at the time of maximum velocity (most rapid recession). Thus, if we picture the star as alternately swelling and shrinking, it is of about the same size when it is brightest as when it is faintest: smallest during the rise to maximum brightness, and largest on the way to minimum. Therefore it is a mistake to imagine that the changes of brightness are a consequence of changes of size. Actually the radius would change by only about 10% if the variations of velocity are seen in terms of pulsation—too little to account for the change of brightness by about a magnitude, even if the star were largest at maximum. Almost the whole of the change in brightness is caused by change of temperature of the star's surface, which is hottest at maximum and coolest at minimum.

These changes of temperature are reflected in the spectrum of a Cepheid, which is of earliest spectral class near maximum and latest near minimum. The spectra vary from Class F to Class G or K. They are not precisely like those of nonvariable stars, especially at maximum.

From the shortest periods to the longest, the Cepheids show a progression of spectrum: those of longest period are the coolest. Maximum spectrum progresses less with period than minimum spectrum. The relation between period and spectrum (the period-spectrum relation) is therefore more marked at minimum than at maximum brightness, and the spectral range is largest for the Cepheids of longest period. This difference is associated, as might be expected, with larger light ranges for long-period Cepheids. At the shortest periods the photographic range is about eight-tenths of a magnitude, and for longer periods, may be as much as two magnitudes.

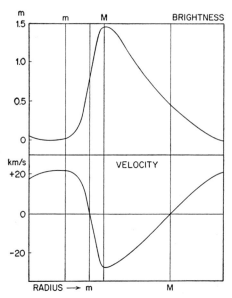

Fig. 14.5. Light-curve and velocity-curve for a typical Cepheid of period about twenty days. Notice that maximum brightness nearly coincides with greatest velocity of approach; minimum brightness, with greatest velocity of recession. The point where the velocity passes through zero, from recession to approach, is the place where the star is smallest; where the velocity passes through zero, from approach to recession, the star is largest. Notice that maximum and minimum size do not coincide in time with maximum and minimum brightness. The star is nearly the same size when it is brightest as when it is faintest.

The motions of the Cepheids are difficult to study, for most of them are too far away for accurate measures of proper motion. The determination of their radial velocities, however, is unaffected by distance. When the radial velocities of Cepheids are analyzed, they show that the Cepheids, as a whole, are moving around the galactic center in nearly circular orbits; thus they share the galactic rotation that is shown by most of the stars in our vicinity.

The Cepheids are all closely confined to the galactic plane. Few, if any, true Cepheids are more than five hundred parsecs from it. This high concentration to the galactic plane is a feature that is shown by all stars that share the motion of galactic rotation (such as the B and O stars, and the galactic clusters). The Cepheids are thus all within the layer of dust and

gas that lies in the central plane of the galaxy. Modern ideas tend to interpret the motions and distribution of the B and O stars, the galactic clusters, and the Cepheids, in the sense that they are young stars, fairly

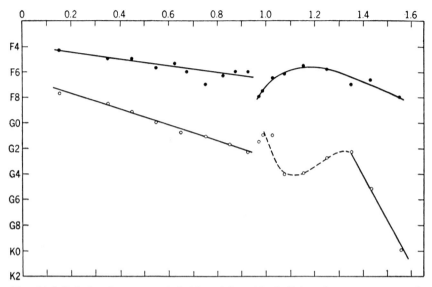

Fig. 14.6. Relation between period (plotted logarithmically) and spectrum at maximum and minimum for Cepheid variables. Compare Figure 14.3.

recently formed from the galactic dust and gas, and still moving in it. There are a few abnormal Cepheids that partake neither of the motion nor of the distribution of the ordinary Cepheids, and as will be seen, their other physical properties differ also. They have evidently had a different history.

Magnitudes, periods, and spectra are given below for several well-known bright Cepheids.

SOME WELL-KNOWN CEPHEID VARIABLES

Name	Magnitude	Period (days)	Spectrum
SU Cassiopeiae *	6.05–6.43	1.95	F2–F9
TU Cassiopeiae	7.90–9.00	2.14	F5–G2
Polaris	2.08–2.17	3.97	F7; also a double star
δ Cephei	3.71–4.43	5.37	F4–G6
β Doradus	4.24–5.69	9.84	F2–F9
ζ Geminorum	3.73–4.10	10.15	F5–G2
X Cygni	6.53–8.09	16.39	F8–K0
l Carinae	3.6 –4.8	35.52	F8–K0
U Carinae	6.30–7.55	38.75	F8–K5
SV Vulpeculae	8.43–9.40	45.13	G2–K5

* The first systematic discoverers of variable stars began to name them by the name of the constellation in which they are found, preceded by the letters R, S, T,

3. THE RR LYRAE STARS

The eclipsing-stars that resemble Algol, and show one sharply marked minimum, were early recognized as double stars that vary because they occult one another periodically. As the discovery of variable stars progressed, some were found that seemed to display exactly the opposite behavior—they showed one sharply marked maximum. For some time these were regarded as a sort of antithesis of the Algol stars, and were named the "antalgol stars." But when the pulsation theory seemed to have provided a satisfactory interpretation of the general behavior of the Cepheids, this second group was recognized as also Cepheid-like in character and was included among the stars whose variations were ascribed to pulsations. They are now generally known as the "RR Lyrae stars," after the first of the group to be thoroughly studied and also one of the brightest of them in apparent magnitude.

Bright RR Lyrae stars are uncommon in our catalogues, and these stars were at first discovered in rather small numbers, as compared, for instance, with the long-period variables. The great impetus to their study came from their discovery in large numbers in some of the globular clusters, and for this reason they are still sometimes known as the "cluster-type variables." This name is rather confusing since other types of variable stars are also found in globular clusters; we shall therefore not use it.

Light-Curves of RR Lyrae Stars. Early in the present century, Solon I. Bailey studied the variable stars in several of the nearer globular clusters, especially ω Centauri (named like a star because it is a hazy naked-eye object) and Nos. 3 and 5 of Messier's Catalogue. He found no such variety of light-curve as we have described for the Cepheids. There are two main types of light-curves for the RR Lyrae stars—a symmetrical curve and a curve with a steep rise in brightness and a slower fall. He recognized two varieties of this latter curve—in one the rise and fall were both very rapid, and in the other, the fall was considerably slower than the rise. Later studies have shown that Bailey's three types of light-curve (which he called, in the order described, types "c," "a," and "b") suffice to describe all the light curves of RR Lyrae stars that have been studied, whether in our galaxy, in globular clusters, or in other systems.

The "c" and "a" types of curve are shown in Figure 14.7. The curves of

---

etc. Thus we have R Ceti, S Ceti, T Ceti, U Ceti, etc. As discoveries progressed it was found necessary to use two-letter combinations, thus: RR, RS, ..., RZ, SS, ST, ..., SZ ..., ZZ. As even these combinations did not suffice, the letters begin again with AA ..., AZ, finishing with QZ. Further variable stars are then distinguished by numbers, preceded by the letter V, as: V 369 Cygni. A few of the brightest variable stars, which had already been named, have retained their original names and have not been given letters and numbers.

type "a" almost always show the characteristic dip just before the rise in brightness.

Periods of RR Lyrae Stars. The RR Lyrae stars are the intrinsic variables of shortest period, with a range from about an hour and a half to

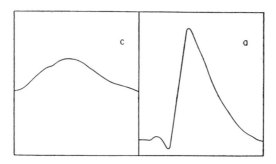

Fig. 14.7. Typical light-curves for RR Lyrae stars of types "c" and "a" (see text).

a little over a day. CY Aquarii, with the shortest period known, runs through its cycle in eighty-nine minutes; but even shorter periods may exist. Such rapidly varying stars are difficult to detect; the fainter ones must be sought by photography, and if the exposure necessary to record the star is of similar length to the star's period, the variation will be blurred out. It is convenient to draw the line between Cepheids and RR Lyrae stars at a period of a day. However, other properties besides period (such as spectrum, distribution, and motion) define the class, and on this basis a few RR Lyrae stars with periods greater than a day are known.

Bailey noticed in the RR Lyrae stars of ω Centauri a tendency for the symmetrical "c" type light-curves to go with the shorter periods; the "a" type light-curves set in suddenly at a period rather less than half a day, and the "b" type curves appear more gradually for the longer periods. This relationship between type of light-curve and period has been found also within other globular clusters, though the point at which the break between "c" and "a" curves appears does not occur at the same period in all clusters. The same tendency is also found among the RR Lyrae stars, within our galaxy, that are not associated with the globular clusters. The break however is not absolute: CY Aquarii (period of 0.06 days) and XX Cygni (period of 0.13 days) have strongly asymmetrical light-curves.

An interesting suggestion has been made about the significance of the differences of light-curve. The pulsating, or vibrating, star may be thought of as analogous to an organ pipe or trumpet which, while sounding a fundamental note, also sounds overtones. The type "a" curves are supposed to be vibrating with the fundamental tone and the type "c" curves with the first overtone, the fundamental being suppressed, for some reason not fully understood. They resemble "overblown" pipes which sound with a note an octave higher than their fundamental note. For a wind instrument, the overtone that gives a note an octave above the fundamental has *exactly* twice the frequency of the fundamental tone. But for the star, conditions are more complicated; here the vibrating body is a sphere of gas that increases in density toward the center, and the frequency of the overtone is not exactly twice that of the fundamental. The calculations that are needed to determine precisely what it must be are difficult and laborious, and require an intelligent guess about how the star's density increases toward the center. The most likely assumptions lead to a ratio of 0.7:1 for the periods (i.e., 1:0.7 for the frequencies) of the overtone and the fundamental. This ratio is not far from the ratio of the periods of the "c" type to the "a" type stars in a cluster, which makes the suggestion seem very plausible.

Spectral and Velocity Variations. The RR Lyrae stars vary in spectrum with the same period as their light changes. As with the Cepheids, the major part of the light variation is a result of the changes of surface temperature, and therefore of surface brightness, that are evinced by the changes of spectrum.

The spectrum of an RR Lyrae star is abnormal—even more so than that of a Cepheid. The average change of spectrum is from about A4 to F4. Unlike the Cepheids, the RR Lyrae stars show no apparent relation between spectrum and period; those of shortest period are no earlier in spectrum, and only a very little bluer, than those of longest period. The spectrum of an RR Lyrae star of longest period is quite different from that of the Cepheids of shortest period, being a whole spectral class earlier at maximum; on the basis of spectrum alone one can tell whether a star with period near a day is a Cepheid or an RR Lyrae star. Stars of either kind are most uncommon at this period.

The RR Lyrae stars show changes of radial velocity with the same period as the changes of brightness.

Distribution of the RR Lyrae Stars. The RR Lyrae stars form a sort of haze of stars that seems to be symmetrical around the galactic plane and around the galactic center. A few are known to lie eight or nine kiloparsecs above the galactic plane, and many more at comparable distances certainly remain to be discovered. They thin out away from the galactic plane and also away from the galactic center, but they thin out in both directions more slowly than any other kind of variable star, perhaps than any other kind of star at all. In this they present a striking contrast to the Cepheids, which are almost all confined to a thin layer about half a kiloparsec deep above and below the plane, within the galactic "dust-sandwich."

The RR Lyrae stars seem to increase in numbers toward the galactic center. In a region about one kiloparsec from the center, half of all the variable stars are RR Lyrae stars—a far higher proportion than in our neighborhood. These RR Lyrae stars are, moreover, of considerably shorter period than the average for all those known in the galaxy (most of the latter represent our own district rather than the central regions). This difference is not at present understood.

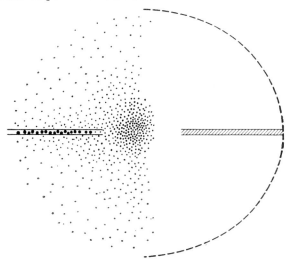

Fig. 14.8. Schematic vertical section through the galaxy, to illustrate the distribution of RR Lyrae stars (small dots) and Cepheids (large dots). Only half the galaxy is sketched; the other half, inaccessible to observation, is probably similar. Notice that the RR Lyrae stars are distributed in a spheroid greatly concentrated toward the galactic center, and somewhat less so toward the galactic plane. The Cepheids are all confined within a narrow slice in the galactic plane (bounded by straight lines) which contains the galactic dust and gas. Probably the layer of dust, and the Cepheids, do not extend through the galactic center.

Motions of the RR Lyrae Stars. The motions of the RR Lyrae stars are of the kind usually associated with a spherical or spheroidal distribution. They may be thought of as representing orbits with a large variety of eccentricities and inclinations; and the average both of eccentricity and inclination is larger than for stars of any other class (except, perhaps, some of the high-velocity dwarf stars). In other words the RR Lyrae stars, as a group, do not share the galactic rotation; though they may have a small average rotation, they are not moving in the same way as the Cepheids, the O and B stars, and the majority of stars in our neighborhood, in nearly circular orbits about the galactic center.

When we discuss the globular clusters, we shall find that these subsystems of the galaxy possess the same properties of distribution and motion as the RR Lyrae stars, perhaps to an even greater degree.

Absolute Magnitude of the RR Lyrae Stars. Within any one globular cluster, little, if any, progression of brightness with period is to be found for the RR Lyrae stars. In ω Centauri there are 135 RR Lyrae stars known with periods from 0.07 days to over 0.8 days (in addition to other variables of longer period that are not RR Lyrae stars). They show no marked progression in brightness throughout this range of period.

In other words, the RR Lyrae stars in individual clusters *show no period-luminosity relation.* The relation of the brightness of the RR Lyrae stars to that of other stars in the clusters, to the total brightness of the clusters, and other properties, make it appear very probable that the brightness of the RR Lyrae stars in all clusters is the same. It is not much more of a step to conclude that the brightness of all RR Lyrae stars is the same, wherever they may be found.

It is easier to decide that all RR Lyrae stars are of the same brightness than to decide what that brightness is. As with the Cepheids, the problem has been solved indirectly by the study of RR Lyrae stars in isolated systems, this time the globular clusters. The details will be made clear in the chapter where globular clusters are described. The absolute magnitudes of the RR Lyrae stars are very close to zero (absolute photographic magnitude $= 0.0$). Thus the RR Lyrae stars are fainter than the Cepheids, even those of shortest period.

Importance of Luminosities of RR Lyrae Stars. A group of stars that are all of the same absolute magnitude, if they are readily recognized, is obviously a powerful tool for surveying our galaxy. We do not even need to know the period, so long as we can be sure that the star is an RR Lyrae star.

The presence of RR Lyrae stars in globular clusters enables us to determine the distances of the clusters. This work has resulted in a revision of our ideas as to the dimensions of our galaxy, as will be described in the next chapter.

If RR Lyrae stars can be found in an extragalactic system they will furnish excellent means of measuring the distance of that system. Unfortunately only a very few galaxies are near enough to us for their RR Lyrae stars to be detectable. Both Messier 31 and Messier 33, the two nearest spiral galaxies, are too distant for the discovery of the RR Lyrae stars which are almost certainly present in them. One of the nearest elliptical galaxies is very rich in them, and a few have been found in a cluster near the Small Magellanic Cloud.

Within our own system the RR Lyrae stars are a powerful surveying tool. About fifteen hundred can at present be placed in our map of the galaxy, but a vast number are still undiscovered, and a still larger number are too remote for discovery, on the opposite side of the galaxy. It has been estimated that our own galactic system contains a hundred thousand RR Lyrae stars.

4. TYPE II CEPHEIDS AND SEMIREGULAR
YELLOW VARIABLES

The Cepheid variables have characteristic variations of light that are related to period, and their spectra and radial velocities vary concurrently. Their motions conform closely to the galactic rotation, and they are closely confined to the galactic plane. The Cepheids that are defined in this way are sometimes known as *"classical Cepheids,"* to distinguish them from some stars of similar period, also pulsating stars, which differ from them in several of these respects. The latter are sometimes known as "W Virginis stars," after one of the brightest known.

The Variable Star W Virginis. There is a variable star in Virgo, of about the ninth magnitude, that lies far from the encircling line of the Milky Way. Its "galactic latitude" (measured up from the galactic circle, exactly as terrestrial latitude is measured from the equator) is $+61°$, the largest known for any Cepheid-like star. This, and its considerable distance from us, means that W Virginis is situated far from the central plane of the galaxy, at a distance of about 1750 parsecs, so it is quite outside the "central sandwich" of dust and gas within which most of the Cepheids lie.

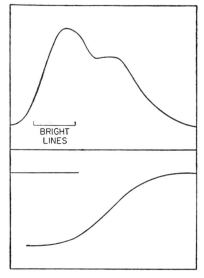

The motion of W Virginis is as unusual as its situation. Its radial velocity places it among the high-velocity stars, such as the RR Lyrae stars, hence it does not share the galactic rotation, but is traveling in an eccentric orbit of large inclination.

The period of W Virginis is 17.27 days. There are many classical Cepheids with periods near to this one, but their light variations are quite different. The light curve of W Virginis is illustrated in the figure.

The spectrum of W Virginis is rather like that of a classical Cepheid of the same period, though it does not vary quite as much during the light cycle. But it shows a striking feature that is quite unknown for a classical

Fig. 14.9. Light-curve and velocity-curve for a W Virginis star (schematic).

Cepheid: bright lines of hydrogen are present during part of the cycle and are especially conspicuous near the maximum. Moreover, as shown by Sanford at Mount Wilson, for a short time during the rise in brightness, *two absorption spectra are seen simultaneously.* These two spectra cannot be interpreted in terms of two stars, as is done for some spectroscopic binaries; they represent two disturbances flowing out through the star's surface. One is the previous surge dying away, and the other is the new surge coming up through the surface. During the interval when both are present, the old surge (at first the stronger) gradually fades, and the new surge (at first the weaker) grows stronger and finally is the only survivor. Seventeen days later it is dying away and another new surge is rising to supplant it. This doubling of the spectral lines for a short interval is shown also by the RR Lyrae stars, though it has never been noted for the classical Cepheids. In RR Lyrae the doubled lines last only about twenty minutes. Thus in spectroscopic behavior, as in velocity and distribution, the RR Lyrae stars have something in common with W Virginis.

W Virginis is a typical specimen of a group of pulsating variables, which we shall call the *Type II Cepheids,* because they are associated with the kind of stellar population that is marked by spherical or spheroidal distribution about the galaxy, and by "high velocity." This is sometimes called the *Type II Population,* as contrasted with the *Type I Population* to which most of the stars near the sun belong (see p. 439).

The Type II Cepheids or W Virginis Stars. W Virginis is an uncommon star, but it is not unique. Thirty or forty galactic Cepheid-like stars are known that lie far from the galactic plane, and that have light-curves like that of W Virginis. Those whose radial velocities have been measured are evidently high-velocity stars, and those whose spectra have been studied tend to show bright-line phenomena like those of W Virginis. They evidently form a real group, similar to, but not identical with, the classical Cepheids of similar period.

All these W Virginis stars have periods greater than ten days, and most of them are concentrated around sixteen or seventeen days; the shortest known period is that of AL Virginis (10.30 days) and the longest, that of TW Capricorni (28.58 days). It is interesting to inquire whether the luminosities of these stars are the same as those of classical Cepheids of similar period. Fortunately there exists a source of independent information about the W Virginis stars; they can be located as members of stellar systems of known distance. A number of them have been found in globular clusters; the light-curves of these stars are of W Virginis type (virtually no typical classical Cepheids are found in globular clusters), and the spectral variations (for stars bright enough to be investigated) show the characteristic bright lines. From the information furnished by globular clusters, we find that the W Virginis stars are between one and two magnitudes fainter than Cepheids of similar period.

When the variable stars that have been found in globular clusters are examined, they will be found to fall into three main classes. A great many of them have not been studied thoroughly enough to reveal their character, but the large majority of those that have been studied are RR Lyrae stars with periods of less than a day. There are 30 or 40 that are periodic, with periods greater than a day, and most of these are of period greater than 10 days. Those with periods between about 10 and 40 days are clearly W Virginis stars, and there are a number more, which behave very like W Virginis stars, with longer periods of up to about 150 days. The group of stars with periods between 40 and 150 days resembles closely a group of galactic variables that have been called "RV Tauri stars," after one of the best known. The RV Tauri stars are evidently an extension of the W Virginis stars into the longer periods, and there does not seem to be any reason (save the accident of history) for making any distinction between them. The RV Tauri stars are less regular in their behavior than the W Virginis stars, their light-curves not repeating so regularly; however, all red stars tend to be less regular and punctual at longer periods.

The RV Tauri Stars as a Group. The presence of RV Tauri and W Virginis stars in globular clusters, whose distances can be inferred from the presence and brightness of RR Lyrae stars, permits us to determine their absolute magnitudes. They are found to show a period-luminosity relation, but it differs from that of the classical Cepheids; that is, it does not ascend steadily toward the longest periods. The W Virginis stars are a little fainter than the classical Cepheids of similar period; the RV Tauri stars of shortest period (65 to 90 days) are the brightest, and those with periods up to 150 days become progressively fainter. From the W Virginis stars through the RV Tauri stars, there is a definite period-spectrum relation, the stars of longest period being of latest spectral class, reddest, and therefore of lowest temperature. The RV Tauri stars, like the RR Lyrae and W Virginis stars, show bright lines, and doubled absorption lines, in similar parts of their light cycles.

The motions and distribution of the RV Tauri stars confirm the conclusion (drawn from their behavior and their presence in globular clusters) that the galactic RV Tauri stars are more closely associated with the RR Lyrae stars than with the classical Cepheids. They are known (from the work of Joy at Mount Wilson) to be high-velocity stars, and many of them lie far from the galactic plane. This is the more remarkable when we remember that they are much nearer in period to the Cepheids than to the RR Lyrae stars. As we shall see in the next section, the semiregular red variables are again nearer to the Cepheids in dynamical properties, and the long-period red variables are intermediate between the RR Lyrae stars and the Cepheids. Arrangement of variable stars in order of period is not an arrangement in order of dynamical spatial properties.

5. THE RED VARIABLE STARS

The stars that vibrate in the longest periods are, as might be expected, those of the smallest density—the large red giants. As we have already seen, large size goes with low surface temperature, and it is not surprising that these low-density stars are of late spectral class. The characteristic of the spectra of cool stars is the presence of molecular bands, and these low-density stars show molecular spectra conspicuously. The atmospheres of some are rich in oxygen, and display the spectral bands of the metallic oxides—titanium oxide, zirconium oxide, and scandium oxide. Others are rich in carbon, and in their spectra the bands of cyanogen (CN), hydrocarbon (CH) and the carbon molecule (C_2) are prominent. In other words, their spectra are of classes M, S, and N.

Whether these low-density variables are rich in oxygen or in carbon, their general behavior is much the same. Their periods range from about 70 to over 700 days. Many of them vary greatly in light (at least to the eye and the photographic plate); some change by as much as eight magnitudes in brightness between maximum and minimum. Some of them, however, have very small light changes, to be detected only with the most refined equipment.

The red variables that have large light ranges (over two and a half magnitudes) are known as the *long-period variables*. Mira Ceti, the "wonder star," sometimes as bright as the second magnitude, is their prototype, and they are sometimes called "Mira stars." Those of smaller range are often called the red semiregular or irregular variables, because their changes are not so punctual as those of the Mira stars. They grade into the nonvariable red giants.

Motions of the Red Variables. The motions of the red variables are diverse and puzzling. Many long-period variables, especially those of shortest period, are high-velocity stars, and like all high-velocity stars, they are found far from the galactic plane and are commoner toward the galactic center than in our district. But those of longer period show less of the high-velocity character, are more crowded towards the galactic plane, and less concentrated (if at all) toward the galactic center. In fact those of the longest periods seem to share the galactic rotation. They are not divided into two distinct groups (as are the Cepheids), but show gradations all the way from the motion and distribution of Population II stars to those of Population I stars. Most of the semiregular and irregular red variables are more nearly of Population I than of Population II, and the same is true of the nonvariable red giant stars. These facts make the long-period variables particularly interesting: they are a sort of stellar "missing link" between the two types of population, and they suggest that an absolute distinction should not be made between Population I and Population II stars.

The Long-period Variables. The large variations of brightness of the long-period variables are neither quite punctual nor quite regular, and the shape of the light-curve varies from time to time. Perhaps this is a result of their large size and low density (they are several hundred times the sun's size and have about a millionth of its density).

The long-period variables show such conspicuous changes that the brightest have been known for many years. Four were known before 1800, and fifty-six had been recorded by 1860. First to be discovered was a star in the constellation Cetus, which was noted by Fabricius on August 13, 1596, and named by him "Mira" (the marvellous). It varies with a period of about 332 days, and at its brightest is of the second magnitude, easily visible with the unaided eye. The next long-period variable to be discovered was Chi Cygni, which varies through a range of eight magnitudes with a period of about four hundred days (see Figure 14.11); it was noted in 1686 by Kirch, and many of the early records are the work of his wife, Margharita, one of the first women to contribute to astronomy.

Many long-period variables thus provide records that go back for more than a century. These long series of observations display not only the underlying regularity of the changes of brightness, but also the small but definite differences from one cycle to the next. The diagram of Chi Cygni leaves no doubt of the basic periodicity, and also gives evidence of subsidiary waves which may have a regularity of their own, although no general rules for these secondary variations have ever been successfully formulated.

Because many long-period variables are fairly bright at their maximum, and because their variations are large and striking, they can profitably be studied by observers with modest equipment. There are few more fascinating hobbies than the study of the slow, rhythmic changes of brightness of these stars with a small telescope or even a pair of binoculars. Amateurs of astronomy in the United States who are interested in making systematic observations of this kind have organized the American Association of Variable Star Observers, which provides finding charts and magnitudes of comparison stars, and suggests programs of observation to its members.

The very large changes of brightness that are observed with eye and photograph for the long-period variables are in a sense illusory. The total brightness varies by less than a magnitude, as may be seen when more of the light is collected with an integrating device such as a bolometer. That the *visual* change is so much larger results from our limited color range, and from the changes of energy distribution in the stars' spectra as their light varies. They are cooler at minimum than at maximum, and the main part of the starlight is at infrared wave lengths, invisible to the eye. We only see the tail end of the spectral energy curve, and even this slips out of sight toward the red as the star cools. Another factor, that conspires to make the range of brightness seem larger than it is, may be found in the obscuring effect of the molecular bands, which become stronger at the lower temperatures of minimum light.

The changes of surface temperature that accompany and enhance the

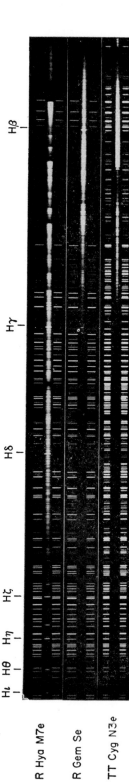

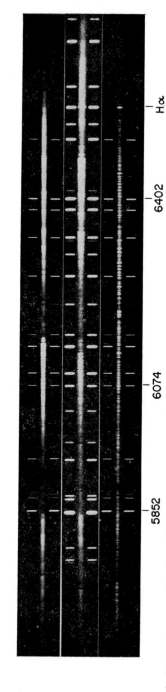

Fig. 14.10. Typical spectra of red variable stars. The upper section shows R Hydrae (M7e), R Geminorum (Se), and TT Cygni (N3e) for wave lengths 3700–5200 A. The positions of the Balmer lines are marked. The lower section shows R Leonis (M8e), T Geminorum (Se), and R Leporis (N6e) for wave lengths 5600–6600 A; the positions of three standard iron lines *in the comparison spectrum* are marked. All the stars except TT Cygni are long-period variables; TT Cygni is semiregular.

The Balmer lines show in *emission*. Notice the partial obliteration of Hβ in R Hydrae by overlying band absorption, and of Hε by the "H" line of Ca II.

The dispersion of these spectra is almost linear. By interpolating between the wave lengths of the marked lines, you can identify the characteristic bands in the spectra of the M, S, and N stars, listed in the footnote table on the facing page.

variation that we see are shown by spectral changes. The long period variables go from perhaps M2 to M8 from maximum to minimum, and the change is largely shown by the varying strength of the bands of titanium oxide. An even more striking fact is the presence of bright lines of hydrogen and some of the metals; the bright lines make their appearance as the star brightens, are very strong at maximum, and fade as the star fades. They show some intricate behavior with which we shall not concern ourselves. We have already seen similar bright-line phenomena for the W Virginis and RV Tauri stars, of which the long-period variables are in a sense a continuation toward longer periods.

One very revealing feature of the bright-line spectrum should, however, be mentioned. In Figure 14.10 several of the lines of hydrogen can be seen, notably Hγ, Hδ, and Hζ. There is little sign of Hβ or Hϵ, though we should expect the former to be the strongest hydrogen line in the figure, and the latter to be intermediate in strength between Hδ and Hζ. Usually the bright lines of hydrogen fall off steadily in intensity from long to short wave length. A glance at the figure will show that Hβ lies within a strong band of titanium oxide, and Hϵ falls within the H line of calcium. It is inferred with great plausibility that the hydrogen lines are produced *under* the layers of titanium oxide and calcium. Spectra taken with powerful instruments actually reveal the bright Hβ pushing up feebly between the intricate pattern of lines that make up the molecular band of titanium oxide, and bright lines of hydrogen even farther to the ultraviolet are seen underneath the pattern of absorption lines of iron. As the star declines, Hϵ appears to show through the H line.

Here is evidence of the layering of an atmosphere, with the bright lines produced near the surface, and the metallic vapors poised high above, apparently floating over the surface of the star in a sort of atomic and molecular umbrella. The bright lines usually found in stellar spectra come from enormously distended atmospheres of stars at high temperature, but those of the long-period variables seem to come from near the surfaces of very cool stars, by a process not yet understood. It seems as though we are seeing through the skin of the star to a hotter region inside.

There is indeed one well known long-period variable where this lower hot layer is occasionally unveiled. The star R Aquarii varies in brightness

	Class M		Class S	Class N
TiO	*V O*		*Zr O*	*C$_2$*
5760	5738		6229	5165
4955			6345	4737
4804	*Ca I*		6473	
4737				
4669	4227			
4584				

* (The spectrum of TT Cygni is by R. F. Sanford, the rest by P. W. Merrill. Mount Wilson Observatory.)

steadily with a period of 387 days. Sometimes, however, the variation seems almost suppressed for several years, and at the same time the star becomes much bluer and shows all the spectroscopic evidences of high temperature. It seems very likely that at such time the hotter internal layers are exposed for some reason, and the outer envelope (which plays the major role in producing the large light range) is temporarily depleted.

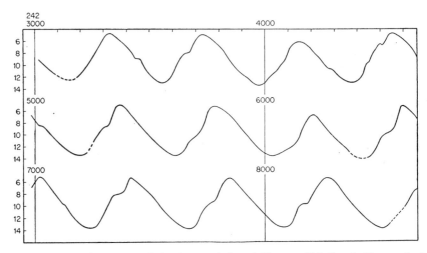

Fig. 14.11. Light-curve of the long-period variable star Chi Cygni. The vertical scale shows magnitudes; the horizontal, Julian days. (From visual observations by the American Association of Variable Star Observers.)

At some time in the past, R Aquarii has seemingly suffered an even greater disturbance, which amounted to an explosion. The star is surrounded by a peculiar cloud of glowing atoms, which is expanding at a steady rate and has evidently been thrown off from R Aquarii in the remote past. Perhaps at that time the star was a nova, and even during the milder disturbances just described, its spectrum had much in common with that of a new star (p. 385). Other stars that behave in rather a similar way are known.

Mira Ceti "the marvellous," true to its name, has a feature that is at present unique for a long-period variable. It is a double star, which was discovered visually by Aitken after Joy had predicted its duplicity from a study of the spectrum. The companion is about as bright as Mira itself at minimum, and seems itself to be a variable star. But it is a very different kind of variable from the long-period variable that it accompanies: the spectrum is of Class B, with peculiar bright lines. The absolute visual magnitude is about +8, so the star lies perhaps ten magnitudes below the main sequence, and is probably an "intermediate," with properties between those of the subdwarfs and white dwarfs. Thus the members of the system of Mira present a contrast as striking as that presented by the components of Sirius (p. 283, 357).

The long-period variables are not as valuable as Cepheids and RR Lyrae stars in measuring distances, because their absolute magnitudes differ and are not known with precision equal to those of the variables of shorter period. The brightest long-period variables have absolute visual magnitudes of about —2.5; the faintest are near zero.

No long-period variables have hitherto been found in any stellar system except our own, chiefly because of their comparative faintness. They are very rare in globular clusters, and this is surprising, because the globular clusters are characteristic Population II environments and many of the long-period variables show Population II motion and distribution. Their general absence from globular clusters is a significant fact for the study of their history, but it is not yet understood.

The Semiregular Red Variables. With shorter periods, in general, and about the same luminosities, semiregular red variables seem to be doing the same things as long-period variables on a smaller scale. Their light-curves are less regular and less periodic, and their spectral changes, though similar, are less marked and the bright lines less conspicuous, if visible at all. Their absolute magnitudes are near zero or —1. The most obvious difference is that the semiregular red variables have rather later spectra, period for period, and are thus probably cooler. Most of them have periods around a hundred days. There are a few semiregular red variables with much longer periods and much greater luminosities; Betelgeuse and Antares are among them, with absolute visual magnitudes between —3 and —4, and correspondingly large dimensions and low densities. It will be remembered that Betelgeuse is one of the few stars whose diameters can be measured with the interferometer (p. 283). Betelgeuse and Antares vary rather irregularly in cycles of several hundred days. The former has been found with the interferometer to vary concurrently in size. These luminous red variables show the Population I character in distribution and motion. They are common in the well-defined Population I environment of the Magellanic Clouds.

Many red variables are almost or quite irregular in their variations. They probably do not differ much physically from the semiregular stars, into which they grade imperceptibly.

Precise studies with the photoelectric cell have shown that all red stars with banded spectra are quite prone to variability. The later the spectral class and the higher the luminosity (i.e., the lower the density), the greater is this tendency. These variations are small and erratic and are probably caused largely by the molecular absorbing bands. Similar erratic variations are no doubt superimposed on the changes of brightness of the more regular red variables and contribute to the differences between individual maxima.

6. NOVAE

The term "new star" or "nova" implies a star that has been newly created; however, modern observation has made it clear that such objects are pre-existing stars that have suffered explosion.

The nova that gives perhaps the best idea as to what happens during the explosion was the one that appeared in 1918 in the constellation Aquila.

Nova Aquilae 1918. That novae are not newly created stars is definitely shown by this one, which can be found on the earliest photographs of the region as a slightly variable star of about the eleventh magnitude. Before the outburst of 1918 it seems to have been a bluish star, perhaps of spectrum A, which was flickering in brightness in an irregular way. Today, more than thirty years after the explosion, it is about the same as it was before, and is still flickering.

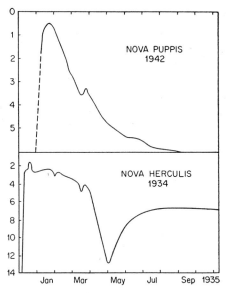

Fig. 14.12. Light-curve of the typical nova, Nova Puppis 1942, and the more unusual nova, Nova Herculis 1934. The vertical scale shows visual magnitudes. The horizontal scale for Nova Puppis is marked in days; that for Nova Herculis, which developed more slowly, in months.

The most striking thing about a nova is the suddenness of the main outburst. The whole range (ten to twelve magnitudes) is attained in a few hours. The light-curves of two novae are shown in the diagram. After the precipitate rise in brightness, Nova Aquilae grew fainter, and then after a short time, it began to undergo semiperiodic fluctuations of light, reminiscent of those of a Cepheid, and suggestive of pulsations. The fluctuations gradually died away, and after a few years the star seemed to have returned to its former status. These post-maximum fluctuations are shown by many novae, but sometimes they are inconspicuous or altogether absent.

Nova Aquilae was observed just before maximum to have a supergiant A2 spectrum and a radial velocity of -1700 km/sec. This was, of course, the velocity with which the surface rose. Apparently the pre-maximum spectrum was not very different from A (at least in energy distribution). It rose from magnitude 11.8 to -1.4, over 13 magnitudes. Its distance has been found to be 1200 light years. Thus it started as a blue star of absolute magnitude $+4.4$ (a subdwarf) and rose to absolute magnitude -8.8, without any great change in superficial energy distribution.

Directly after maximum, a profound change immediately took place in the spectrum. The single spectrum of absorption lines, all of the same radial velocity, was replaced by several different absorption spectra, all with different radial velocities. At the same time the whole spectrum was suffused by bright lines, faint at first, but soon dominating the whole picture. We may imagine that the surface of the star burst at the time of maximum, and that the rising material began to fly out into space in a series of sprays and jets. The material that lay between us and the core of the star produced the observed absorption lines, and the material that surrounded the core produced the bright lines as it spread out into space. The spectroscopic details are too intricate to be described here.

Just after maximum, the original absorption-line spectrum (that of the original "skin") could be traced, and the bright lines that appeared were the same as the absorption lines that were first seen. In other words, the material was absorbing where it was in front of the stellar core and producing bright lines elsewhere. The spectrum of the material in front of the core, which was coming toward us with the full explosion speed of 1700 km/sec, was displaced toward the violet (short wave length) by the Doppler effect, and the material giving the bright lines was moving with the full range of speeds from -1700 km/sec. on the side of the star toward us, to $+1700$ km/sec. on the far side of the star. The bright-line envelope was nearly or quite transparent to its own light, therefore each stellar line consisted of a broad, bright line, with an absorption line at its violet edge.

The spectrum runs rapidly through a series of profound changes as the light of a nova declines. The bright-line spectrum, which at first duplicates the original absorption spectrum, soon develops lines that go with higher temperatures (He I, He II, O II, etc.), and the secondary absorption spectra commonly show these lines also. After a short time, some bright lines appear that are not accompanied by absorption lines. The first to do so are so-called "auroral lines," "forbidden lines" with relatively high transition probabilities. After a short time, "nebular" bright lines begin to appear, the most conspicuous among them being "forbidden" lines of

O III, which we have already seen as typical of gaseous nebulae. The spectrum of a nova runs through successive stages that may be described as those of an absorption-line star, a bright-line star, an aurora, and a gaseous nebula. Several of these spectra may be present at the same time; then they begin to die away in the same order in which they appeared. A late stage of the nova shows the nebular-type spectrum only. Still later, even the nebular lines die away, and we are left with a weak bright-line spectrum that suggests a Wolf-Rayet star. In the end, all the bright lines

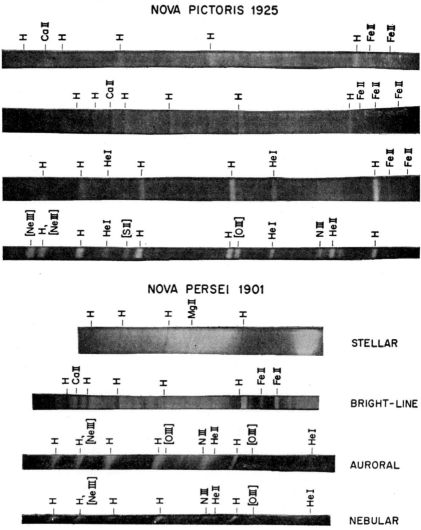

Fig. 14.13. Spectral development of Nova Pictoris and Nova Persei. The first photograph of Nova Persei is the historic picture mentioned in the text, that first proved that a nova can give a starlike spectrum.

seem to die away altogether, but this stage is not reached for years, or even for decades.

The series of spectra may be understood in a general way in terms of the rapidly decreasing density of the expanding materials around the star. Reference to page 337 will recall that the nebular spectrum appears only under conditions of very low density, which are attained first in the outermost regions of the expanding envelope.

An exploding nova is not as simple as the preceding descriptions would suggest. It is not so much like a bursting balloon as like a Roman candle. Many novae throw off large jets or blobs of material, and in addition there is probably a general spraying of material from the surface. Details differ in each case; many novae show some sort of axial symmetry, but some, like Nova Persei, seem to blow off on one side. Nova Aquilae is an especially well-observed case, and here, by the study of the structure of the bright and dark lines from the first day onward, we can form a definite picture of how the material was thrown out. There were two large jets which came out at about the time of maximum, one directed nearly toward us, the other nearly away from us; the one going away from us

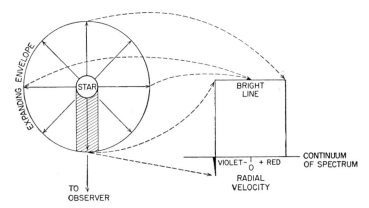

Fig. 14.14. Diagram to illustrate the production of the bright and dark lines in the spectrum of a nova by an expanding envelope of gas. The envelope is supposed to expand with uniform speed in all directions. In the direction of the observer it is seen silhouetted against the star; in other directions it appears as a luminous sphere of gas.

The star and expanding envelope are shown on the left; on the right is the spectrum line produced by one kind of atom. The cylinder of approaching gas, silhouetted against the star, produces an absorption line, displaced with *minus* the speed of the expanding envelope, relative to the velocity zero. The sphere of luminous gas builds up a broad bright line, with velocity ranging from − to + the speed of expansion. Broken arrows connect the parts of the envelope with the parts of the bright line to which they contribute. The gas that is receding from the observer gives rise to the red edge of the bright line; the gas traveling across the line of sight, with zero displacement relative to the observer, contributes to the middle of the bright line. The violet edge of the bright line is produced by the advancing atoms that are just outside the cylinder of absorbing gas.

was the bigger and brighter of the two. A number of lesser jets built up a series of "smoke rings," which were all ejected within six hours.

A final confirmation of the actual ejection picture is provided by photographs taken of the nova at intervals since 1918. It is seen to be surrounded by an expanding nebular envelope, which grows larger with time. During the past few years one of the two main jets just described has been seen to move clear of the expanding nebular disc, and to appear as an entity. The nebulous material is still expanding with the same speed as it had at the time of the original explosion. Because it is moving through a vacuum, there is nothing to retard it; it left the effective gravitational field within a few hours, and in spite of its low density, it obeys the first law of motion.

Fig. 14.15. The expanding nebulosity around Nova Persei 1901, photographed about fifty years after the outburst was seen. The elliptical shell seems to be concentrated on one side of the star. Probably the expanding gases are colliding with, and exciting, nebular material that lies in space near the star. (Palomar Observatory)

Some Other Bright Novae. Nova Persei has already been mentioned. It showed similar postmaximum fluctuations, and rather similar spectral changes, to Nova Aquilae. Its spectrum was notable for the great strength of the "forbidden lines" of Ne III and Ne V. The most interesting feature of Nova Persei was, perhaps, the rapidly expanding nebulous ring that was seen around it during the first few days. This ring could scarcely be in physical expansion (it was traveling much too fast), and it was early interpreted as the passage of a "light pulse," carrying the news of the maximum brightness through some surrounding nebulosity, previously invisible. The illuminated nebulosity had a recognizable structure. Today, fifty years after the original outburst, *the same structure has reappeared,* probably this time as the result of the arrival of the exploding materials from the nova and the production of shock-waves in the diffuse nebulosity through which they are passing.

Nova Herculis 1934 is an example of a nova that went through its development much more slowly than Nova Aquilae and Nova Persei. It remained near maximum brightness for three months, and the temperature rose but little during that time. It was one of the few novae to display the "forbidden lines" of ionized iron—the reason being that the density fell low enough to permit these lines to appear, while the temperature was still low enough for appreciable amounts of Fe II to remain in the envelope. Nova Herculis was not unique; an almost identical nova, in light curve and spectral changes, appeared in Auriga in 1891.

Nova Pictoris 1925 was a nova that developed even more slowly. Although it appeared over twenty-five years ago, it is still on the decline, and had for decades an interesting bright-line spectrum. It showed strong "forbidden lines" of ionized sulphur, and a remarkable succession of the successive spectra of iron: "forbidden lines" of Fe II, Fe III, Fe V, Fe VI, and Fe VII have been observed as it developed.

Nova Serpentis and *Eta Carinae* are novae that have developed still more slowly. The latter, which was bright in 1843, has returned little more than half-way to its original brightness in over a hundred years.

7. RECURRENT NOVAE

Most novae have been seen in historic times to explode once only. But members of a small group undergo semiperiodic explosions every few decades. Well-known examples are T Coronae Borealis, which brightened in 1866 and in 1946, and T Pyxidis, which has had four recorded appearances, the first in 1890. These recurrent novae are of special interest: they have rather smaller ranges (about eight magnitudes) than most novae; and several of them have shown the *coronal lines* of enormous excitation in their spectra, which ordinary novae have never been seen to do. T Coronae Borealis seems to be a red star, apparently a giant, at minimum, and perhaps the recurrent novae represent explosions of stars of a different type from ordinary novae, which are generally thought to originate from blue subdwarfs. But our direct knowledge of the prenova stages for normal novae is almost confined to Nova Aquilae.

There is a link between the recurrent novae and the red variables in such stars as Z *Andromedae* and *AX Persei,* which are basically red giant stars, but which undergo outbursts that resemble nova outbursts (except in scale) at periodic intervals of about nine hundred days. They show the spectroscopic properties of a nova, with bright lines and the continuum of a high-temperature star, in the course of each outburst. A very similar set of phenomena is presented by *R Aquarii,* the peculiar long-period variable described on page 382, which shows nova-like spectral features at times and is the site of a bright nebular envelope that seems to derive from a nova outburst of perhaps a thousand years ago.

8. SUPERNOVAE

In the year 1885 a nova appeared in the central region of the great spiral in Andromeda. It was of the seventh magnitude—about one-tenth as bright as the whole galaxy in which it appeared. It can be seen on photographs of the galaxy taken that year, but unfortunately no photographs were taken of its spectrum. If only the Andromeda spiral were twenty light years farther from us (less than one-ten-thousandth of its distance), the light of the nova would not have arrived until photographic spectroscopy was well established, and we should know far more about supernovae than we do.

The light curve of the supernova, which received the name S Andromedae, was not unlike those of other novae, and although the visual examination of the spectrum suggested that it was far from typical, the star was not at first regarded as very different from the numerous novae that had been observed in our galaxy. If it had resembled them, the absolute magnitude would have been about -7, which would have placed the great spiral at a distance of less than 10 kiloparsecs (about the same distance as that of the sun from the galactic center, as known at the present time).

When Cepheid variables were discovered in the Andromeda spiral in 1925 by Hubble, it became clear that the distance is much greater than 10 kiloparsecs. Therefore the nova that appeared in 1885 must have been much brighter than the novae of the galaxy. In addition to the Cepheids, Hubble has discovered over a hundred novae of the ordinary type in the Andromeda spiral, which are of the sixteenth or seventeenth apparent magnitude. S Andromedae was therefore about 10 magnitudes brighter than an ordinary nova. Stars of this sort are called *supernovae.*

About fifty supernovae have been discovered in distant galaxies. Some of them have been brighter than the whole galaxy in which they appeared. Three are known to have appeared in our own galaxy: Tycho's nova of 1572, Kepler's nova of 1604, and the nova of 1054, which will be discussed later. Supernovae seem to occur most commonly in spiral galaxies like Messier 33. It has been estimated that there is on the average about one supernova per galaxy in 359 years; but three supernovae appeared in a single galaxy within a couple of decades.

Supernovae are not all similar. There seem to be two types, which differ in brightness, in light variations, and especially in spectrum. The brighter type (called Group I), to which S Andromedae belonged, attain an

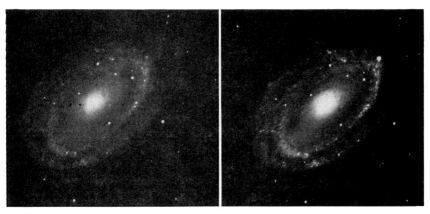

Fig. 14.16. A supernova in the galaxy NGC 4725. The photograph on the left was taken May 20, 1931; that one the right, May 10, 1940. The identification of the supernova is left as an exercise for the reader. (Mount Wilson Observatory)

absolute magnitude of about −16 at maximum; their spectra are enigmatic. The supernovae of Group II are over 2 magnitudes fainter; they differ from ordinary novae only in scale, and their spectra are like those of ordinary novae, except that they give evidence of greater velocities of expansion, which in turn produce broader emission lines. The Group II supernovae seem to exceed those of Group I in a ratio of about six to one, but the sample from which these figures are deduced is very small.

A supernova that is observed in a distant galaxy provides a method for determining the distance, exactly similar to the methods based on the brightness of RR Lyrae stars and Cepheid variables. It is necessary to determine whether the supernova is of the brighter Group I, or the fainter Group II, kind, and this can be done if the light-curve is well observed, because the light-curves of the two kinds differ.

The spectra of the supernovae of Type I present a very interesting problem; they are somewhat featureless and suggest the coalescence (as a consequence of large velocities of expansion and huge Doppler broadening) of a number of bright lines. Another possibility, however, is that the spectra consist of absorption bands with the detail obliterated by Doppler broadening. About the only thing on which there is general agreement is that the spectra show little evidence of the lines of hydrogen, which are generally the most prominent feature of a bright-line spectrum of a nova, at least in early stages. The spectra suggest, indeed, that supernovae are abnormally poor in hydrogen. The explosion of a supernova is associated with a vast amount of energy (about 10^{50} ergs in visual light) and might well produce a spectrum that has no parallel under other conditions.

The three galactic supernovae mentioned above were brilliant naked-eye objects. Tycho's nova of 1572 had apparent magnitude of about -4 and was visible in daylight. Kepler's nova of 1604 rose to -2.5, and the supernova of 1054 was reported to be as bright as Venus, so it must have rivaled Tycho's nova. We do not know the magnitude range of supernovae, but attempts to identify the remains of Tycho's and Kepler's novae with stars have failed. Probably each remaining star is of the fifteenth magnitude or fainter, which would make the range at least eighteen magnitudes, more than five magnitudes greater than those of ordinary novae. The remains of the supernova of 1054 are visible as a faint star, probably of about the sixteenth magnitude, so that the original range must have been about twenty magnitudes. The distance of this star has been measured (as will presently be described), and we conclude that it was of absolute magnitude -16.5 at maximum and is about $+3.5$ at minimum (i.e., at the present time). We cannot, however, conclude that it has returned to its original state, as ordinary novae seem to do, and it may have been considerably brighter before its outburst than it is now—perhaps a giant star.

The Crab Nebula. The faint star that is the remains of the supernova of 1054 would not have attracted attention if it were not surrounded by the most remarkable bright nebula in the galaxy. Lord Rosse, the Irish astronomer who made the first studies of spiral galaxies with his 60-inch reflector, observed and named it in 1844. Photographs show it as a complex filamentary structure with a faint blue star at its center. No other bright nebula looks like it.

Photographs taken at intervals during the last thirty years show that the Crab Nebula is expanding in all directions away from the central star. Spectroscopic observations show that the nebula has a radial speed of expansion about the central star. As radial velocities are given directly by the Doppler displacement, and are independent of distance, we know the true rate of expansion from the spectroscopic observations. The photographs of the nebula give the angular rate of expansion. If we assume that the motion across the line of sight (measured on the photographs) is equal to the motion in the line of sight (measured from the spectra), the unknown distance factor in the cross motions can be determined and the distance of the Crab Nebula may be calculated. It is found by this method to be about five thousand light years, and the absolute magnitude at maximum follows from the fact that, at its brightest, the star was as bright as Venus. The process is the inverse of that by which the tangential velocity of a star is calculated from its proper motion and its distance (p. 286):

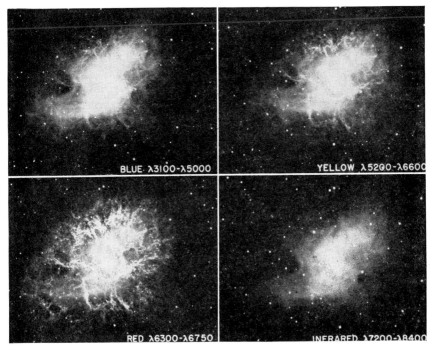

Fig. 14.17. The expanding nebulosity (the "Crab Nebula") that is the debris of the supernova observed in A.D. 1054. The photograph, in light of four colors, shows the different distribution of radiation from different atoms. Notice the extraordinary filamentary structure in red light. The remains of the exploding star are visible as one of the faint objects, of about the sixteenth photographic magnitude, near the center. The star is of fantastically high temperature, but much of the brightness of the nebula is probably due to the mutual collisions of the gaseous filaments. (Mount Wilson Observatory)

$$\text{tangential velocity} = 4.74\mu/\pi \quad (\text{km/sec})$$

or

$$\pi = 4.74\mu/V,$$

where V is the radial velocity.

Since the distance of the Crab Nebula is known, its present size can be calculated from its angular size, and this, combined with a knowledge of the velocity of expansion, enables us to calculate back and deduce the time at which the explosion took place. The conclusion is that the Crab Nebula has taken about nine hundred years to reach its present dimensions. Thus the light of the original outburst must have reached the earth between 1000 and 1100 A.D. The actual outburst, of course, took place some five thousand years before that, but its light took that time to reach us.

As a result of these calculations, a search was made of the early records to find whether there had been a bright new star in the position of the Crab Nebula (in the constellation Taurus, near the tip of one of the horns of the imaginary Bull) near that date. Oddly enough, there were no records of such an event in the Western Hemisphere—astronomy was not an active science in Europe at the time—but the star was recorded both in Japanese and Chinese documents as having been observed about July 4, 1054. It is considered practically certain, therefore, that the Crab Nebula represents the debris of the explosion of a supernova, nine hundred years after the outburst.

The physical study of the Crab Nebula is of the greatest interest. The outer filamentary parts show a bright-line spectrum of "forbidden lines" and of hydrogen—the latter unusually weak, and suggesting a low abundance of hydrogen. The inner part of the nebula shows a continuous spectrum, which probably comes from a cloud of electrons at high temperature. The central star itself seems to be a white dwarf with a temperature of perhaps half a million degrees at the surface and about one fiftieth of the size of the sun. We cannot determine its mass, but if it is as massive as the sun, it has a hundred thousand times the solar density. The mass of the nebula, on the other hand, can be roughly determined from the amount of light it gives out, which can be used as a measure of the number of atoms it contains. The nebula is estimated to have a mass of about fifteen suns. If all this mass was combined in the original star before the outburst, the Crab Nebula originated as a giant star; masses of fifteen suns or more are found among the brightest main-sequence stars and the yellow and red supergiants. Our ideas of stellar development must provide for the origin of supernovae from one of these stellar types.

The Crab Nebula may be radiating the transformed light of the central star. Another possibility has recently been suggested—that the outer filamentary regions, which are in violent motion, are exciting the gas by their rapid internal motions, and producing "shock-waves" by the collisions of nebulous masses.

The Crab Nebula radiates not only in visible and photographic light; it is one of the strongest radio sources in the sky and is thus emitting intense radiation of very long wave length.

Cause of Nova and Supernova Explosions. The explosion of an ordinary nova, spectacular though it is, affects the star only superficially. Novae return after a few months or years to essentially their former state, and the amount of matter lost is between one thousandth and one ten-thousandth of the mass of the star. The loss from recurrent novae seems to be even less. It is calculated that if ordinary novae explode about once in a million years, they lose in the long run about the same amount of matter and energy as a recurrent nova over a similar interval of time. A supernova, however, is fundamentally affected by its outburst; it loses an amount of energy comparable to the whole energy of an average star and apparently throws off the major part of its mass. If we are to judge by the Crab Nebula, a supernova does not return to its former state after explosion. Therefore it is probable that novae and supernovae present different problems.

Three main possibilities may be considered in accounting for stellar explosions: collisions of stars (with one another, with planets, or with interstellar material); sudden release of energy within a single star, which survives essentially unaltered; or sudden internal transformation of a single star into a star of different structure.

That the energy radiated in a nova explosion is the *mechanical* energy of collisions between stars is unlikely, because ordinary novae all have nearly the same absolute magnitude at maximum, whereas mechanical energy of collision between stars might have any value, depending on their masses and relative speeds. Collisions between stars and nebulae, or between nebulae or clouds of solid particles, are an even less probable cause for novae. The former would scarcely be sudden, and the latter, besides being gradual, could not produce the necessary energy. Moreover, recurrent novae are hard to explain in terms of collisions, for the chance of a single star suffering a collision every few decades is very low. If a recurrent nova were the result of periodic approach of the components of a double system, it would be strictly periodic, whereas the recurrent novae are only roughly periodic. However, it is not impossible that the collision of a star with another star, planet, or nebula, might act as a trigger to cause a sudden release of internal energy—the second hypothesis suggested.

As a cause for the ordinary nova, sudden release of internal energy seems a plausible idea. Modern theories (very like those that have been worked out concerning the propagation of underwater explosions) have shown that a sudden explosion inside a star would probably have many of the features shown by the nova outburst—the high speed, the smaller outbursts following the first, the sharply rising temperature. Opinion inclines to the idea that the nova outburst originates deep within the star, perhaps near the center, but the nature of the explosion, and what sets it off, are not yet understood. A gradual change of internal conditions—e.g., of temperature—might finally precipitate a nuclear process that released a large quantity of energy suddenly. In an ordinary nova, this energy (though large) is only a small fraction of the total energy of the star; its release may be supposed to relieve the condition that produced it, and the star, after throwing it off, returns to its original state.

A supernova does not, seemingly, return to the pre-explosion stage. For these stars, the third possibility—a radical internal transformation—seems the most plausible suggestion. Perhaps they represent an internal rearrangement by which a red supergiant turns into a white dwarf.

9. DWARF NOVAE, SHELLS AND FLARES

The supernovae have absolute magnitudes at maximum of about —16 and —14. The ordinary novae are of about absolute magnitude —7.5 at maximum. These are not the only stars that are observed to undergo explosive ejections. There is a group of stars whose members undergo explosions on a far smaller scale, and they may conveniently be called the *dwarf novae*. They are sometimes subdivided into the "U Geminorum stars" and the "Z Camelopardalis stars," after two of the best known examples.

U Geminorum and SS Cygni. These two stars are the best known members of the first subdivision. SS Cygni is perhaps the most thoroughly studied of all stars, for it has been followed continuously by visual observers all around the world during some of its outbursts.

The light-curve is that of a nova in miniature, with a very rapid rise in brightness, a brief maximum of varying duration, and an abrupt fall in light, always of the same steepness and not as rapid as the rise. Outbursts are not strictly periodic, but they follow a definite cycle, recurring at intervals of a few weeks, or months at the most. The range of brightness is about 4 magnitudes. Their absolute luminosities are rather hard to determine, but are generally supposed to be about +8 to +10 at minimum, so that even at maximum the U Geminorum stars have the low absolute magnitudes +4 to +6. Between outbursts their minimum brightness is nearly or quite constant.

The spectra of the U Geminorum stars give definite evidence that the outburst is nova-like. At maximum, SS Cygni looks like an A star with very wide, hazy lines; the spectrum of U Geminorum is nearly featureless and suggests very high temperature. Toward minimum there is a bright-line spectrum, suggesting that of an early-stage nova, but no "forbidden lines" appear. During their protracted minima, these stars are somewhat redder than at their maxima, and one or two have line spectra like those of G dwarfs.

It has been suggested that some U Geminorum stars are actually double, and large variations of radial velocity within a short interval have been thought to provide additional evidence of this. But detailed proof is lacking, and, as we have seen, there are some stars (such as R Aquarii) that seem to combine the spectra of matter under incompatible conditions.

Z Camelopardalis. The stars of the second subdivision of the dwarf novae have more frequent outbursts than SS Cygni or U Geminorum (every two or three weeks), and such outbursts are smaller, with a range of about two and a half magnitudes. The maxima last so long that there is often no constant minimum; however, sometimes the variation dies out completely for a time, and the star remains constant in brightness at a magnitude about intermediate between maximum and minimum. The spectral changes are like those of the previous group and point to nova-like outbursts on a small scale.

Shell Stars. Ejection of material is observed also from stars of very different type. The supernovae may originate from giant stars (perhaps red giants); the novae seem to be subdwarfs at minimum, and the dwarf novae also are apparently subdwarfs, perhaps further down the subdwarf sequence. But bright main-sequence stars may also be subject to ejections. Many B stars have bright-line spectra, which we have learned to associate (at least for hot stars) with extensive envelopes. These bright-line B stars are much commoner among the early B stars (of highest temperature) than among later ones, and for A and later stars they are almost unknown.

The occurrence of bright lines in the spectra of the hottest stars is very often associated with very broad, shallow absorption lines. Such broad, shallow lines can be interpreted on the supposition that the stars are in rapid rotation, so that the absorption line is broadened by Doppler effect (one edge of the star is approaching us, the other receding). Altair (α Aquilae) has broad lines of this kind, which are seen as evidence that the star is rotating with an equatorial velocity of about 250 km/sec. Many of the brighter Pleiades have an equally rapid rotation and also show bright lines.

The fact that rapidly rotating stars tend to have bright lines, and therefore extensive gaseous envelopes, suggests that their rotational speeds are great enough to spill material off. Many close eclipsing-stars (components near together) have very short periods of less than a day. It is certain that the tidal distortion of these stars causes the tidal bulges to be aligned, so that the individual stars are rotating in the same period as their revolution. Their spectra, as would be expected, show broad and shallow lines. Many of them, also, show bright lines, and careful studies have revealed in several cases that there is actually a luminous ring of gas around at least one of the component stars. This loss of material by a rapidly rotating and revolving double star may be an important factor in the development of such pairs.

Luminous rings of gas around rapidly rotating stars are not, however, the only evidence for the loss of material from the surfaces of bright main-sequence stars. Some of them undergo explosions that more nearly suggest those of novae. Two well-known examples of such explosions are furnished by γ Cassiopeiae and by Pleione.

γ Cassiopeiae. About ten years ago the bright star γ Cassiopeiae increased in brightness by nearly a magnitude. Usually its spectrum shows extremely broad and hazy absorption lines (signs of rapid rotation) with bright lines superimposed upon them (signs of the shedding of atoms into space). But soon after the star brightened, another spectrum appeared superimposed upon these—a spectrum consisting of very sharp lines, whose strengths pointed to the fact that they must be formed in a region of lower density, and less intense radiation, than the main spectrum of the star. Here was evidence that a shell of atoms had been thrown off (much as in a nova, but with smaller speed) and was traveling slowly away from the star. After a few months the shell dissipated into space, and the usual spectrum of the star once more took control.

Pleione. One of the brighter Pleiades, Pleione seems to be slightly variable in brightness and has indeed received the name BU Tauri. There is an old classical legend that the Pleiades, the daughters of Atlas, were fleeing in terror from Orion, and were changed into a flock of doves in the sky; however, one was lost. This may, possibly, represent an early variable star observation, indicating a change of brightness of one of the Pleiades in the past!

Pleione is probably the most variable of the bright Pleiades at present. It varies not only in brightness, but in spectrum. In about 1888 it had, in addition to the broad hazy lines indicative of rotation, a bright-line spectrum. But the bright lines disappeared at the beginning of the twentieth century. About a decade ago they appeared again, and soon afterward, Pleione was observed to throw off a shell, rather like the one thrown off by γ Cassiopeiae. The shell remained visible for a year or so, and then dissipated into space. Pleione evidently suffered a very small nova-like outburst. Such behavior seems to be frequent among the B stars, especially those that show bright lines and are rotating rapidly.

"Symbiotic Variables." The star AG Pegasi first attracted attention as a shell star of spectrum B; it seems to have a period of about eight hundred days, at which interval it suffers ejections rather like those just described. But in addition it has an underlying spectrum of Class M! A number of other stars show an equally startling combination of spectra: Z Andromedae, a periodic nova (about nine hundred days), has an underlying M spectrum; and several other stars like it are known. We have already met R Aquarii, a long-period variable with an underlying, variable B spectrum.

When these stars were first discovered, they were regarded as intimate combinations of two stars that were profoundly affecting one another, comparable to the "symbiotic" combinations, such as lichens, that are found in the domain of biology. But the conviction is gaining ground that these objects are really units, one star that displays, on occasion, a wide variety of conditions. The long-period variables, with their late-type spectra (which indicate low temperature) and their bright lines of hydrogen (only to be produced at much higher temperature) should have prepared us for this possibility. So should the sun, which runs the gamut from the sunspot spectrum (temperature perhaps 4500°) to the corona (temperature corresponding to 1,000,000°), and which is without doubt a single object! Even more relevant is the tendency of the surface of the sun, on rather rare occasions, to exhibit "flares," intensely bright regions whose spectra show evidence of far higher temperature than the surroundings. It is very likely that other stars may also display flares, which can be thought of as localized nova-like outbursts (probably very superficial ones). The discovery of such stars is a very recent event.

Flare Stars. The stars that are subject to flares seem to include, if not to comprise, yellow or red dwarfs of low luminosity. Their brightness suddenly increases for a very short time and is sometimes more than doubled. Because of the short duration of the flares, most of those that have been accurately observed have been found by accident, for example, during continuous photoelectric study of a red dwarf. A survey of old photographs has given evidence that a number of red dwarfs are subject to flares, though at present we can hardly form an opinion as to how often they occur. The study of these stars, which may have the effect of linking solar phenomena with the shell stars and the novae, is still in its infancy; it holds promise of interesting results.

10. VARIABLE STARS IN NEBULOSITY

The galactic plane is pervaded by a layer of dust and gas, and even from ordinary photographs, it is obvious that the distribution of the part of this nonstellar material that forms absorbing clouds is very nonuniform. Some well-known dark spots are the Coal Sack near the Southern Cross, and the broad streaks of obscuration in Taurus and in Ophiuchus and Scorpio.

The distance of these large and obvious dark nebulae is easily determined by comparing the numbers of faint stars that are visible within them and outside their boundaries. The point at which the numbers of fainter stars begin to fall off within the bounds of the dark nebula marks the distance at which the obscuration begins. If the absolute magnitude corresponding to this brightness is known, the distance of the stars can be determined, and this is also the distance of the obscuring cloud. Most of the large and obvious dark nebulae are rather near to us—within a few hundred parsecs. More distant dark clouds are too small in angular dimensions to cut off identifiable regions, and their effect appears in the cumulative dimming of more distant stars.

The volume of a dark nebula is perhaps of the order of 1000 cubic parsecs, and a few stars would be expected to lie *within* such a dark cloud. On this basis there should be, for example, about seventeen stars within the Taurus nebulosities. It is of interest to examine whether these stars are affected by their situation, and differ in any way from stars that are not inside a dense cloud of dust and gas. They do, in fact, differ from "free" stars in many ways. One of the most obvious is that they tend to vary in brightness; their spectra are also of most unusual kinds.

The stars in the Taurus nebulosity were studied by Joy of Mount Wilson. He noticed that these stars have absorption spectra of class G, K, or sometimes M. The distance of the cloud is known by the method mentioned above and shows that the stars within the cloud have the luminosities of dwarf stars. However they have a feature that is not usually associated with normal dwarf stars—their spectra show bright lines, especially the H and K lines of ionized calcium. Moreover, many of the stars seem to be connected with bright nebulous wisps, which have similar spectra to the emission spectra of the stars, additional evidence that the stars are really *in* the nebula. Furthermore, most, if not all, these imbedded stars are slightly or considerably variable in brightness—so much so that it is impossible to locate nonvariable stars in the immediate neighborhood for purposes of comparison. Evidently the nebula is having an influence on the stars within it.

A further remarkable fact is that, although only seventeen stars were to be expected (on the assumption of uniform distribution inside and outside the Taurus cloud), nearly two and a half times as many were located by Joy within the cloud. This lends color to the idea that not only are the imbedded stars affected by the nebula, but that they are also probably being produced in it.

Very similar results were later obtained by studies of Ophiuchus-Scorpio nebulosity by Struve and his collaborators. Here again stars of peculiar spectrum, some of them variable, were found, and in larger numbers than would have been expected for ordinary regions.

Even more striking is the region that contains the Orion Nebula. This nebula is most conspicuous for its bright wisps and filaments, but they are clearly associated with dark streaks; and the whole region is pervaded by both dust and gas. In fact, long-exposure photographs of the region show that the whole constellation of Orion is embedded in a huge nebula. Many years ago Miss Leavitt discovered several dozens of faint variable stars within the boundaries of the Orion Nebula. But all attempts to find periods for these stars have failed, and they must be classed as completely irregular in behavior. Moreover, they have absolute magnitudes from +4 downward, so they are much fainter absolutely than most intrinsic variables. They, also, show late-type spectra, often with emission lines, very like those described for the Taurus Nebula. There are more than a hundred such stars within the Orion nebular region, whereas only ten could have been expected. Here again we have a strong suggestion that the nebula is not only affecting, but actually forming, the stars within it.

There are a couple of hundred stars, late-type dwarfs (G-M), that show similar symptoms, especially bright lines of ionized calcium, although not all of them are within obvious dark nebulosities. Nonetheless, they tend to be crowded in regions where accumulations of dust are also found, and very likely all or most of them are being affected by the dust in much the same way as the stars within the dense, dark nebulae.

Some of the stars within the nebulae, though they may be variable in brightness, do not show emission lines, and these tend to be the stars of earlier spectrum and, therefore, apparently of higher temperature. A star of high temperature, which is pouring out radiation intensely, exerts a "radiation pressure" on the matter that may surround it, and probably the hottest stars in nebulae are literally blowing the nebulosity and dust away from themselves and thus preventing the formation of an envelope that can produce the bright lines. A few of the other variables in nebulae have a continuum that seems to point to a higher temperature than their spectra (G or thereabouts) would lead us to expect, and it has been suggested that these stars are causing fluorescence in solid particles that surround them— particles, perhaps, resembling meteoric or cometary matter. The spectra of the variable stars in nebulosity are still not completely understood.

The stars that have just been described are often called the "T Tauri variables," after a well-known member of the group. This star, which varies in a spasmodic manner, has an underlying spectrum of Class F, with peculiar emission. The most striking thing about T Tauri is that it lies at the tip of a nebula of funnel-shaped form, and the nebula varies as much as, or a good deal more than, the star does. It is quite possible that the variation that is noted is, in fact, very largely that of the nebula, which forms a luminous envelope round the star.

Several other stars are known to be associated with funnel-shaped nebulae that vary in similar striking fashion. Two of the best known are R Monocerotis and R Coronae Austrinae. In each case the star is at the tip of the funnel. R Monocerotis does not vary much, but the nebula itself varies greatly. Photographs taken at short intervals show an effect as though shadows were coursing across the nebulosity, and probably these really are shadows, cast by nebulosity drifting past the star. Although the spot near R Monocerotis looks like an eighth-magnitude star, careful studies suggest that the actual star, when seen, is not brighter than the fifteenth magnitude, and that the rest of the light comes from the nebulous envelope. There are dark Balmer lines in the spectrum, but they are probably produced in the envelope, not characteristic of the star. Exactly how the light of the star operates on the envelope is not known. There are a number of faint variable stars near R Monocerotis, clearly in the same nebula, and their absolute magnitudes are from $+3$ to $+6$; hence they are very like the groups described before. The spectrum and behavior of R Coronae Austrinae are very like those of R Monocerotis, and here, also, a group of faint irregular variables is associated with the dark cloud.

11. R CORONAE BOREALIS STARS

There is another well-defined type of variable star, not yet mentioned, that may be classed with the nebular, or at least, with the extrinsic variables.

R Coronae Borealis is the best-known example, and the group is named after this star.

The variations of R Coronae Borealis are very conspicuous and quite irregular. It remains bright for long intervals, sometimes weeks or months —once it remained of constant brightness for over ten years. It then undergoes a sudden dimming, by as much as eight magnitudes; it does not remain extremely faint for long, but returns gradually and rather spasmodically to its former brightness. About a dozen well-observed R Coronae Borealis stars are known. They have one thing in common: all of them show evidence of being unusually rich in carbon. Several stars are of Class R at maximum, and R Coronae Borealis itself, classed as G0p (peculiar), though it does not display carbon bands (probably because the temperature is too high), has unusually strong lines of carbon. The spectra show the marks of high luminosity.

As the star changes in brightness during the fall to minimum, the spectrum does not change in the way in which those of intrinsic variables change as their brightness diminishes. For the first few magnitudes of dimming, the spectrum is essentially unaltered, and as the star becomes still fainter, the dark lines are replaced by bright lines, especially those of ionized calcium. It is difficult to believe that a change of temperature is responsible for these stars becoming faint—rather, a progressive obscuration is suggested. Furthermore, no change of radial velocity is seen to accompany the change of brightness, as would be the case if the precipitate drop were caused by a change in dimensions. The cause of R Coronae Borealis variations is as yet incompletely understood.

STELLAR GROUPS

Many a night I saw the Pleiads, rising thro' the mellow shade,
Glitter like a swarm of fireflies, tangled in a silver braid.

TENNYSON, Locksley Hall

The preceding sections have introduced the stars as individuals. Physical properties, geometrical properties, and variations have revealed the enormous range within the stellar population. The sections now to come will consider the stars as members of the stellar communities in which they are found. A good deal may be found out about a star, considered as an isolated object. But much of this information is of the nature of a momentary snapshot of the star as we see it now. In order to learn something about how stars have developed, and are likely to develop, they must be considered in relation to their neighbors. Stars are to be understood by the company they keep.

If stars are found in close association, it may safely be assumed that they have always been in close association. There are perhaps 10^{11} stars in our own galactic system, and comparable (though often smaller) numbers in other galaxies. But stellar systems are so vast that the distances between individual stars are very great. It has been shown that during the probable age of the stellar system (less than 10^{10} years) close encounters between stars are so very unlikely that only a negligible fraction of the pairs that are now connected can have been produced by mutual capture. And for groups of stars that contain more than two members (multiple stars and star clusters), the chance that they can have been collected together by capture is effectively nil. In other words, when two or more stars form a physical system, we can assume that the matter from which they came has been close together from the first (whatever that loose term may mean). The simplest groups of stars from which to start are double and multiple stars.

1. DOUBLE STARS

We have already studied double stars under the heads of visual binaries (p. 293), spectroscopic binaries (p. 299), and eclipsing binaries (p. 304).

All these groups, taken together, will give some idea of the types of star that are found associated together. Binary stars give a very good picture of stellar associations in general, because all the evidence points to the probability that a large proportion, perhaps a majority, of the stars in our neighborhood are members of double or multiple systems. This is not necessarily true of all neighborhoods; for example, in the globular clusters or the regions of the galactic center double stars may be less common.

Visual binaries are found because they are far enough apart to be seen separately. The easiest visual binaries to discover are those that are nearly equal in brightness, like the two brighter components of α Centauri (a G0 and a K5 dwarf, of absolute magnitudes 4.7 and 6.1), or those of 42 Comae (F5 dwarfs, both of absolute magnitude 4.6). If one component of a visual binary is very much fainter than the other, it will be more diffi-

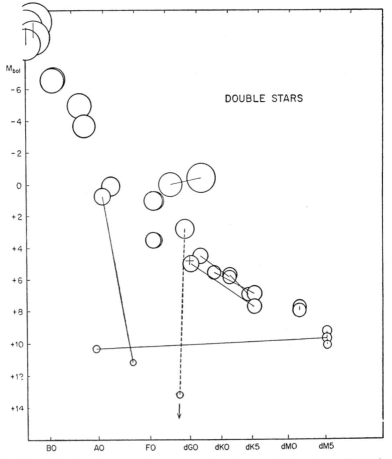

Fig. 15.1. Relations between the components of visual double stars. Spectral class is plotted against absolute bolometric magnitude. Size is on a conventional scale.

cult to discover unless the pair is very nearby. Nevertheless, there are many well-known visual binaries that have a large disparity in brightness. The α Centauri system has a third component, of absolute magnitude 14.9 ("Proxima Centauri," the nearest of the stars, a red dwarf). The system of o² Eridani is even more remarkable than that of α Centauri. It consists of two main-sequence stars and one white dwarf * (A = G5, absolute magnitude 6.2; C = M6, 12.9; B = A0, 11.4). The triple star μ Herculis consists of one rather bright G dwarf and two M dwarfs (A = G5, absolute magnitude 3.7; B = M2, 10.2; C = M2, 11.2). One of the most disparate pairs is that of Sirius (A = A0, +1.3; B = F0, +11.3). A star that almost qualifies as a visual binary, though observable only with the interferometer as double, is Capella (G0, −0.2; F5, +0.1). If these stars are

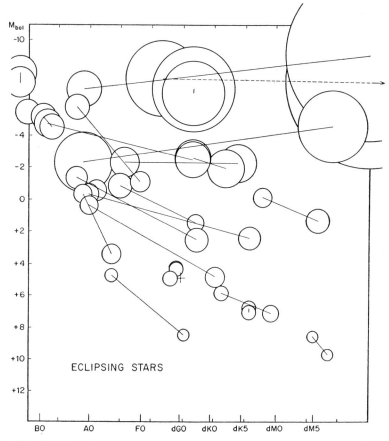

Fig. 15.2. Relations between the components of eclipsing-stars, presented as in Figure 15.1.

---

* The components of a multiple star are denoted, in order of brightness, by the letters A, B, C, etc.

displayed on a spectrum-luminosity diagram, the span of possible associations is seen to be wide.

The data are precise about these stars, and among them we find associations of main-sequence with main-sequence star, main-sequence with white dwarf, and giant with giant. The picture is amplified and the conclusions are strengthened when information is pooled for a larger number of visual binaries. The members of associated pairs are by no means always alike; in fact, they seem more prone to be unlike, despite the fact that like, or nearly like, pairs are the easiest to see.

The division of binary stars into visual, spectroscopic, and eclipsing pairs is an arbitrary one, which depends on method of detection. Wide pairs tend to be seen as visual binaries; close pairs, as spectroscopic and eclipsing binaries. For the two latter classes, also, it is possible to assemble the known facts about a number of well-observed pairs. They bear out the story told by the visual binaries: associated stars may be alike, but they may also differ greatly. The picture that follows contains information for visual and eclipsing binaries.

The variety of *size* of the stars in Figures 15.1 and 15.2 can only be illustrated schematically. The actual sizes are in a ratio of more than a thousand to one. The sizes in these and the following diagrams are on consistent scales.

The display of associated stars, drawn from the best-known visual and eclipsing binaries, shows a great catholicity in the choice of company. The visual binaries and some of the eclipsing-stars mark out the main sequence by a number of "twin systems," consisting of stars similar in brightness and color. But we also see main-sequence stars connected with giants, subgiants, and white dwarfs. There are also associations between pairs of giants (usually dissimilar) and pairs of supergiants (also usually dissimilar). There is one pair of subdwarfs, and one pair of white dwarfs is also known, but uncertainty of color and exact luminosity have excluded it from the diagram. Many white dwarfs are known as members of dissimilar pairs; their most common association is apparently with red dwarfs, though white dwarfs are coupled with the main-sequence stars Sirius and Procyon.

There are no recorded cases of M giant paired with M dwarf, but such pairs, if they did exist, would be the most difficult to find. A pair of M stars that differed by more than ten magnitudes would be practically undetectable, whether as a visual, spectroscopic, or eclipsing binary. It is not at all impossible that such pairs may exist. In fact, when the picture is surveyed, the conclusion that can be drawn is that *no association is definitely ruled out.* Therefore pairs of stars of almost every degree of diversity may represent objects of the same age.

One particular type of association is to be noted among the eclipsing-stars. A number of pairs consist of a main-sequence star and a subgiant.

These binaries appear as the "Algol stars," with one deep and one shallow minimum. The subgiant members of many of these pairs do not comply with the mass luminosity law (p. 311): they appear to be more luminous than other stars of equal mass. It has been suggested that these subgiant companions of main-sequence eclipsing-stars were formed (by splitting of a single, rotating star) at a fairly recent time (cosmically speaking) and have not yet had time to settle down completely.

The pairs of stars that are joined by the lines in the diagram may be regarded as of common origin and similar age. It does not follow that all the pairs themselves are of similar age. However, when we pass in the next sections to systems that contain larger numbers of members, we shall find arrays of stars, with a great variety of properties, that must also be supposed to be of common origin and similar age. These arrays are quite similar in many respects to the array shown in the previous diagrams. But there are differences too, not only between these diagrams and the cluster arrays, but also between individual cluster arrays. These differences offer our first clue to the fascinating and still controversial problem of the evolution of stars.

2. STAR CLUSTERS

Our study of stellar associations moves from the small groups of double and multiple stars to much larger groups, the clusters of stars. Two main kinds of star cluster are recognized within our galaxy, and their counterparts are also found in other galaxies near enough to permit detection of such groups. They are known as the *galactic* (or open) *clusters* and the *globular clusters*. The globular clusters are, of course, galactic objects too. The galactic clusters are so called because they occur only near the galactic plane, whereas the globular clusters are seen at all galactic latitudes. Some of them are the farthest known galactic objects from the Milky Way plane.

Galactic clusters and globular clusters differ sharply from one another. Galactic clusters contain, at most, a few hundred stars; globular clusters, hundreds of thousands. Galactic clusters occur in the galactic plane and share the galactic rotation. Globular clusters form an almost spherical system about the galactic center and are high-velocity objects. Moreover, as we shall see, the stellar populations of the two kinds of clusters show striking differences that provide a springboard for speculations on the development of stars.

3. GALACTIC CLUSTERS

About three hundred galactic clusters are known, and probably there are five or ten times as many in the parts of the galaxy that we cannot see. Many bright galactic clusters are seen in the nearest galaxies, the Magellanic

Clouds. In the spiral galaxies where they can be detected, they follow the outlines of the spiral arms.

Some galactic clusters are well-known naked-eye objects. The Pleiades and the Hyades are probably the best known; we may also mention the great double cluster in Perseus, the naked-eye group Coma Berenices, and ϰ Crucis, beside the Southern Cross. The Hyades cluster is not only an obvious group in the sky, but is observed to be drifting through space as a unit. A well-known group of stars, not so obvious as a cluster to the eye, but identified by common drifting motion, is the "Ursa Major cluster," which comprises most of the stars of the Big Dipper, and also the bright star Sirius. Some of the galactic clusters that contain bright, hot stars (such as the Pleiades) are embedded in nebulosity, a thing never observed for globular clusters.

The components of double and multiple stars must have had a common origin and must therefore be coeval. The same may be asserted, a fortiori,

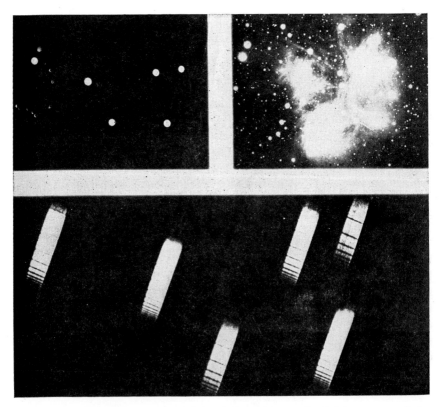

Fig. 15.3. The Pleiades. The left-hand figure shows the brighter stars of the cluster; the right figure, the surrounding nebulosity (compare Figure 13.14); below are shown the spectra of the brighter members of the cluster, all of Class B. (Yerkes Observatory)

for the members of clusters. Therefore it is of great interest to examine the display of stars of different kinds that is shown by a given cluster. Just as for double stars, we find a broad range of associations, but because clusters are more populous than multiple star systems, the relationships and trends are placed on a more solid observational basis. Recent work goes to show that the nature of the population of a cluster is very restricted.

The Pleiades and the Hyades. Even a superficial glance at the two bright clusters in Taurus shows that their makeup differs. The brightest stars in the Pleiades are blue, medium B stars, and the fainter stars have progressively later spectra and therefore lower surface temperatures. About two hundred and fifty stars are recognized in the cluster, and their absolute visual magnitudes range from about —3.5 to +10.5. The cluster Praesepe is rather like Pleiades, in that the brightest stars are the bluest, and the fainter stars are progressively redder; however, the brightest stars are not earlier in spectral type than A. In the Hyades, on the other hand, the brightest stars are giants of Class K, and there is also a Praesepe-like sequence, with fainter members of lower temperature, beginning at Class A. These clusters are all of about the same size, diameter about 10 parsecs; the brightest stars are crowded in the middle, so that the first impression is of rather smaller dimensions.

Fig. 15.4. Messier 8 (left) and the Trifid Nebula, in Sagittarius. Notice the cluster of bright stars within Messier 8. (Harvard Observatory)

When spectrum and luminosity are compared, we obtain a number
of sequences for different galactic clusters, which may be compared with
the points, joined by straight lines, that were shown in the diagram that
accompanied the previous section. Such a picture is shown below in
schematic form. The "main sequences" of the clusters break off in the
picture, not because of the absence of fainter stars, but because these
stars are too faint for their spectra to have been studied.

Photoelectric Studies of Magnitude and Color. Until a few years ago
the information on the population of clusters rested on estimates of spectral class, which are difficult for the fainter stars. Recently, however, the
spectrum has been replaced by a more accurate parameter—the color,
as measured with the photoelectric photometer. These data on colors and
magnitudes in galactic clusters have refined and clarified our information
about the physical condition of the members.

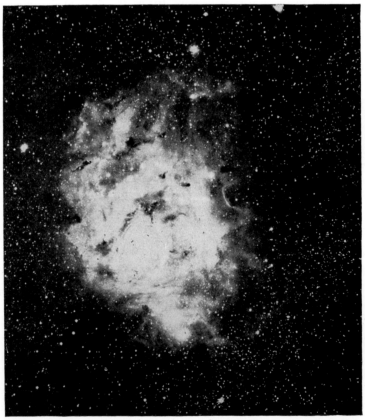

Fig. 15.5. Messier 8 on a larger scale. Notice the dark variegations on the face of
the bright nebula, and the dark bay on the right. The bright, high-temperature stars
that are responsible for the bright nebulosity are not visible in this photograph because of the brightness of the nebula. (Mount Wilson Observatory)

Within a single cluster we always find a sequence of stars whose brightness and color are closely related. For the stars of the Pleiades, for example, we find a sequence that runs from bright blue stars in a narrow sequence down to comparatively faint reddish stars. There is only a small spread in luminosity for stars of a given color.

The stars of the Hyades cluster and the cluster in Coma Berenices shows similar narrow sequences of stars, in which the bluest are the brightest; but these stars are neither so blue nor so bright as the most conspicuous of the Pleiades. And in addition, each of these clusters contains some red giant stars, about as bright as the brightest blue stars in the same cluster. All galactic clusters that have been studied show similar patterns of color and luminosity. All have a main backbone of stars that run from brighter blue to fainter red stars, and some also contain red stars about as bright as the brightest blue stars in the cluster. All galactic clusters seem to follow a definite pattern, but some of them fill out different parts of the pattern from others.

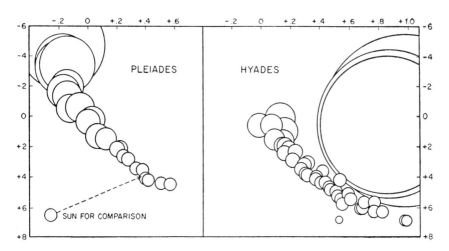

Fig. 15.6. Relation between absolute bolometric magnitude and color for the Pleiades and the Hyades. (Diagram based on photoelectric measures by Eggen.)

Color-luminosity Pattern for Nearby Stars. An even more striking result has emerged from a comparison of the color-luminosity arrays for galactic clusters with that for nearby stars (not members of multiple systems), whose parallaxes are well known (Figure 15.7). These stars are found to fall into much the same pattern as the cluster members, so that we can make the fairly safe assumption that they are similar physically to the stars in the clusters. Thus we can use the ideas that result from the supposition that the cluster stars have to have a common origin, to speculate on the origin of the non-cluster stars in our neighborhood.

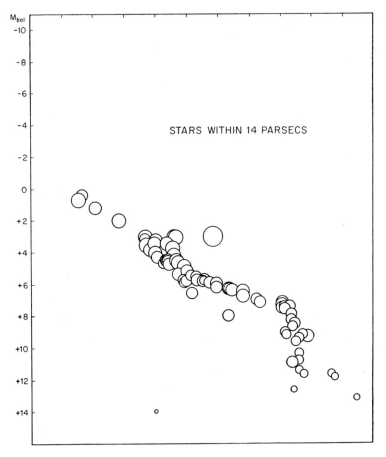

Fig. 15.7. Relation between absolute bolometric magnitude and color for a number of nearby stars. (Based on photoelectric measures by Eggen.)

We notice that no data for bright red or yellow supergiants are contained in the diagram. Further work is needed to determine whether they, too, fit a restricted pattern. There are galactic clusters (such as the double cluster in Perseus) that contain both red and blue supergiants, and these will furnish information on the matter when accurate colors and magnitudes have been measured.

The striking similarity between the color-luminosity arrays for galactic clusters and nearby, non-cluster stars is balanced by an equally striking difference. The same kinds of stars are represented in the two groups, *but not in the same proportions.* We have seen that among the stars in our neighborhood, the numbers increase rapidly as we go down the main sequence, so that the commonest kind of star is the red main-sequence star, somewhat cooler and fainter than the sun. But the fainter main-sequence stars do not mount up nearly so rapidly in numbers within a galactic cluster; in fact we receive the impression that (except at the bright end of the sequence, where the stars thin out) all luminosities are represented in about equal numbers. The contrast suggests a difference of history. Less than one star in a hundred thousand in our galaxy is now a member of a galactic cluster, and from what we know of the stability of galactic clusters and the age of the galaxy, it seems likely that the vast majority of galactic stars have never been members of galactic clusters in the past.

4. GLOBULAR CLUSTERS

The globular clusters are the most striking and most populous stellar groups within our galaxy. They are named from their shape, which suggests a compact ball of stars. They show a beautiful and striking symmetry; many look as though they are spherical, but some are slightly flattened.

In contrast to galactic clusters, which consist of hundreds of stars, globular clusters contain many thousands. It is difficult to say how many stars there are in a globular cluster, because we can observe only the

Fig. 15.8. The globular cluster ω Centauri. (Boyden Station, Harvard Observatory)

brightest; but the total brightness of the whole cluster gives some information. In our neighborhood of the galaxy, the number of stars continues to increase with decreasing absolute brightness down to perhaps absolute magnitude $+15$ or $+16$. No globular cluster has yet been examined for stars fainter than $+5$, and if the number of fainter stars goes on increasing as rapidly as in our neighborhood, the next ten fainter magnitudes may account for a very large number of stars indeed. Estimates of the population of a rich globular cluster range from fifty thousand to fifty million; perhaps the best guess places it as of the order of a hundred thousand. There are some globular clusters that seem to be less populous than others; there may be a factor of ten or more in total content, but perhaps only the brightest stars are deficient.

Over a hundred globular clusters are known in our galaxy, and probably these represent at least half the total number. Globular clusters, therefore, are much rarer than galactic clusters, of which the galaxy may contain about a thousand. Two of them (ω Centauri and 47 Tucanae) are hazy naked-eye objects of the fifth magnitude; these are among the nearest globular clusters, but others may be even nearer to us and less bright because of fainter absolute magnitude or obscuration.

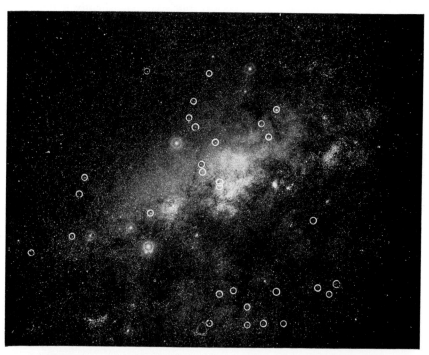

Fig. 15.9. The region of the galactic center. Circles are drawn around the globular clusters. Thirty (almost one-third of all those known) are visible in this one photograph. (Harvard Observatory)

The distribution of the globular clusters is remarkable. Most of them are concentrated on one side of the sky, toward Sagittarius and Scorpio. At the time when the sun was thought to be at the center of the galactic system this lopsided distribution was puzzling. Thirty years ago, however,

Shapley realized that these clusters can be considered to represent a sort of spherical exoskeleton for the galaxy. The new view resulted in pushing the solar neighborhood toward the edge of the galactic system; the unsymmetrical distribution of the clusters then appeared as a natural consequence of the fact that we are viewing a cosmic continent from a point near the shores; naturally most of the major features lie on one side of us.

Not only did Shapley's work displace the sun from a central position. It also revised completely the contemporary ideas as to the size of the galaxy. Many globular clusters contain RR Lyrae variables (p. 370), and the presence of these stars of known luminosity made it possible to measure their distances. The galaxy turned out to be ten times as large as had been thought before, and our distance from its central regions to be about nine kiloparsecs.* As we can see stars about five kiloparsecs from us in the direction opposite to the galactic center, it follows that the radius of the galaxy is about fourteen kiloparsecs, and its diameter nearly thirty. The symmetry of the system of globular clusters placed the central regions of the galaxy in the constellations of Scorpio, Sagittarius and Ophiuchus, the region in which we see most of them.

It may seem surprising that our eccentric position in the galaxy is not made obvious by a striking asymmetry in the brightness of the Milky Way. The light from the central regions is, however, impeded by the clouds of dark material that lie in the galactic plane and cut off the brightest districts completely. The region of the galactic center is difficult to study by photographic means; the nearest region to it that has yet been analyzed lies a thousand parsecs from the nucleus, above the galactic plane.

Radio observations, however, show an intense maximum at the position of the galactic center: microwaves penetrate the obscuring matter much better than the shorter light waves. If there were no obscuring matter in the Milky Way, the brilliance of the central regions would be very striking. Unfortunately for us the sun lies near the central plane of the galaxy. The solar orbit seems to be slightly inclined to the plane, and a million years ago we may have been outside the dense central "sandwich filling." Perhaps some of the primitive inhabitants of our globe saw the nuclear regions in their full brilliance.

The distances of a number of globular clusters were measured by means of their RR Lyrae stars. From these clusters Shapley devised other measures of distance, based on apparent diameters, bright nonvariable stars, and so forth. The distances of most globular clusters are now determined with fair certainty. Their total luminosities can thus be determined from their apparent magnitudes. The absolute magnitudes of globular clusters have a large range: the brightest (ω Centauri and perhaps 47 Tucanae) have absolute photographic magnitudes about −10; for the faintest, the absolute magnitudes are as low as −5. Probably the brightest

* 1 kiloparsec = 1000 parsecs (see Appendix I).

globular clusters are the most populous, at least in the brighter stars. As we shall see, the faintest galaxies seem to be about as luminous and as populous as the brightest globular clusters, but their dimensions are very much greater. No intermediate objects have been found.

The motions of globular clusters are as remarkable as, and closely related to, their distribution. Radial velocities for about fifty have been measured by N. U. Mayall at the Lick Observatory (those accessible from the Northern Hemisphere). They range from $+290$ to -360 km/sec. In other words, *the globular clusters belong in the high velocity group.* Unhappily, they are so remote that proper motions cannot be determined. However, from their large radial velocities and their almost spherical distribution, it is clear that they travel in orbits of high eccentricity and high inclination about the galactic center; probably both in eccentricity and in inclination they display the extremes of orbits encountered in the galaxy. The stars that come nearest to the globular clusters in these respects are the RR Lyrae stars; and it is significant that such stars are particularly common in globular clusters. RV Tauri stars and Type II Cepheids (W Virginis stars), which are also high-velocity objects, are likewise well represented there.

The stellar population of individual globular clusters is quite distinctive; unlike the Pleiades-like clusters, they show a distribution in which *the brightest stars are the reddest.* They contain supergiants, the reddest being at most of absolute visual magnitude -2. There are no published data about stars in globular clusters of absolute magnitude fainter than $+4$, but with modern equipment it will be possible to extend the information

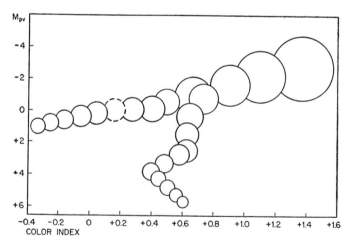

Fig. 15.10. Relation of absolute photovisual magnitude to color index for the globular cluster Messier 3; sizes of stars are shown on a conventional scale; the ratio in diameter from largest to smallest is actually 100 to 1. The broken circle denotes the RR Lyrae stars of the cluster. (Based on the work of A. H. Sandage.)

to about $+5$ or $+6$. With the exception of a minority of stars, all globular clusters show essentially the same "family portrait," which is sketched in the diagrams. It should be compared with the very different arrays for galactic clusters (Figure 15.6) and for the nearby stars (Figure 15.7).

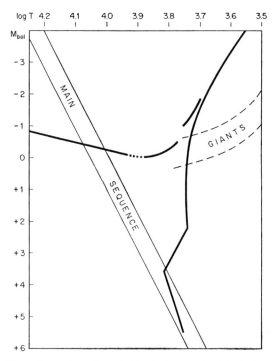

Fig. 15.11. Heavy lines show the relation between absolute bolometric magnitude and logarithm of effective temperature for the observed stars of the globular cluster Messier 3. The dots show the "gap" in which the RR Lyrae variables are found. The course of the main sequence and the location of the giants of Population I are indicated (compare Figure 11.6). (Based on the work of A. H. Sandage.)

This very striking diagram (still largely conjectural) shows, at least, that the globular cluster population is quite unlike the galactic cluster population, and resembles in rough outline the subgiant-subdwarf sequences for the nearby stars, only sparsely represented in the nearby galactic population, and associated in our neighborhood with high velocity. We may perhaps regard the globular cluster population as the "family portrait" of a group of stars of advanced age—older, at any rate, than the galactic-cluster arrays. The family of stars that lies near the galactic plane, shares the galactic rotation, and displays the family portrait outlined by the galactic clusters, is called "Population I"; the globular-cluster population, associated with eccentric, high-inclination orbits and stars that do not share the galactic rotation, is known as "Population II."

Probably they represent extremes of stellar development, and we may expect to observe transitions from one to the other, unless we are to admit that our galaxy and other galaxies consist of two interpenetrating and concentric systems that differ completely in origin—a most unpalatable suggestion.

The globular clusters contain many RR Lyrae stars, and RV Tauri and W Virginis stars are well represented. There are few true long-period variables (some astronomers think that there are none) and only a few red semiregular stars like those near to us. One nova has been observed in a globular cluster and one well-known planetary nebula; both these types of object are generally regarded today as belonging to "Population II," because of their distribution within the galaxy and their motions (insofar as these are known).

Individual globular clusters differ in their variable star population: some contain over a hundred RR Lyrae stars, some have not been found to contain any. And the lengths of period, as well as relation of period and form of light-curve, differ from cluster to cluster without obvious relationship to other cluster properties. These differences are related to differences in color-luminosity array. They must be of great significance in the problem of the history of globular clusters.

The conditions within a globular cluster are extreme; nowhere else, save perhaps at the centers of some galaxies, are the stars so close together as at the center of a dense cluster, or distributed so completely at random. The globular clusters seem to contain no interstellar dust, and no evidence exists that they contain interstellar gas. In this way they are very different from galactic clusters.

The stars in globular clusters are undoubtedly all moving in orbits about the center of gravity of the cluster. Many of the orbits must have high inclinations, or the clusters would not be so nearly spherical; probably they also have a large proportion of high eccentricities. The theoretical study of motions within a globular cluster is difficult, and the observation of the motions of single stars is even more so. The few stars whose radial velocities have been measured have not very large speeds relative to the center of the cluster; but this would be expected because the clusters, though populous, are not very massive; the mass of a typical cluster is perhaps one millionth of the mass of our galaxy.

When our galaxy was recognized as being surrounded by an aura of globular clusters, a search was begun for globular clusters associated with other spiral galaxies. Only the nearby galaxies give any promise of success; but about two hundred globular clusters are known to be associated with the great spiral in Andromeda. Globular clusters have also been found in the nearest dwarf galaxies.

Clusters that look like globular clusters have been observed in the Magellanic Clouds. However, the colors and "family portraits" of some of them seem (from present evidence) to be more like those of galactic clusters than of our own globular clusters. Evidently a cluster can be globular in form and galactic in population. In our own galaxy, the properties of galactic clusters associate them with Population I, the flattened system that partakes of the galactic rotation; those of globular clusters link them with Population II, with motions in orbits of high eccentricity and high inclination.

Careful photoelectric study of the clusters near the two Magellanic Clouds has revealed that some have the integrated colors typical of galactic clusters, but that others resemble globular clusters in color. That at least one of these, near the Small Magellanic Cloud, has many of the properties of a globular cluster is shown by the fact that several RR Lyrae stars have been found within it. The clusters of the Magellanic Clouds are of extreme interest because they are associated with galaxies that are in many ways very unlike our own, as will appear in the next chapter.

CHAPTER XVI

STELLAR SYSTEMS

O vast Rondure, swimming in space,
Cover'd all over with visible power and beauty,
Alternate night and day and the teeming spiritual darkness,
Unspeakable high processions of sun and moon
    and countless stars above . . .
Now first it seems my thought begins to span thee.
WALT WHITMAN

The previous section has dealt with stellar groupings, has discussed the types of stars that are found grouped together within the galaxy, and has introduced the subject of stellar populations. We are now concerned with stellar systems, or *galaxies,* which are more or less complex collections of stellar groupings, isolated from one another in space (though often found grouped in clusters of galaxies).

Galaxies, like stars, display wide variety. They range from the uniform and symmetrical, through complex symmetrical types, to complex forms without structure or obvious symmetry. They can be arranged in a rough sequence, from elliptical galaxies, through spirals of various shapes, to irregular galaxies, and this sequence probably has an evolutionary significance. We shall begin by considering the irregular galaxies, generally thought at the present time to be the least advanced. We say "least advanced" rather than "youngest," because the stage of development of a galaxy probably does not depend on years alone, but on years and mass, or years and dimensions, or years and composition.

1.   THE MAGELLANIC CLOUDS,
IRREGULAR GALAXIES

The Magellanic Clouds, the two galaxies nearest to our own, are good examples of the irregular type. They were first recorded by the expedition headed by Magellan, that circumnavigated the globe in 1519, and they

420

have been named after its leader. They lie in the far southern sky, in the constellations Dorado and Tucana. They are faintly visible to the naked eye as bright clouds that look like isolated scraps of the Milky Way; they have even fooled unwary observers from the north into the belief that the sky was clouding up! One of them is larger and brighter than the other, and they have received the names Large Magellanic Cloud and Small Magellanic Cloud, or the Nubecula [little cloud] Major and Nubecula Minor.

Fig. 16.1. The Large Magellanic Cloud. (Boyden Station, Harvard Observatory)

The Large Cloud is a galaxy much above the average size and brightness; the Small Cloud, though fainter, is nevertheless brighter than the average. The total absolute magnitudes of galaxies have the large range from about -19 to about -11, with the average about -14. The Large and Small Clouds have, respectively, absolute magnitudes -17.5 and -16.

Both the Magellanic Clouds are roughly circular in outline and both show concentrations of brightness toward the center. The Large Cloud, in fact, has a well marked central "axis," and suggests an embryonic barred spiral. Both are well resolved, i.e., many individual stars are photographed within them. The brightest stars are of about the ninth magnitude, and fainter stars are detected in increasing numbers down to the eighteenth magnitude; there are certainly many fainter stars which cannot be reached with existing telescopes. Hundreds of thousands of individual stars can

Fig. 16.2. The Small Magellanic Cloud. Two galactic globular clusters are seen in the foreground; the bright one to the right is 47 Tucanae (compare Figure 11.4). (Boyden Station, Harvard Observatory)

be studied in the two Clouds, more than in any other galaxy except our own. So we can form a good idea of the nature of their population.

The two Clouds are at about the same distance from us; the "modulus" (the quantity $m - M$ in the important formula, p. 274, that connects absolute magnitude, apparent magnitude, and distance) is about 19, so the brightest stars observed in them are very luminous supergiants of absolute magnitude -10, and stars can be observed a little fainter than absolute magnitude zero.

The supergiants in the Clouds are both blue and red stars. There are a number of O, B, and W stars, and also some luminous K stars. The distribution of color among these high-luminosity stars suggests Population I, represented by some galactic clusters, such as the Double Cluster in Perseus (Pleiades and Hyades do not attain such high luminosities). The colors of the fainter stars in the Clouds have not been studied at present; they will be most important and significant in comparison with the data for the colors of galactic stars.

The presence of dust and gas can be determined by the occurrence of obscuration and of bright nebulosity. Bright nebulae are numerous and conspicuous in the Large Cloud. The "Tarantula Nebula," 30 Doradus (Figure 16.3), is one of the largest and most brilliant bright nebulae known. In the Small Cloud bright nebulae are less common. Dark nebulae are detected in the Large Cloud by irregularities in apparent stellar distribution, but such evidence is absent for the Small Cloud. An even more sensitive way of detecting dark material is to count the number of faint, distant galaxies that can be seen *through* the two Clouds; the Large Cloud cuts down the number appreciably, but the Small Cloud does not. Hence we can conclude that the Large Cloud contains considerable interstellar dust, and the Small Cloud, comparatively little.

Fig. 16.3. The "Tarantula Nebula," one of the largest bright nebulae known, a member of the Large Magellanic Cloud. Notice that it contains a large cluster of stars which must, from their apparent brightness, be extremely luminous. (Boyden Station, Harvard Observatory)

The radial velocities of the two Clouds are readily measured by means of the well-defined bright-line spectra of the nebulae within them. Both appear to be receding from our galaxy at high speed (273 km/sec for the Large Cloud, 170 km/sec for the Small). However, this high speed is almost entirely a reflection of the galactic-rotation motion of the solar neighborhood, and when allowance is made for its probable speed, the two Clouds are inferred to be describing almost circular orbits round our galaxy. They are unfortunately too distant for their proper motions to be detectable, even against the background of faint and distant galaxies.

The best-known property of the Magellanic Clouds is the presence of large numbers of Cepheid variables. The relation of apparent (and

Fig. 16.4. A region of the Large Magellanic Cloud, showing luminous stars, nebulosity, star clusters, and irregular obscuration. (Boyden Station, Harvard Observatory)

therefore absolute) magnitude to period for these stars led to the discovery of the period-luminosity law (p. 365), which forms the basis of the study of cosmic distances. The relation between period and brightness is not an exact one; at any one period, the stars show a range of about a magnitude in brightness. Even if allowance is made for inaccuracies of measurement and of standards, and for the possibility that some of the variables may be unresolved double stars, the magnitudes still show an appreciable spread. In the Large Cloud irregular obscuration may be responsible, but as the Small Cloud contains little dust, the effect must be small there. If the Clouds were thick systems, some of the variables would be appreciably farther from us than others, and their moduli correspondingly different. But even when probable allowances have been made for all these factors, there still seems to be real dispersion in brightness at a given period. Possibly the period-luminosity curve is actually compounded of a number of overlapping, roughly parallel curves.

Several thousand Cepheid variables are known in the two Clouds—more than have been found in our whole galaxy. At least for the Small Cloud, discoveries of Cepheid variables are probably almost complete, whereas in our galaxy not a tenth of the Cepheids may have been found. Cepheids are at least as common, probably much commoner (in proportion to the total number of stars), in the Clouds than in our own system. Even if the total numbers of Cepheids should prove to be the same in the Clouds as in our galaxy, the proportion would be higher in the former, since our galaxy contains far more stars.

Fig. 16.5. A region of the Large Magellanic Cloud, showing several star clusters. S Doradus, the star marked in the upper center, is probably the brightest individual star known. It is of high temperature, and has an absolute photographic magnitude about −9; it is possibly an eclipsing-star of very long period. (Boyden Station, Harvard University)

Although it is generally (and probably correctly) assumed that the course of the period-luminosity curve is the same in the Clouds as in our galaxy, the distribution of the periods of the stars is very different. The Small Cloud, in particular, displays large numbers of Cepheids with periods of two and three days—periods that are quite uncommon in our galaxy. The Large Cloud shows a period distribution nearer to that of our own system; in both Clouds, the type of light-curve that goes with a given period seems to differ very slightly from that found in our galaxy for the same period, although the general progression of light curve with period seems to be similar.

The fact that the Magellanic Clouds, particularly the Large Cloud, are very rich in luminous blue stars, nebulosity and dust, and are populated by Cepheid variables, all of which are well-known members of Population I in our galaxy, has suggested that in the Clouds we may see "pure" specimens of that population. However, both Clouds contain clusters that look like globular clusters; some of these have colors typical of globular clusters, and one at least contains RR Lyrae stars. Moreover, each of the Magellanic Clouds has been the site of three known novae, and novae are probably members of Population II. Probably, therefore, the Clouds do contain some Population II stars. It seems likely that most stars in the Large Cloud are of Population I, and perhaps the populations in the Small Cloud are divided about equally. In any case, our galaxy has a much larger proportion of Population II stars (about 90%) than either Cloud.

Fig. 16.6. The small irregular galaxy I.C. 1613. (Mount Wilson Observatory)

Other Irregular Systems. In the immediate neighborhood of our galaxy there are two other irregular galaxies that are somewhat like the Magellanic Clouds: No. 6822 of the *New General Catalogue* and No. 1613 of the *Index Catalogue* (NGC 6822 and IC 1613). Both are more distant than the Clouds. Both contain nebulous material. Both are rich in Cepheids, which conform to the period-luminosity relation, and both are easily resolved into stars and evidently contain many supergiants.

The irregular galaxies, typified by the Small Magellanic Cloud, are conglomerations of stars, nebulae, and some dust, without obvious symmetry, although they are not structureless in the sense in which a globular cluster is structureless. They are a well-defined galactic type, comparatively rare relative to elliptical galaxies, or even to spirals. They can be placed with confidence at one end of a sequence of stellar systems. Today there is a tendency to regard them as galaxies in the "early stages," though this does not necessarily imply that their age, *in years,* is smaller than that of some "more advanced systems." The rate at which a galaxy develops depends on several factors, of which the elapsed time is only one.

The Magellanic Clouds are large irregular galaxies, with diameters about 4.5 kiloparsecs and 3 kiloparsecs respectively (perhaps a fifth and a seventh of the size of our own galaxy). The diameters of IC 1613 and NGC 6822 are even smaller than that of the Small Cloud, about 1.1 and 0.9 kiloparsecs. Roughly speaking, the smaller irregular galaxies tend to be more structureless than the larger ones, and indeed the Large Cloud suggests a barred spiral form.

Spiral galaxies are commoner among stellar systems than irregular galaxies, and tend on the whole to be larger. The smaller spirals may be 2 or 3 kiloparsecs across, the largest ones, eight or ten times that size.

3. THE SPIRAL IN TRIANGULUM, MESSIER 33

The beautiful spiral in Triangulum (No. 33 of Messier's catalogue) furnishes an introduction to spiral galaxies. It is a fairly small specimen, well known to us because it is one of the two nearest, the other being the great Andromeda spiral, Messier 31. Messier 33 is a loosely wound spiral, fairly symmetrical in shape. In appearance it suggests a pinwheel, and a pinwheel it is, turning on its nucleus at considerable speed. Unfortunately it is impossible to decide in which direction this particular spiral is turning; with others we have been more fortunate.

About twenty-five years ago the distance of Messier 33 was first successfully measured when Hubble found a number of Cepheid variables in it. The *modulus* $(m - M)$ is nearly 24 magnitudes for this system, according to the best measures now available. Thus the distance is about 500 kiloparsecs or 1,750,000 light years (p. 274). The distance is rather uncertain for three reasons. First, we do not know exactly how much allowance to make for cutting down of brightness by nearby obscuring matter within our own galaxy: guesses can be made by counting very faint and distant galaxies in the same direction, which would share the obscuration; however, galaxies (even faint and distant ones) are not quite evenly distributed, and such counts may lead one astray. Secondly, the measured magnitudes may not be quite correct: to determine a scale of magnitudes that goes all the way from the first to the twenty-second (a light ratio of more than a hundred million) is a very difficult observational problem. Photoelectric measures are removing our doubts (and errors) from the magnitude scale, and revisions may well amount to several tenths of a magnitude. Thirdly, the zero point of the adopted scale of absolute magnitudes for the Cepheids may not be quite correct. These three sources of uncertainty might conspire to double or halve our estimate of the distance of the spiral in Triangulum.

When the distance of the spiral is known, the angular size tells us its true size immediately. The obvious outlines of the system, as seen on a photograph, are one degree of arc by half a degree, or 7.6 by 3.8 kiloparsecs. The system looks elliptical, but long-exposure plates, analyzed with a sensitive photometric instrument, show that it is surrounded by a less elliptical haze, $1°.5 \times 1°$, or 11.4 by 7.6 kiloparsecs. Probably we are looking at an object that is essentially spheroidal, and would look circular if seen face-on. We can visualize Messier 33 as a flattened system, tipped toward us so that its central plane makes an angle of about 60° with the central plane of our galaxy. We shall see later that there is direct evidence for this tipping. However, we cannot tell which edge of the spiral is tipped toward us, which away from us.

The bright inner body, readily visible on a photograph, consists of the spiral arms, and is about 7.6 kiloparsecs across. It is probably quite flattened. The spiral arms are imbedded in a larger region, which looks hazy

Fig. 16.7. The galaxy in Triangulum, Messier 33. (Mount Wilson Observatory)

on our photographs; it is probably spheroidal, and is about 11.4 kiloparsecs across. Such a structure is probably typical of all spirals: a flattened central region that contains the spiral arms and a larger, less flattened substratum.

The most conspicuous features of Messier 33 are its bright nucleus, and the luminous knots that define the course of the spiral arms. These luminous knots are not generally stars, but star clusters and bright nebulae, the larger ones comparable perhaps to the whole constellation of Orion, the smaller ones to the Pleiades. Not many individual stars are to be seen: all those that are visible on the photographs are within the spiral arms. They are either blue stars, or Cepheid variables, or novae. Since the modulus of the spiral is nearly 24 magnitudes, and since stars fainter than magnitude 23 cannot be photographed, all these stars are brighter than about absolute photographic magnitude —1, and most of them are true supergiants.

It is safe to say that the stars seen in the spiral arms of Messier 33 are like those that are found in the galactic clusters of our own system, or in the parts of the galaxy that lie near the galactic plane and share the galactic rotation. Along with the bright nebulae and the star clusters we can trace streaks of dark material. Probably the gas and dust are confined closely to one plane, as in our own galaxy.

There must be vast numbers of stars too faint to be photographed in Messier 33. The total absolute magnitude of the system is about —17.5 (intermediate between those of the Large and Small Magellanic Clouds), and probably most of the light comes, not only from stars too faint to be photographed, but from stars fainter than the sun, which would be *more than six magnitudes below the limit of observation.* The same is true of our own galactic system: stars fainter than the sun form the vast majority.

The light from the spiral arms is only a small part of the total light of Messier 33: there is much more light in the continuous substratum, which mounts up steadily in brightness from the faintly seen edge to an extremely brilliant center. The substratum is *unresolved,* or in other words, the individual stars are fainter than absolute magnitude —1; it is also redder, on the whole, than the arms. It seems to contain little nebulosity or dust. Probably it consists of stars of Population II, such as we have seen making up the personnel of globular clusters. The arms, on the other hand, represent Population I. Probably at least three-quarters of the light of Messier 33 comes from the substratum, with its unseen stars, and less than a quarter from the arms.

The variable stars that were found in the arms gave the first clue to the distance of Messier 33. Cepheid variables (apparently just like the galactic ones of similar period) are rather common there. A very few novae have been found, but novae are only a little more common in this spiral than in the Magellanic Clouds. No RR Lyrae stars are known in Messier 33, nor are any long-period variables. But as these stars would be of about the twenty-fourth magnitude, their nondiscovery is not surprising.

All spiral galaxies look like pinwheels, and it was natural that attempts should be made to estimate their rotational speeds. When the distances of galaxies were not known, hopes might reasonably have been entertained that they were near enough for the internal motions to be directly measurable. Thirty years ago, indeed, some observers believed that they had detected very small rotational motions in several of the spirals, including Messier 33. However, when Cepheid variables were found, the galaxies were realized to be so distant that the suspected rotational speeds were physically implausible, being greater than the speed of light. How the very careful measures led to a result that is now generally considered spurious is not thoroughly understood; perhaps the measures resulted from a combination of unconscious prejudice and differences in plate quality between the earlier and later photographs.

A better method of measuring the rotation of galaxies gives results independent of distance. The spectrum of the system was photographed from edge to edge. The opposite edges showed velocity differences that were clearly Doppler shifts, so that the galaxy was seen to be turning, with one edge approaching the observer, one receding from him.

In order that the rotation be measurable by Doppler motion, the system must of course be somewhat tipped; all parts of a system that was seen face-on would be moving across the line of sight if it were rotating, and no Doppler motion would be observed. In fact, the greater the tilt, the more does the Doppler effect appear, and it would be a maximum for a system that was presented exactly edge-on. Messier 33 shows a marked rotational effect, so it must have an appreciable tilt; and the tilt of about 60°, suggested by its apparent shape, can be used to deduce the real rotational motions from the projected ones that are observed.

The system is rotating in a very remarkable way. It does not turn like a wheel; neither does it behave like a planetary system, with the parts nearer the center showing the largest speeds. The actual motion is, indeed, a compromise between the two. Near the center the system turns most like a wheel, but the farther we go from the center, the more "planetary" does the rotation become. If the system turned like a wheel, it would remain always the same in shape and structure; however, since there are departures from the wheel-like (or "rigid") rotation, the form of the spiral arms must be changing continually.

Although we can say so much about the rotation of Messier 33, we do not know which way the spiral is turning. Is it going arms forward (*leading*) or arms trailing (*following*)? Look at the picture and try to visualize first one edge, and then the other, as the near edge. There is no reason to choose, for this system, between the two orientations. Fortunately some other galaxies are kinder, and for many of them we can say with some confidence which edge is inclined toward us, and which away from us.

The velocity with which Messier 33 is turning (from 50 to 150 km/sec, depending on location) would carry it around in from 20 million to 200 million years. On simple assumptions as to mass-distribution, one can calculate from the laws of motion that the whole galaxy Messier 33 must have a mass more than 3000 million times that of the sun. This mass, though large, is little more than one-hundredth of the mass of our own galaxy or of the Andromeda spiral.

4. THE ANDROMEDA SPIRAL, MESSIER 31

The great spiral in Andromeda is probably the best known of all spiral galaxies, largely through the work of Baade and of Hubble. It is comparatively near to us, so that a good deal of detail can be studied in its structure. It was No. 31 in Messier's famous catalogue.

The distance of Messier 31 is about 450 kiloparsecs. Our knowledge of the distance depends on three data: the actual measured magnitudes, the absolute magnitudes of recognizable members, and the amount of light

Fig. 16.8. The great galaxy in Andromeda, Messier 31. Notice the two companion galaxies. The ring around the bright star is a photographic effect. (Lick Observatory)

Fig. 16.9. The central region on Messier 31. (Mount Wilson Observatory)

lost by nearby absorption. All three may be somewhat revised in the future, especially the first, and perhaps the second. So the distance may be rather different from the figure mentioned; it is perhaps more likely to be greater than less, but the difference will not be very large.

Our knowledge of the distance of Messier 31 dates from the discovery of Cepheid variables in the system by Hubble. Before that time, as with Messier 33, it was thought that actual rotational velocities could be directly measured, but the motions thus suspected are now considered to have been spurious; at the distance of nearly 2 million light years, rotational motions would only be detectable on a baseline of centuries. The radial velocities, however, show that the spiral is rotating, and rotating in a complex fashion, not like a wheel. The total luminosity of the system is about −19.5.

Structure of Messier 31. The system is elliptical in outline and is evidently much tilted; the angle with the line of sight is about 15°. This is more fortunate than unfortunate, for it means that rotational speeds are readily measured by radial velocities; it also makes possible a decision as to which edge of the system is nearer to us, so that we can determine which way the spiral is turning. This was not possible for Messier 33.

The total diameter of the spiral, as outlined by the stars that can be readily photographed, is about 2°. With delicate photoelectric measurements, the system can be traced to about twice that distance. The total diameter is about 40 kiloparsecs, a little larger than we now suppose our own galaxy to be. The readily observable portion is about 25 kiloparsecs in diameter.

The system consists of an obvious bright nucleus, about which two symmetrical systems of arms make several coils. Four arms can be traced on one side of the nucleus, five on the other, so each arm makes between two and three complete turns. In Messier 33, on the other hand, one can trace the arms through less than one complete turn. So the Andromeda spiral is more "tightly wound" than Messier 33.

The Nucleus and Substratum. Careful measures of the distribution of light in the system show a steady fall in brightness from a very bright

Fig. 16.10. The southern region of Messier 31, showing the bright Type I stars in the arm. (Mount Wilson Observatory)

nucelus toward a faint edge. Upon this uniform distribution the spiral arms stand out as a slight excess of brightness. The smooth central distribution may be called the "substratum"; it carries most of the brightness of the system, though its individual stars are only just observable. The spiral arms, where the stars can be readily seen, contribute a smaller part to the total light, probably about one-sixth. As Baade has expressed it: the body of the system is the cake; the arms are just the frosting—or they might be likened to the filling of a layer cake.

The substratum seems to be uniform and structureless. It contains little or no dust, and if there is gas, it does not appear in the form of bright nebulae. The substratum is, in fact, almost transparent; galaxies can be seen through it (*between the arms*) almost to the center of the system. It consists of great numbers of faint stars, the brightest of which can just be observed as a uniform haze, both in the nucleus and between the arms. They are between the twentieth and twenty-first magnitude; therefore, they are of about absolute photographic magnitude -1 or -2, very like the brightest stars observed in globular clusters. The substratum is spherical, or at least spheroidal, in form. In addition to the individual stars, it contains a number of globular clusters, which will be mentioned in more detail presently.

The Spiral Arms. The spiral arms are outlined by bright nebulosity and dust (visible because it obscures part of the system). They contain bright blue stars, Cepheids, star clusters that resemble our own galactic clusters, and emission nebulae, of which a thousand have been charted. Everything seen in or through the arms is slightly reddened by the dust.

The most interesting denizens of the arms are the numerous Cepheid variables. About fifty were observed by Hubble, with periods between ten and forty-eight days, and of magnitudes from 19.3 to 18.1. Their apparent magnitudes show that they fit the period-luminosity curve, and it was by means of these stars that the distance of the system was first successfully measured. Their light-curves are very similar to those of galactic Cepheids of like period. There are no doubt many fainter Cepheids, of shorter period, in the system, and these will doubtless be discovered when they are sought on photographs that go not only to the twentieth but to the twenty-first or twenty-second magnitude, made with the 200-inch telescope. The very brightest stars in the arms have absolute magnitudes of -7 or -8, comparable with the brightest stars known in our galaxy. Many of these seem to be supergiants of Class F, something like Canopus; a few are variable stars, and one has the exceptionally large period of about six years.

The spiral arms of Messier 31 contain, in fact, stars similar to those found in the galactic plane of our own system, that share the galactic rotation. They are probably closely confined to a plane in Messier 31.

The Globular Clusters. About two hundred globular clusters can be observed around the Andromeda spiral. The edges of these clusters are just resolved; they are seen to consist of stars very like those of the substratum. In fact, they are simply tight condensations of substratum stars No RR Lyrae stars have been found in them as yet, though we may expect that such stars are common, both in the globular clusters and in the substratum of the spiral. A search with the 200-inch telescope shows that they are below the limit of observability.

The absolute magnitudes of the globular clusters around Messier 31 are quite like those of the galactic globular clusters. These globular clusters have furnished one very important piece of information. Baade has used them to decide which edge of the system is the nearer to us. This is done on the assumption that they are arranged in a spherical or spheroidal way. If, then, there is obscuring matter in the arms, those that lie behind this obscuration will be dimmed, and the numbers of globular clusters of a given brightness will be cut down for the half of the sphere that is behind the central plane. Baade has concluded that the south side is nearer to us than the other.

A naïve look at the system leads to the same conclusion. The south side of the nucleus is crossed by a much stronger obscuration than the other. If we think of the nucleus as a uniform bright ball, we can visualize the central layer of obscuration lying in front of it and thus picture which way the whole system is oriented.

Over a hundred classical novae have been observed in Messier 31, in addition to the supernova of 1885 (S Andromedae). They seem to be very similar in every way to the novae of our galaxy, and appear in the Andromeda system about as freely as in our own. In Messier 33 they are much rarer, and in the Magellanic Clouds they are very rare indeed. The way in which the novae are distributed is rather like the distribution of the globular clusters and suggests that they are not confined to the plane of the system, but have a somewhat spheroidal distribution. They are evidently not confined to the arms; for instance, they occur in the structureless nucleus, where Cepheids, bright B stars, and nebulosity are not found.

Roughly speaking, we may consider that the Andromeda spiral consists of a uniform substratum (the "cake"), of Type II stars, including novae and globular clusters, and the spiral arms (the "frosting"), of Type I stars, such as bright blue stars, yellow supergiants, Cepheids, dust, and bright nebulosity.

Rotation of Messier 31. The large tilt of the system makes a determination of the rotational speed rather easy. It is in nonuniform rotation; the circular speed is small in the center and increases outward to about 5 kiloparsecs, and it decreases still farther out. In other words, the rotation is more wheel-like ("rigid") toward the center, more planetary toward the edges. The speed is about 225 km/sec at its greatest. From this speed, and from the known dimensions of the system, we conclude that the mass of Messier 31 is about 2×10^{11} suns, comparable to that of our own galaxy.

A combination of the rotational speed with the knowledge of which edge of the spiral is the nearer to us enables us to say which way the system is turning. If the conclusions about tilt, drawn from globular clusters, are correct, this system is turning in such a way that *the arms are trailing,* an idea expressed many years ago by V. M. Slipher.

There are a few other galaxies for which rotational speeds have been measured, and for which it is possible, from the distribution of absorbing material across the nucleus, to say which edge is nearer to us. All these systems, according to Hubble, are turning in such a way that their arms are trailing.

5. OUR GALAXY

As twixt the poles, with lesser lights and great
Patterned, the Galaxy so whitely glows
That thereof sages question and debate.
DANTE,
Paradiso, Canto XIV, 97–99

The shape and structure of the great spirals in Andromeda and Triangulum are readily traceable and their sizes easily measured, as their distances are known. Our own galaxy is much more difficult to survey: we cannot see the forest for the trees.

The problem of surveying our own galaxy may be likened to the problem of drawing a map of New York city on the basis of observations made from the intersection of 125th Street and Park Avenue. Although it would be clear to an observer at this spot that the city is a big one, any statement as to its extent and layout would clearly be impossible. London would offer an even better analogy, for the neighborhood is not only congested but foggy.

The astronomical difficulties of surveying the galactic system are, in brief, the results of being inside it, so that the apparent brightness of its stellar denizens depends on their distance from us, and on local obscuration as well. In the distant spirals, all the stars are effectually equidistant from us, and their relative brightness can at once be inferred from their apparent brightness. But within the galaxy we can determine the distance of an object only if (a) it is near enough for parallax measures, or (b) it is of known absolute magnitude. The distance of *groups* of stars can be derived from parallactic drift, but this is only an indirect approach to individuals. All stars of measurable parallax are within less than a thousand parsecs, less than one-fiftieth of the whole diameter of the galaxy. For a survey on a broader scale we are driven back upon stars of known absolute luminosity.

A difficulty even harder to evade is the local fogginess and dustiness, which has the effect of dimming the stars, and therefore of making them appear more distant than they actually are. The effect is greatest in directions where the dust lies thickest—i.e., in the galactic plane and regions near its central layer. Since the sun happens to lie very nearly in the galactic plane, we encounter the full effect of this difficulty. The amount of obscuration that affects any star must be measured by determining the amount by which its light is reddened (which is proportional to the total obscuration), or less satisfactorily, by indirect methods. Sometimes the intensity of the interstellar (stationary) absorption lines that are superimposed on the spectrum of a star will permit us to infer how distant it is.

The best that can be done in surveying the galaxy is therefore to select stars of known luminosity, correct their distances as best we can for the effects of dust and fog, and use the resulting map as sort of skeleton for our plan of the whole system.

The "Milky Way." Even to the naked eye the heavens are seen to be encircled by a hazy band of light, which has from time immemorial been

Fig. 16.11. Part of the Milky Way in the Southern Hemisphere. Notice the bright band of stars and the obscuration that edges it. The same region may be seen, on a smaller scale, in Figure 13.10. (Harvard Observatory)

known as the "Milky Way." It has figured in the primitive legends of many countries, as (rather oddly) the path in which the sun rides across the heavens—though of course it does not coincide with the ecliptic—or as the underside of the celestial cow that constitutes the heavens, or in other ways. Even to the naked eye, the Milky Way seems to be somewhat variegated, and there are some very conspicuous dark spots on it, such as the Coal Sack near the Southern Cross.

The telescopic or photographic view of the Milky Way shows that it consists of large numbers of faint stars and is even more variegated than the naked-eye view suggests. It also shows that there are a number of bright nebulae along the band, and that most of them encompass bright blue stars.

Neither the visual nor the telescopic view of the Milky Way shows the extreme contrast between the center (in Sagittarius) and the anti-center (in Taurus) that would be expected from what we now know about our eccentric position in it. The Sagittarius "clouds," though broader than those on the opposite side of the sky, are so much cut up by obscurations that their superior brilliance is masked. Microwave observations (radio), however, are of such long wave length that they penetrate the dark clouds better than the eye or the photograph, and they show a great concentration in the direction where we know from other sources that the central regions of our galaxy lie.

Even though our view is impeded by dark clouds, much information about the arrangement of the stars in the Milky Way, and outside it, can be obtained by the simple but laborious process of counting faint stars in various parts of the sky, and by making approximate allowances for the obscurations. These studies of the numbers of faint stars in various directions show that the enormous preponderance of stars lies near the galactic plane, and thins out progressively above and below it. Star counts, however, give only general information, because we do not usually know the distance of any particular star, and must draw our conclusions from assumptions as to the distribution of absolute magnitudes among the stars counted.

Surveys of Stars of Known Luminosities. For survey purposes, the most valuable stars are the identifiable ones of very high luminosity, which can be seen at great distance. Such stars are luminous O and B stars, both individuals and members of galactic clusters. These are especially valuable in allowing us to sound the galaxy in the direction away from the center. They are concentrated in the Milky Way (the galactic plane), and none has been found in the *anticenter* direction at a distance greater than 4 kiloparsecs.

Another valuable sounding-line is provided by the classical Cepheids. Here again, we find none toward the anticenter farther off than 4 kiloparsecs. An especially valuable region is the constellation Canis Major, in this direction, and here all the faint Cepheids prove to be of short period and therefore not very luminous. The most luminous supergiant stars, identifiable by their spectra, are all found to be in the galactic plane; some of them are F stars like Canopus, and some are low-temperature red stars of spectrum S (the latter may ultimately prove to be the best of all sounding-lines for distant regions, because they are easily photographed on plates sensitive to the infrared).

Besides the stars of known brightness that are confined to the galactic plane, the bright nebulae (excited by hot stars) and the obscuring clouds are similarly restricted. The obscuring clouds are not only recognized by the spotty appearance of the Milky Way (evidence only of the very nearest clouds), and by the general obscuration and reddening in low latitudes, but—most striking of all—by the fact that they cut off completely the external galaxies in a zone that coincides with the Milky Way. This zone, known as the "zone of avoidance," because galaxies seem to avoid it, provides evidence of the "sandwich" of dark matter that cuts through the center of our galaxy. The same type of "dark sandwich" can be seen excellently in the photographs of many distant galaxies that are viewed nearly on edge.

We have mentioned hitherto only the stars of known luminosity that are confined to the galactic plane: the O and B stars; the Cepheid variables; the luminous supergiants; the galactic clusters, as well as the bright nebulae that are associated with some of these; and the dark clouds that lie among the stars. All these objects belong to the population that was noted in the Magellanic Clouds; all seem to share the galactic rotation; and all may be identified with Population I.

It is difficult to draw a picture of the *structure* of the galactic system from these Population I objects that lie in the plane; the individual distances are too much affected by the irregular obscurations. From counts of stars and the distribution of the Cepheids, we can say that there is one stellar condensation about 1 kiloparsec toward the center of the galaxy, as seen from us, and another less dense one about 3 kiloparsecs in the other direction. The inference is that we lie near the outer edge of one arm, and that there is at least one other arm outside our position. How many other arms lie between us and the center has not been determined by observation, but we may guess that there are some at least. In the Andromeda spiral the arms seem to be about 1000 parsecs across, and the distance between them (at least those best observed) to be about 2000 parsecs, and those of the galaxy do not seem to be very different.

The course of the arms is more difficult to trace than the place where our line of sight, toward and away from the galactic center, intersects them. They seem to run at an angle of very nearly 90° from the galactic center, from the bright star clouds of Carina (especially rich in hot stars and Cepheids) to the Cygnus clouds. Toward the edge of the Cygnus

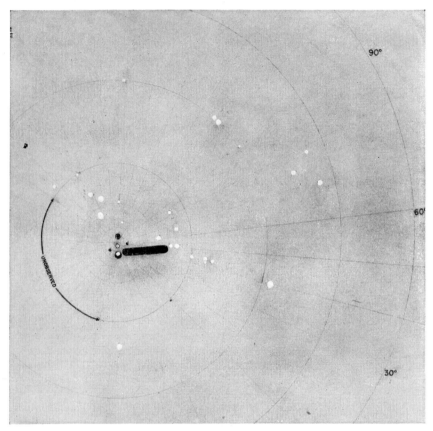

Fig. 16.12. The spiral arms of our galaxy, as outlined by bright nebulosity. The galactic center is beyond the lower rim of the figure. Nebulosities in three arms can be seen. The long, dark cigar is a schematic representation of the Milky Way rift. The two larger black dots are the Taurus-Perseus and Scorpius-Ophiuchus dark nebulae, the two smallest are the Northern and Southern Coal Sacks (in Cygnus and Crux). (W. W. Morgan, Yerkes Observatory)

cloud we see a dust-free lane, where we evidently look along a line that lies between arms.

Stars of known luminosity that are not confined to the galactic plane are found in the very numerous RR Lyrae stars, which give an idea of the size and shape of the galactic *substratum*. They are commonest in the galactic plane, and fall off very slowly (in comparison to the Type I stars) above and below the plane. They seem, in fact, to constitute a sort of spheroidal corona about the galactic center. Even more striking is the spherical distribution of the globular clusters, which (as already mentioned) outline the galactic system, and which provide what was the first, and is still the best, determination of our distance from the galactic center

—rather less than 10 kiloparsecs. These are the standard examples of Population II.

Other examples of Population II, the subdwarfs, the high-velocity dwarfs, and to some extent the individual subgiants, are not detectable at distances great enough for us to determine how their numbers fall off as we leave the galactic plane. However, their motions (and especially the inclination of their galactic orbits) show that they must share the spatial properties of the corona of stars defined by the globular clusters and the RR Lyrae stars. This corona does not share the galactic rotation that is shown by the population of the "arms," but it may have a small net rotation (otherwise the proportion of retrograde galactic orbits shown by this class of stars would be greater than it is).

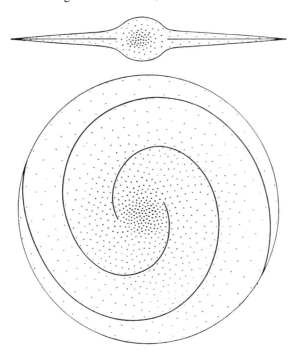

Fig. 16.13. Schematic diagram of the galaxy.

Another very important group of objects that belong to the galactic substratum contains the planetary nebulae—more or less symmetrical nebulous envelopes around hot stars that lie below the main sequence in luminosity. There are about four hundred planetary nebulae known in the galaxy, and their absolute magnitudes, on the average, are about −1. They share the prime Population II characteristics: they are greatly concentrated toward the galactic center, and their radial velocities show a large *dispersion,* from −250 to +250 km/sec. Planetary nebulae may be likened to very mild novae; and very likely the novae, at least those of normal or classical type, are also Population II objects, for they, too, tend to concentrate about the galactic center. Determination of their velocities, however, is a difficult problem; most of them are too far away for measures of proper motion, and their true radial velocities, complicated by the speeds of the exploding material, are indeterminate.

Although an accurate map cannot at present be drawn of the galaxy, the available observations, pieced together with Messier 31 and Messier 33 in view, suggest that it also is a spiral system that consists of substratum and arms. The arms are in orderly rotation, but the substratum does not share the rotational motion. The arms are confined to a plane; the substratum is a spheroidal system of stars. The galactic center is between 6 and 8 kiloparsecs from us, the outer edge, about 4 kiloparsecs; thus the overall diameter, deduced from observable objects, is about 25 kiloparsecs, somewhat smaller than that of Messier 31. In making this statement we assume that the galaxy is symmetrical, though with a few exceptions we have observed nothing on the far side of the galactic center.

The Direction of Rotation. We know that the motion of stars in the solar neighborhood is in the direction of the concentrations of stars in Cygnus. An intense radio maximum is also observed in that direction. This at first suggested that the arms of our galaxy are *leading* rather than *trailing*. More recently, however, the course of the spiral arms has been traced around most of a complete turn within the galaxy, by means of the radio waves sent out by the atomic hydrogen concentrated in the arms. The consequent map of the course of the arms shows without ambiguity that they are *trailing* like those of other well-observed spiral systems. The intense radio maximum toward Cygnus is a local peculiarity, and does not indicate the course of the main structure.

The Galactic Center. The actual galactic center lies behind heavy obscuration and has not been successfully observed photographically. The radio observations, which place the center just at the position indicated by the symmetry of the system of globular clusters, cannot reach individual stars; they record only the integrated radiation of a cloud of stars. Infrared photography with the most powerful instruments might conceivably reach stars at the galactic center, but we have at present to be content with looking for "windows" in the obscuring clouds that allow us to come fairly near the center. One such window, when used to survey stars of about the twentieth magnitude, shows a district about 1000 parsecs from the center. It has been studied by Baade and S. Gaposchkin. If the galaxy is anything like Messier 31, this place is well within the nucleus, and therefore its population is of great interest.

The distribution of the star colors in this central region is quite similar to that in a globular cluster, and there is little evidence of dust or bright nebulosity. Most interesting of all, however, is the variable-star population.

More than half the very numerous variables are RR Lyrae stars of exceptionally short period (average, less than a third of a day, as compared with half a day in our own neighborhood); there are also a smaller number of long-period variables and a few semiregular red stars, but no classical Cepheids. All this shows that the population of the region near the galactic center is characteristic of Population II. In the future we may hope to see surveys made of regions even nearer to the center. The evidence available at present suggests that the galactic center has the population properties of a globular cluster; perhaps the whole nucleus is a sort of gigantic globular cluster. The nuclei of some other galaxies give exactly this impression.

In our own neighborhood, there is about as much material in the form of dust and gas as in the form of stars. This is not necessarily true of the regions far between the arms, or of the central regions, which, if we can reason by analogy with Messier 31, may be rather free of dust and gas. Nor is it true of regions far from the galactic plane; dust and gas should be trapped in the plane, but not so the stars.

The speed of the solar revolution about the galactic center is about 200 km/sec; in conjunction with our distance from the center, this leads to a mass of about 10^{11} suns for the whole system, comparable to that of Messier 31, and probably slightly smaller.

6. ELLIPTICAL GALAXIES

The Andromeda spiral is accompanied by two (perhaps four) well-defined smaller objects. They are elliptical in form, and do not show any structure. Indeed until a few years ago they had not been resolved into stars, although their spectra showed that they are essentially stellar and not nebular. We see Messier 32 superposed upon the spiral arm of the larger galaxy, and NGC 205 is quite near to it. These small galaxies are evidently grouped in space with the Andromeda spiral.

The successful photography of individual stars in these two systems was achieved by Baade at Mount Wilson. Even with the 100-inch telescope, no definite stellar images could be recorded on photographic plates sensitive to blue light. Baade reasoned that if the brightest stars in these systems are red, they will be more than a magnitude brighter in red light than in blue, and he attempted to record them on red-sensitive plates. The result was one of the most remarkable observational achievements of our time; by dint of the most careful focusing and guiding, and a wise adjustment of exposure time (since too long exposures fog the plates by the light of the "auroral spectrum" of the night sky), he succeeded in photographing the individual stars of the two elliptical galaxies.

Messier 32, which lies on the arm of Messier 31, proved to be of amorphous structure. In the middle the photograph was "burned out" (i.e., the images ran together), but Baade described the outer part as being

Fig. 16.14. NGC 204, the satellites of the Andromeda galaxy, photographed on a red plate. Note the resolution into innumerable stars. It is visible on a small scale in Figure 16.8. (Baade, Palomar Observatory)

broken up into an "unbelievable mass of the faintest stellar images." In the words of an innocent bystander, the negative looks like a heap of pepper. The edges fade into the background of faint stars that make up the body of Messier 31, so the actual size is not easily determined.

NGC 205, which lies free of the great spiral, proved easier to resolve, and the photographs showed stars up to the center. It evidently contains fewer stars than Messier 32.

The nucleus of Messier 31 looks like another Messier 32, and it proved to be similarly resolvable with the same means. The center consists of "a dense sheet of extremely faint stars," and "there is not the slightest doubt that . . . the Andromeda nebula is resolvable into stars right up to the very nucleus." We can regard Messier 32, NGC 205, and the nucleus of Messier 31 as *three similar systems of stars.*

The brightest stars in all three systems have the same apparent magnitude, and stars appear in very great numbers at this magnitude in all three. Since all three are to be presumed at the same distance, the absolute magnitudes of the brightest stars follow at once from the distance of Messier 31, as determined from the Cepheids in the arms. They prove to be of absolute magnitude (photographic) -1.1. The color index $(pg - pv)$ for the brightest stars in NGC 204 is found to be $+1.3$, so the absolute visual magnitudes of these stars are -2.4.

The brightest stars in our three elliptical galaxies are of the same brightness as the brightest stars in globular clusters ($Mpg - 1.3$), and also of the same color. We may infer that the stellar makeup of these systems is similar to that of a globular cluster, which we have called the population of Type II. They contain no *hot* main-sequence stars of high luminosity (characteristic of Population I), another point of similarity with globular clusters.

Since the distance is the same as that of Messier 31, the sizes and total magnitudes can immediately be determined. NGC 205 has an absolute magnitude -14.0 and a diameter (to the outermost identifiable members) of 1.6 kiloparsecs. Messier 32 has an absolute magnitude -13.8, and as mentioned earlier, its size is indeterminate, probably about 2 kiloparsecs. Luminosity and size of the nucleus of Messier 31 are harder to estimate; presumably we must consider that it really embraces the whole substratum, in which case it is much larger and brighter than either of the companions. According to Holmberg the substratum contains 5/6 of the light of Messier 31. Both Messier 32 and NGC 205 are below the average in absolute luminosity; the average galaxy is probably a little brighter than -14.0.

NGC 147 and NGC 185. Two more elliptical galaxies, still fainter than the preceding, were studied by Baade at the same time, and these also he successfully resolved into stars which seem to be quite similar to those of the two bright companions of the Andromeda spiral. They are about at the same distance (if they were much farther away, the resolution would have been impossible) and probably also are satellites of Messier 31, because they are only 40,000 parsecs from it.

NGC 185, the brighter of the two, has an absolute photographic magnitude -13.7 and a diameter of 1600 parsecs; NGC 147 is of absolute magnitude -13.4, and its diameter is about 2 kiloparsecs. They are thus about four magnitudes brighter than the brightest known globular clusters, but have about twenty times the diameter.

The Sculptor and Fornax Systems. Two other elliptical galaxies, much less centrally condensed, have been found in the constellations of Sculptor and Fornax. Both are nearer to us than Messier 31. The Fornax system is at a distance of 140,000 parsecs, is 2000 parsecs across, and of absolute magnitude -11.9. The Sculptor system is only 68,000 parsecs from us and is our nearest galaxy, except for the Magellanic Clouds. It is 900 parsecs in diameter, and has an absolute magnitude -10.6. It is three quarters of a degree in apparent diameter, and counts of faint, distant galaxies show that in spite of the concentration of stars *it is quite transparent;* hence it presumably contains no dust. The same probably holds for the other elliptical systems. The Fornax system has been shown to be accompanied by some globular clusters, very few in comparison to our galaxy, but quite comparable to the galactic ones.

The most interesting information about the Sculptor system comes from discovery of a great many RR Lyrae stars within it. First noted at Mount Wilson, these stars have been studied by Thackeray at the Radcliffe Observatory in Pretoria (South Africa). Over two hundred RR Lyrae stars have been found, and it is estimated that there are perhaps seven hundred altogether in the system. In luminosity, in light-curves, and in periods, they are very similar to those in the globular clusters of our galaxy. A few of period greater than a day have been noted, and these stars are very like the Type II Cepheids that are found in some globular clusters, such as ω Centauri. Probably all the similar elliptical galaxies also contain RR Lyrae stars, but the Sculptor system is the only one near enough to have been studied at present. Almost one in a thousand of its stars must be a variable star.

Giant Elliptical Systems. The six elliptical systems hitherto discussed are relatively small and faint galaxies; their diameters are about 1000 parsecs and their luminosities between −10 and −14, below the average. Some of the very brightest galaxies, however, are elliptical systems. Messier 87, for example, is of absolute magnitude about −15.5. It is probably a gigantic globe of Population II stars, quite amorphous in structure and

Fig. 16.15. The great globular galaxy Messier 87. Notice the aura of globular clusters that surrounds it. (Palomar Observatory)

presumably free of interstellar dust. It is surrounded by an aura of bright objects (absolute magnitudes perhaps −6 to −8), which are almost certainly globular clusters. They are too far away, if so, to be completely resolved into the stars. Perhaps Messier 87 is comparable to the nucleus and substratum of Messier 51; it has about the same absolute magnitude. It is not unique; however, the great majority of the elliptical galaxies seem to be smaller and fainter ones, and these are probably even commoner than present-day discoveries indicate. They can be recognized only when they are comparatively close to us.

7. CLASSIFICATION OF GALAXIES

Previous sections have introduced the three main types of galaxy: irregular, spiral, and elliptical. Each of these types embraces a broad variety which will now be described in more detail. A simple scheme of classification has been adopted for galaxies, with the object of arranging them, if possible, in a coherent sequence. Such an arrangement turns out to be possible, although individual galaxies have idiosyncrasies that make the classes less specific than those that we have encountered for stars.

Fig. 16.16. The spiral galaxy, NGC 4594, in red light. Notice the layer of dark material that cuts through the middle. (Mount Wilson Observatory)

Irregular galaxies can hardly be sub-classified; at best one can describe degree of concentration, resolution, and symmetry, and mention dimensions and luminosity for those of known distance.

Spiral galaxies, however, display forms that run all the way from tightly wound systems, mostly nucleus, to loosely coiled shapes with small nuclei. In some, the arms seem to spring from a nucleus that is circular in plan (or that would presumably look circular if the system were not seen in projection). In others, the arms spring from the ends of a straight bar. The former are called simply "spirals," and designated by an S; the latter are known as "barred spirals," and designated SB. Within each type three main forms are recognized and indicated by the letters a, b, c, which represent degrees of closeness in winding, "a" being the tightest and "c" the most open. It must be remembered that many spirals are seen nearly on edge, and almost all display some effect of projection, so that even though a system is clearly spiral, it may be difficult to assign to a class.

All spirals (i.e., galaxies that show spiral structure) have some features in common: star clusters, individual bright stars, nebulosity, and obscuring matter are characteristic of the arms. The nuclei are not resolved, save in the nearest systems; they and the general substratum consist of stars that occur in large numbers at absolute magnitude between -1 and -2 and are yellow-red in color. Dust is nearly unknown in the nuclei, and evidence of nebulosity very rare there.

A few spirals with very bright condensed nuclei show bright "nebular" lines of O II and O III in the nuclear spectrum. These lines (which show that the nuclei contain considerable numbers of gaseous atoms) are of extreme interest, and have not hitherto been satisfactorily interpreted.

In describing the three classes of spirals, Sa, Sb, Sc, Hubble originally gave them the alternative names of "early," "intermediate," and "late" spirals, and these terms are still sometimes used. However, since they imply an order of development that has not been substantiated, it seems best not to employ them.

Fig. 16.17. The spiral galaxy Messier 81. (Palomar Observatory)

Fig. 16.18. The spiral galaxy Messier 83. (Harvard Observatory)

449

Fig. 16.19. The edgewise spiral NGC 4565. (Mount Wilson Observatory)

Fig. 16.20. The spiral galaxy NGC 628, seen almost face-on. (Mount Wilson Observatory)

Fig. 16.21. A galaxy with a bright nucleus and faint arms. (Harvard Observatory)

Fig. 16.22. The Sd galaxy NGC 7793.

Fig. 16.23. The barred spiral NGC 1300 (Palomar Observatory)

Elliptical galaxies, which are always symmetrical, structureless, and innocent of obscuration, are classified by their apparent shape: E0 designates a circular disc, and degrees of flattening are recognized as far as E7. Of course, the observed flattening of the images is the least that the actual galaxy can have; if seen face-on, even an E7 would look like, and be classed as, E0. The frequency of apparent shapes makes it likely that many elliptical galaxies are really spherical; there are too many round images to be the result of chance orientation.

Hubble has suggested that the elliptical galaxies are connected with the spiral systems through a transitional type, which he calls S0, in which spiral structure is on the verge of visibility. His scheme of classification is connected as follows:

$$
\begin{array}{c}
\hspace{4cm} \text{Sa—Sb—Sc} \\
\hspace{3.5cm} \diagup \\
\text{E0—E3—E7—S0} \\
\hspace{3.5cm} \diagdown \\
\hspace{4cm} \text{SBa—SBb—SBc}
\end{array}
$$

It will be noted that Hubble's scheme does not embrace the irregular systems.

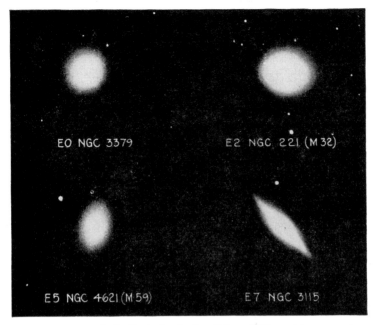

EO NGC 3379

E2 NGC 221 (M 32)

E5 NGC 4621 (M 59)

E7 NGC 3115

Fig. 16.24. Typical elliptical galaxies. (Mount Wilson Observatory)

Fig. 16.25. A peculiar galaxy, difficult to class. (Harvard Observatory)

Shapley has suggested a slightly different scheme (which does not include the barred spirals):

I—Sd—Sc—Sb—Sa—E7—E0

The implications of these classification systems will be discussed when the development of galaxies is considered (p. 473).

8. THE BRIGHTEST GALAXIES

Three galaxies (the Magellanic Clouds and the Andromeda spiral) can be seen with the naked eye. Several dozen are brighter than the tenth apparent magnitude. The twenty brightest galaxies are summarized in Table 16.1, which is based on a tabulation by Shapley. A few of the data have been changed to be consistent with Table 17.1.

TABLE 16.1. THE TWENTY GALAXIES OF BRIGHTEST APPARENT MAGNITUDE

Galaxy	Type	Apparent Photographic Magnitude	Right Ascension 1950		Declination 1950		Galactic Latitude	Figure
			h	m	°	′	°	
* Large Magellanic Cloud	I–SB?	1.2	5	26	−69		−33	16.1, 16.3, 16.4, 16.5
* Small Magellanic Cloud	I	2.8	0	50	−73		−45	16.2
* Andromeda galaxy Messier 31 (NGC 224)	Sb	4.3	0	40.0	+41	0	−20	16.8, 16.9, 16.10
* Triangulum galaxy Messier 33 (NGC 598)	Sc	6.2	1	31.1	+30	24	−31	16.7
NGC 253	Sc	7.6	0	45.1	−25	34	−88	. . .
NGC 55	Scp	7.8	0	12.5	−39	30	−77	. . .
Messier 81 (NGC 3031)	Sb	7.8	9	51.5	+69	18	+42	16.17
Messier 83 (NGC 5236)	Sc	8.0	13	34.3	−29	37	+32	16.18
Messier 101 (NGC 5457)	Sc	8.2	14	1.4	+54	35	+60	. . .
NGC 4594	Sa	8.6	12	37.3	−11	21	+51	16.16
Messier 64 (NGC 4826)	Sb	8.7	12	54.3	+21	47	+82	. . .
* Sculptor system	Ep	8.8	0	58.9	−33	50	−83	. . .
NGC 2403	Sc	8.8	7	32.0	+65	43	+30	. . .
Messier 51 (NGC 5194)	Sc	8.9	13	27.8	+47	27	+68	11.3
NGC 205	Ep	8.9	0	37.6	+41	25	−21	16.14
Messier 94 (NGC 4736)	Sb	9.0	12	48.6	+41	23	+76	. . .
* Fornax system	Ep	9.1	2	39.4	−34	34	−64	. . .
Messier 82 (NGC 3034)	I	9.2	9	51.9	+69	56	+42	. . .
NGC 4945	Sbp	9.2	13	2.4	−49	1	+12	. . .
Messier 32 (NGC 221)	E2p	9.2	0	40.0	+40	36	−22	16.24

This list of the apparently brightest galaxies is an interesting contrast to the list of the nearest galaxies (Table 17.1) that constitute the "local system"; members of this system are marked with asterisks in Table 16.1. Here we see an example of the tendency for lists of apparently bright objects to overrepresent the numbers of absolutely bright objects. We noted the same tendency in the list of the apparently brightest stars.

CHAPTER
XVII
SYSTEMS OF GALAXIES

Where the wheeling systems darken
And our benumbed conceiving soars.
Francis Thompson

Galaxies, like stars, are found in associations. We saw that, at least in our neighborhood, a large number (perhaps the majority) of stars are members of double or multiple systems. Stellar multiplicity runs all the way from double stars, through small groups, to the galactic clusters that contain hundreds of members and the globular clusters that contain hundreds of thousands.

Our knowledge of groups of galaxies is not as extensive as our knowledge of groups of stars, but they, too, display associations that run all the way from double systems to clusters with hundreds of members. Among the

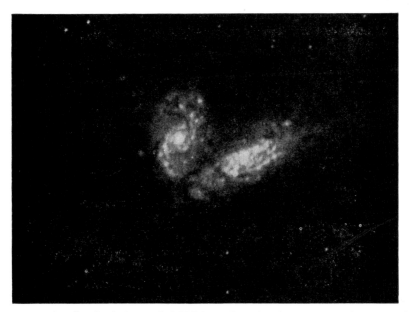

Fig. 17.1. A pair of spiral galaxies (NGC 4567, 4568) that appear to be almost in contact. (Mount Wilson Observatory)

455

galaxies we see analogies with double stars, multiple stars, and galactic clusters; however, there are no aggregations of galaxies that correspond to globular clusters. And as among double and multiple stars we noted associations between very disparate objects, the same is found to be true of galaxies. There are pairs of spirals and pairs of elliptical galaxies; but also there are pairs that consist of one spiral and one or more elliptical systems (Messier 31 and its companions), or one spiral and one or more irregular systems (our own galaxy and the Magellanic Clouds). And in large clusters of galaxies all forms may be represented.

Fig. 17.2. A dissimilar pair of galaxies, NGC 4647, 4649 (Messier 60). They are members of the Virgo cluster. The former is a tightly wound spiral; the latter, an elliptical system. Messier 60 is much the brighter of the two; if they are the same distance it has the higher luminosity. Such pairs have an important bearing on the development of galaxies. (Mount Wilson Observatory)

1. THE LOCAL GROUP OF GALAXIES

Our own system is a member of a small cluster of galaxies, usually known as the "local group"—though this "local" region is more than a million light years across! Although this group of galaxies is by far the best known, we cannot yet be certain that all its members have been found. Table 17.1 contains the known members, with the exception of two faint elliptical systems for which no complete data exist. There are three additional suspects (IC 10, IC 342, and NGC 6946), all of which are large in apparent size but faint in apparent brightness. They are respectively 3°,

11°, and 11° from the galactic circle and are probably greatly obscured. All are of type Sc. Very likely there are other low latitude members of the local group that are hidden by obscuration; these spirals are no doubt above the average in luminosity, and fainter galaxies are likely to have been missed.

TABLE 17.1. THE LOCAL GROUP OF GALAXIES

System	Type	Modulus ($m - M$) Observed	Modulus ($m - M$) Corrected	Apparent Magnitude	Absolute Magnitude	Distance (kpc)	Diameter (kpc)
Messier 31	Sb	23.9	23.3	4.3	−19.6	457	26.2
Our galaxy	Sb–c	. . .	. . .	. . .	−18.6	. . .	24.
Messier 33	Sc	23.8	23.4	6.2	−17.6	479	11.6
Large Cloud	I–SB?	18.6	18.2	1.2	−17.4	44	9.2
Small Cloud	I	18.8	18.5	2.8	−16.0	50	7.0
NGC 205	Ep	23.9	23.3	8.9	−14.0	457	3.4
NGC 6822	I	23.1	22.5	9.2	−13.9	316	1.8
NGC 221	E2	23.9	23.3	9.1	−13.8	457	1.6
NGC 185	Ep	23.9	23.0	10.2	−13.7	398	1.6
IC 1613	I	23.5	23.3	10.0	−13.5	457	3.0
NGC 147	Ep	23.9	23.0	10.5	−13.4	398	2.0
Wolf-Lundmark system	I	23.8	23.5	11.1	−12.7	501	2.0
Fornax system	Ep	21.0	20.8	9.1	−11.9	144	2.1
Sculptor system	Ep	19.4	19.2	8.8	−10.6	69	1.0

The local group of galaxies is probably a typical cluster of galaxies, though a small one. It is the only group that we can study in real detail. The list has been arranged in order of absolute magnitude. Roughly speaking, the sizes fall in the same order, but the faint Fornax system is larger than several systems that are considerably brighter. The brightest galaxies, also, tend to be the most massive: the masses of four galaxies in the local group, deduced from their rotational motions, are given in Table 17.2.

TABLE 17.2. MASSES OF GALAXIES IN THE LOCAL GROUP

System	Type	Absolute Magnitude	Mass (in solar masses)
Messier 31	Sb	−19.6	2×10^{11}
Our galaxy	Sb–c	−18.6	10^{11}
Messier 33	Sc	−17.6	3×10^9
Large Cloud	I–SB?	−17.4	2×10^9

We should note that the three spirals (Messier 31, our galaxy, and Messier 33) are near the top of the list in brightness and in size, with the quasi-spiral Large Magellanic Cloud not far behind. The local group contains no spirals less than 8 kiloparsecs in diameter. This suggests a tendency for spirals to be large and luminous galaxies; in some other nearby groups of galaxies there is also a tendency for the spiral members to be the largest and most luminous, though not always as large as 8 kiloparsecs in diameter. One should not, however, conclude that all highly luminous, large galaxies are spirals; we have already noted the giant elliptical system Messier 87, and there are many other high-luminosity elliptical galaxies.

Nearly all the galaxies at the bottom of the list (including the two name-less galaxies for which no data are tabulated) are elliptical. Clearly there is a tendency for the galaxies of smallest size, smallest luminosity and (pre-sumably) smallest mass to be elliptical. In the intermediate luminosities and sizes, there is a mixture of elliptical and irregular systems. The average luminosity of all the members of the Local System is a little brighter than —13, and this may be taken as the luminosity of an average galaxy. At and below this brightness the majority of systems seem to be elliptical.

From his early study of the brighter galaxies, Hubble concluded that the irregular, spiral, and elliptical galaxies occur in the proportion 5:75:20, but this suggestion that spirals are more than three times as common as elliptical systems is not borne out by the population of the Local System. It was the result of the preferential discovery of systems of highest absolute magnitude—mainly spirals, which can be seen at greater distances down to a given apparent magnitude. Very likely the preponderance of faint elliptical galaxies is a general phenomenon. The same tendency is shown by stars; supergiants are prominent in our catalogues, but the commonest stars are faint dwarfs.

Within the Local System, as we have seen, there are some even more intimate clusterings; our own galaxy is close to the two Magellanic Clouds, and Messier 31 has two, perhaps four, elliptical companions.

With few exceptions, the galaxies of the Local System are the only ones that have been resolved into stars (a few novae and Cepheids have been found in Messier 52, 81, and 101, and a number of quite remote galaxies have shown supernovae). Therefore our knowledge of individual stars re-poses upon them; and upon studies of them and their population hangs our whole scheme for measuring distances in remoter space.

2.   CLUSTERS OF GALAXIES
AND THE RED-SHIFT

The Local System of galaxies is one of the smaller clusters; it contains fifteen recognized and three more possible members. Associations of galaxies range from the rather numerous pairs, through small groups such as "Stephan's quintet" and larger groups such as the Local System, to vast clusters with several hundred members.

Between forty and fifty such clusters of galaxies have been recognized, and the number grows steadily as more distant objects are surveyed with the new powerful instruments. The members of a cluster of galaxies, like the members of a cluster of stars, furnish more powerful methods for dis-tance measurement than isolated objects. Just as within a globular star cluster one can assume that the brightest dozen or so stars will usually be similar, so in a cluster of galaxies one can assume that the few brightest galaxies will be roughly similar in brightness or in size, and thus arrive at conclusions as to the distance of the whole group. One can, moreover, draw some conclusions from the relative brightness of galaxies of different types within one cluster, and notice, for instance, that the spiral galaxies are always large and of high luminosity.

Fig. 17.3. Five galaxies in a compact group (the "Stephan Quintet"). (Mount Wilson Observatory)

With the exception of our own Local System, the cluster of galaxies in Virgo is the nearest one to us. It contains perhaps five hundred galaxies and is about 2.5 million parsecs (2.5 megaparsecs) from us. More than a hundred of its members are brighter than the thirteenth magnitude, and many of them are even found in the Messier catalogue.

About 75% of the recognized members of the Virgo cluster are spirals, and most of the rest are elliptical; there are very few irregular galaxies in the cluster. Many of the spirals are of type Sc. Spiral and elliptical galaxies are found alike among bright and faint members; and all seem to be roughly of the same size.

Many of the component galaxies are bright enough for measurement of their radial velocities. The cluster as a whole is apparently receding from us with a speed of 1200 km/sec. Moreover, the individual galaxies in it are moving with *relative* speeds of about 500 km/sec. From this distribution of speeds, and the over-all dimensions of the cluster, it is possible to obtain an estimate of its total mass. The result is the enormous figure of 10^{14} solar masses, or (if there are 500 galaxies in the group) about 2×10^{11} solar masses per galaxy. This seems an excessive value, for the most massive known galaxies have masses no greater, and possibly there is something in the dynamical theory that has not been taken into account; perhaps there is some nonstellar, nonluminous matter between the galaxies (though the amount required seems improbably large).

The Virgo cluster of galaxies provides a convenient liaison between the bright and well-observed members of the Local System, whose distances are accessible through their variable stars, and the members of more distant clusters which can be studied only through the media of total brightness and apparent size and form. The members of the Virgo cluster include some that can be resolved into stars (the very brightest supergiants), and they can be used to set up standards of comparison for remoter groups.

From the well-observed and extensively studied Virgo cluster we pass to the more distant clusters of galaxies, located on our cosmic map by means of brightness and size of their more conspicuous members. A list of some of the better-known is given in the next table.

Besides these northern objects, primarily studied at Mount Wilson for radial velocity, between 20 and 30 southern clusters of galaxies have been found and studied on Harvard plates. Their populations range from about 10 to nearly 500 galaxies. Two especially large and rich clusters in Centaurus, at distances of 29 and 38 megaparsecs, have shown a great variety

TABLE 17.3. CLUSTERS OF GALAXIES AND RED-SHIFT

Cluster	Galactic Latitude °	Galactic Longitude °	Number of Galaxies (approx.)	Distance (mps)	Radial Velocity (km/sec)
Virgo	+74	256	500	5	+1,200
Pegasus	−49	55	100	15	+3,800
Pisces	−30	96	25	14	+4,360
Cancer	+30	170	150	18	+4,800
Perseus	−13	118	500	22	+5,200
Coma	+87	26	800	28	+7,500
Ursa Major I	+59	106	300	52	+11,800
Leo	+54	201	300	64	+19,600
Gemini	+20	150	200	82	+23,000
Bootes	+66	17	150	140	+39,000
Ursa Major II	+55	115	200	144	+42,000
Hydra	+30	194	over 200	?	+61,000

of population, and each may contain a thousand members. The largest and brightest members of these systems are both spiral and elliptical; some even exceed Messier 31 in dimensions and are accordingly among the largest individual galaxies known.

The clusters in the table that has been given are arranged in order of distance from us. It will be noticed that those of measured velocity show larger and larger recessional velocities, the greater their distances. This phenomenon, known as the "red-shift of the galaxies," has been the occasion for as much theoretical speculation as any other observed fact; nevertheless, there is no unanimity in interpreting it.

The interpretation of the red-shift in terms of a Doppler displacement, such that the distant galaxies are actually receding from us faster than the nearer ones, is only a schematic expedient. Various other interpretations hinge on the theory of relativity, which is beyond the scope of an elementary text in astronomy. Both distance and velocity must be specified in terms independent of the situation of the observer before conclusions can be drawn from correlations between these quantities. Observation shows that the "velocity of recession" is linearly related to the distance of the galaxy concerned. The theory of relativity regards the red-shift as an effect of the "curvature of space."

The theory shows that if the galaxies are uniformly distributed (on the large scale), the apparent distribution will begin to deviate from "uniformity" for very faint galaxies (probably at about the limit we are now able to reach). The nature of the deviation will indicate what are the properties of space: it may be hyperbolic and infinite in extent, or spherical and finite (though of very large radius). Counts of very faint galaxies, combined with accurate measures of their colors, must be related with the velocities of recession to resolve this question. The balance of opinion at present seems to be in favor of hyperbolic space; but the observations cannot be used to determine the time scale.

An alternative way of looking at the facts, developed by Milne and known as "kinematical theory," departs from the procedures of general relativity, and divorces the interpretation of gravitation (one of the chief concerns of general relativity) from the theory of the red-shift. This approach has received less general acceptance among astronomers than the other.

Even though the red-shift is not theoretically understood to general satisfaction, the well-established empirical fact that it is correlated with distance provides a valuable method of determining distances for isolated galaxies, and therefore of measuring their brightness and dimensions. So completely is the empirical relation accepted at present, that distances derived from red-shifts are regarded as the best determinations for galaxies that are not members of clusters.

3. COLLISIONS OF GALAXIES

The Coma cluster (see table) is one of the richest known, and also one of the most compact. It and the similar Corona Borealis cluster share the remarkable feature that, although they contain members of all shapes (from spherical to greatly flattened), there is no evidence of spiral arms in any, and no sign of obscuring material. In other words, stars of Type I and the invariably accompanying interstellar dust seem to be absent from these compact clusters. Spitzer and Baade have recently suggested that the compactness and the population peculiarity may be related.

From the motions of the individual galaxies, and the (small) distances that separate them, it is found that in 3×10^9 years, each galaxy, on the average, will have collided with at least twenty other galaxies. It might be thought that a collision between galaxies would greatly distort the protagonists; but the stars within a galaxy are so far apart that the galaxies seem to pass through one another with relatively few stellar collisions, and not much resultant distortion. On the other hand, any gas or dust would be swept completely out of the colliding galaxies (or at least interpenetrating sections of them), and would congregate in intergalactic space. It might remain there, invisible, or it might have condensed long ago into stars— either individuals or small galaxies, which never become spirals. In the absence of interstellar matter within them, the cleaned-out galaxies would not continue to build Population I stars and would have become the pure Population II systems that are observed. This fascinating idea opens up a new field for speculation about the factors that affect the development of a galaxy. It cannot, of course, provide interpretation of *all* Population II systems; but perhaps it may suggest that Messier 32 and NGC 205 have lost their interstellar matter on a trip through the much larger Messier 31. By the same token, however, the Magellanic Clouds, as such, could never have passed through our galactic plane; the Large Cloud, at least, contains far too much gas and dust.

CHAPTER

XVIII

EVOLUTION

Then gin I think on that which Nature said,
Of that same time when no more change shall be,
But steadfast rest of all things firmly stayed
Upon the pillars of Eternity,
That is contrayr to Mutability.
For all that moveth doth in change delight:
But thenceforth all shall rest eternally
With him that is the God of Sabaoth hight:
O that great Sabaoth God, grant me that Sabaoth's sight.

EDMUND SPENSER
Faery Queen, VII, Canto viii *

The problem of the "age of the Universe" is basic to the consideration of evolutionary problems. During the past half century it has gone through great vicissitudes.

1. THE TIME SCALE

The time is long past when scientific men accepted the date given for the Creation by Archbishop Ussher, who arrived (by the naïve method of adding up the ages of the patriarchs) at the figure 4004 B.C. But during the last fifty years, estimates of as much as 10^{12} years, and as little as 10^9 years, have been successively (and even simultaneously) in vogue. Opinion tends, at the present time, toward the second value. Eddington wrote, in 1914: "Perhaps there is no harm in having some such figure as 10^{10} years at the back of our minds in thinking of these questions." It is impossible to give more than a bare outline of various lines of argument that have led to this conclusion.

Ages of Atoms. It is likely that the atoms are as old as the universe. The known rate of disintegration of radioactive atoms will give some

---

* Spenser might have been describing the nineteenth-century view of the fate of the universe, the so-called "heat-death," when all matter would be stationary at zero temperature, and (in Eddington's words), "nothing whatever can happen however long we wait." Modern views do not anticipate such a heat-death, at least for the substance of stars such as white dwarfs, which are in a degenerate condition. (See p. 357).

idea of how great is the interval since atoms of these sorts were formed. There are, for example, an inappreciable number of atoms of neptunium found in nature—our knowledge of this atomic species depends on *making* it artificially. The half life of neptunium (the time during which half of an assemblage of such atoms go to pieces) is about 10^5 years. The virtual absence of such atoms in nature shows that the age of the universe (so measured) must be much greater than 100,000 years. Both $_{92}U^{235}$ and $_{92}U^{238}$ isotopes of uranium disintegrate spontaneously, forming lead as a stable end-product, at a known rate. If we assume that *all* such lead in the universe has been thus formed, we arrive at 4.5 billion years as the interval during which they have been disintegrating. Again, these two kinds of uranium disintegrate at different rates, and are now found in different proportions; if we assume that they started out equally common, we calculate that the disintegration has been going on for about 6 billion years. Very similar ages are derived for other radioactive species, as summarized below.

Age of:	From	Age in Billion Years
Atoms:	U^{238}, U^{235}	6
Atoms (lead):	Pb^{207}, Pb^{206}	4.5
Meteors *:	U, He	5–7
Earth's crust:	Pb	3

When the broad assumptions are considered, the agreement of these estimates in order of magnitude is surprising. The earth's crust seems a little younger than the atoms; and when allowance is made for the effect of cosmic rays, the ages of meteors are compatible with that of the earth's crust. None of the estimates is greater than Eddington's guess of 10^{10} years.

Age of the Earth. The study of the population of geological strata points to a great age, not only for the earth, but for the earth under conditions not very different from those of today. If there were living creatures on the globe, we can at least infer that the temperature was not above the boiling point, or greatly below the freezing point, of water. An approximate geological timetable is summarized below:

Era	Interval since Era (million years)
Cenozoic (age of mammals):	5–70
Mesozoic (age of reptiles):	110–190
Palaeozoic (age of ancient life):	220–500
Pre-Cambrian:	up to 1800

There is, perhaps, a tendency among geologists to reduce these estimates somewhat, but probably not by more than a factor of two.

There are other ways of estimating the age of the earth approximately: the amount of salt in the sea, and the rate at which it is being brought

* Estimate perhaps vitiated by cosmic-ray bombardment.

down by rivers, indicate an age of about 1500 million years for the oceans. The rate of laying down of sedimentary rocks is not so specific, but points to an interval much greater than 100 million years. The age of the earth's crust, from radioactive evidence, was mentioned in the previous section to be about 3000 million years—understandably greater than the age of the oceans.

Age of the Sun. The problem of the age of the sun was a *cause célèbre* of the closing years of the nineteenth century. It seemed incompatible with the recognized age of the earth. Lord Kelvin showed that if the sun derived its heat from combustion, it could not last more than a few thousand years. He suggested that it shone by means of the release of gravitational energy derived from slow contraction; but even this source would only be adequate for 50 million years. The discovery of radioactivity suggested that the liberation of subatomic energy would suffice to stretch the available time. For a while it was supposed that the sun could convert the *whole* of its material substance into energy, and in that case it could last millions of millions of years. However, the more recent studies of the possible nuclear processes that may sustain the sun (the carbon cycle and the proton-proton reaction, mentioned on p. 353) show that only a small fraction of the total mass can be so converted, and give a possible time span of between 10^9 and 10^{10} years, not much larger, be it noted, than the age assigned to the earth.

The idea that the earth was formed from a long pre-existing sun has given place today to the feeling that earth and sun may have been formed as parts of the same process of development, and even leave the possibility that the earth, as such, may actually be somewhat older than the sun as known today (p. 253).

Ages of Stars. The luminosity of a star measures the amount of energy that is being produced by nuclear processes inside it. Thus the more luminous a star is, the more rapidly it is using up its available mass and energy, and the shorter the time that it can last. Even though the most luminous stars are also the most massive, and have therefore the greatest store on which to draw, they are nevertheless expending their resources so prodigally that they cannot survive as long as less luminous ones. Struve has given a table of possible ages for a series of main-sequence stars.

Type	Luminosity (suns)	Mass (suns)	Lifetime (million years)
O	100,000	14	10
B	1,000	10	400
A	10	2.5	5,000
F	2.5	1.3	20,000
G	1	1	40,000

The remarkable thing about these data is the fact that the most luminous stars must be much younger than the sun, and indeed much younger than the earth, even more recent than the Age of Reptiles!

Ages of Double Stars. It is possible to calculate how long it would take to disrupt double-star systems by encounters with other stars in the course of cosmic history. These very difficult calculations, applied to the known common occurrence of double systems, show that the *upper* limit to the age of our stellar system must be about 10^{10} years (10,000 million years).

Ages of Star Clusters. Clusters of stars must ultimately be sheared apart and dissolved by the differential rotation of the galactic system, as was first pointed out by Bok; the "life" of the Pleiades, on this basis, would be about 3000 million years. The more populous, more compact clusters could survive for much longer.

Ages of the Arms of Galaxies. If the stars in the arms of galaxies are in rotation at a known speed about the nuclei, it is possible to calculate how long the motion must have gone on to produce the degree of "winding-up" that is observed. The resulting intervals are between 10 million and 100 million years.

Ages of Clusters of Galaxies. Like star clusters, clusters of galaxies can be regarded as subject to gradual dissolution; for the Virgo cluster an appreciable change would take about 10^{11} years.

Differences between Near and Distant Galaxies. There seems to be a real difference, definite though small, between the true color of nearby and very distant galaxies. If the observations on which this statement is based are substantiated (they will be critically tested in the next few years) we may consider that an interval of 10^{10} years is an appreciable fraction of the age of a galaxy.

Conclusions drawn from the "Red-Shift." It is sometimes suggested that the red-shift of the galaxies can be interpreted in terms of an interval of about 3000 million years since the beginning of the recession. This conclusion, however, is most uncertain, and the data are susceptible of reconcilement with a large range of intervals.

The results of the various age criteria may be summarized:

Criterion	*Age* (years)
Radioactive atoms	4 to 7 $\times 10^9$
Earth	up to 3 $\times 10^9$
Planets	order of 10^9
Sun	less than 8 $\times 10^9$
Stars	10^7 to 10^{10}
Double stars	less than 10^{10}
Star clusters (galactic)	3 $\times 10^{10}$
Arms of galaxies	10^7 to 10^8
Clusters of galaxies	less than 10^{11}
Difference, near and distant galaxies	less than 10^{10}?
Red-shift (very uncertain)	3 $\times 10^9$

The criteria seem to concur in placing a limit not far from 10^{10} years; this we may regard as the interval of appreciable over-all change—the interval during which we may legitimately speak of development process in the cosmos.

Two groups of data give low ages: the form of the arms of galaxies, and the luminous supergiant stars. It is significant that the latter populate and define the former. We may well suppose that our own galaxy, and the known spirals, have changed appreciably since the Carboniferous Age, when the Coal Measures were laid down, perhaps 280 million years ago.

It must be emphasized that the dates given in the last table are in no sense "dates of Creation." They simply state the limits of our horizon. "The solution of the problem of the time scale will not permit us (not at any rate in the first instance) either to 'date' the present epoch in a 'fundamental calendar' or to forecast with definiteness the 'end.' What it would allow us, however, is to specify an interval of time in which the various aspects of the astronomical universe may be expected to change appreciably." (Chandrasekhar)

2. DEVELOPMENT OF GALAXIES

The variety of galactic types has been described in some detail (p. 448). The very interesting question concerning the relationship between the various types has not yet been discussed. It is one that is far from solution at present, but some possibilities can be pointed out.

The fact that galaxies can be arranged in some sort of sequence makes it seem likely that a process of development is traceable. The classification scheme given by Hubble:

$$
\begin{array}{c}
\text{Sa—Sb—Sc} \\
\diagup \\
\text{E0—E3—E7—S0} \\
\diagdown \\
\text{SBa—SBb—SBc}
\end{array}
$$

does not allow for irregular systems; the spirals are arranged in significant order, and the insertion of S0 (greatly flattened systems which barely show spiral form) between elliptical and spiral galaxies does not, perhaps, fit strictly into the pattern. The existence of two sequences of spirals: normal and barred, is recognized but not elucidated. Hubble's classification was intended to be empirical, without theoretical presuppositions. If the order is reversed, it is in close agreement with modern ideas of the direction of development.

Shapley's scheme:

$$
\text{I—Sd—Sc—Sb—Sa—E7—E0}
$$

was designed to represent an evolutionary sequence. It does not include either barred spirals or S0 galaxies. Present-day work on stellar types and stellar populations is well reflected in such a plan, which contemplates the development of a stellar system from one rich in dust, gas, and supergiants to one that contains none of any of these, and has attained a high degree of uniformity both in population and structure.

The idea of Spitzer and Baade concerning collisions between galaxies allows for the occurrence of S0 galaxies in a sequence that bypasses the normal spirals. Perhaps it might run thus: I—S0—E, although the authors have not committed themselves to any such scheme; they merely consider that a galaxy can reach the Type II population and the concomitant symmetry and uniformity, without passing through intermediate spiral forms.

Development Scale of Galaxies. A most interesting and attractive theory of the evolution of galactic systems has been proposed by the German astrophysicist C. F. von Weizsäcker. It is based on the relatively new and very difficult theory of gaseous turbulence, and many of its conclusions must still be critically tested. It does, however, give a more general view of the course of development of many types of systems than has been attempted by anyone else. The author's own summary of his ideas is as follows:

"All objects are assigned to three morphological groups: spheres, figures of rotation, and clouds. The occurrence of clouds leads to the conclusion that a state of turbulent motion exists. This state is assumed for the original material of all the objects . . . . Figures of rotation form themselves from portions of a turbulent gas, which, as a result of the friction within them that results from turbulence, resolve themselves into a nucleus of small rotation and a residuum that passes out into space, carrying the greater part of the angular momentum . . . . Rotating systems must be young . . . and non-rotating ones, old. Large systems, other things being equal, develop more slowly than small systems. Spiral galaxies must, on this basis, be genetically young, elliptical galaxies and globular clusters genetically (but not necessarily absolutely) older. The irregular galaxies must be the youngest. The distribution of various types of stars and of interstellar material supports this supposition. Spiral structure appears as the natural consequence of the interaction of rotation and turbulent cloud-formation. The O, B, and A stars are thought to be young because of their rapid rotation and other criteria. The other main-sequence stars are old. The transition between these two types results from the radial loss of angular momentum and of mass. The giants do not rotate and are therefore old. It is supposed that they represent a late stage of massive stars with greatly concentrated nuclei, which sets in as a result of lack of hydrogen."

On the basis of these ideas, von Weizsäcker has calculated ages for spiral galaxies, elliptical galaxies, and globular clusters as 7×10^9 years, 1.5×10^9 years, and 10^8 years, respectively. He thus considers the spiral galaxies as genetically the youngest of the three types; a type not younger in years, but one that has developed more slowly because of its size.

If the theory outlined by von Weizsäcker is carried a little further, and the spiral arms are regarded as the *loci* of streaks of material that condense into stars while moving differentially in orbits round the center of gravity of a stellar system, it is possible both to calculate the expected forms of the arms and to estimate their ages. In 10^9 years the whole streak would be "wound up like a ball of twine," and the small number of turns actually observed for spiral arms suggests that they are actually an evanescent feature of the systems which are continually dissolving and re-forming. Their ages, thus deduced, are between 10^7 and 10^8 years—of the same order as those of the supergiant stars that populate them preferentially. The arms of such a system would inevitably trail.

3. DEVELOPMENT OF STARS

The problem of stellar evolution has never been satisfactorily solved. For half a century, theory has succeeded theory, and an account of the changes of thought would have the qualities of a kaleidoscope.

There are no direct evidences of steady change among the stars. Changes must be taking place because stars are pouring out energy, and they are probably consuming their own substance in order to be able to do so. But these changes are so slow that they cannot be detected in the interval over which the stars have been observed. The bright stars listed by Hipparchus over two thousand years ago are still to be seen. The Cepheid variables, their periods accurately timed to their densities, have not shown steady changes of period that point to steady changes of density (though a change of one in a million could be detected). Changes of period there are, but they are quite random: there is probably no steady trend. The novae might be thought to provide evidence of real changes, but even their spectacular outbursts are little more than a skin eruption, which passes off and leaves the star apparently as before. The supernovae stand alone in displaying changes on the large scale, but even for them we cannot be sure that we are not witnessing a cosmic accident, rather than the results of an inevitable process.

We must resign ourselves to the view that we see no more than a snapshot of the cosmic scene. Our main source of information about stellar development is the observed variety among the stars. We have become familiar with supergiants and giants, subgiants, main-sequence stars, and subdwarfs and white dwarfs. We have met the Type I and Type II population. Can they all be fitted into one scheme of development?

Before this question can even be approached, we must recognize not only the great variety of the stars, but also the different proportions in which they occur. Any interpretation of stellar development must keep in mind that the *enormous majority of stars belong to the narrow main sequence,* and that within the main sequence the fainter, smaller, cooler stars are in the greatest numbers. Even the sun is an uncommon kind of star compared with the red dwarfs. Second after the main sequence, in point of numbers, come the white dwarfs. Giants and subgiants are rare; supergiants are statistically negligible. Moreover in different districts the relative numbers are not the same; there are no supergiants at all, for example, in Population II.

One may suppose, either that main-sequence stars are common because they were produced in the greatest numbers, or that they preponderate because they last a long time without alteration. Both factors probably play a part; perhaps the latter is the more potent.

Fifty years ago, contraction was the only process that held out hope of accounting for the continued radiation of the sun and the other stars. Therefore it was natural to think that the stars began their careers large and became gradually smaller and ultimately fainter. Sir Norman Lockyer suggested an evolutionary course from the large, cool stars (like Betelgeuse), which first contracted and grew hotter, passing (so to speak) up the spectral sequence, and then, while still contracting, grew cooler, became red dwarfs and ended by disappearing altogether by becoming too cool to shine. This theory was the more remarkable because it was suggested before the recognition of the "giant" and "dwarf" stars.

In 1913, Henry Norris Russell formulated the theory of stellar evolution that held the stage for a couple of decades. He, too, supposed that stars develop from red giants to blue main-sequence stars and then "slide down the main sequence" to ultimate extinction. In the interval, however, the contraction theory had been abandoned because the discovery of radioactivity had shown that nuclear energy could be liberated and might be invoked as a means for the maintenance of stellar radiation. Though the operation of the process was not understood, it was generally felt that the stars were subsisting on nuclear energy, and it was even thought that a star could ultimately convert the *whole* of its substance into radiation. This *volte face* of opinion expanded the possible time scale of the sun (for example) from the 50 million years allowed by the contraction theory to more than a million million years, the so-called "long time scale."

This theory of stellar development, however, faced a real difficulty in trying to explain how the stars could "turn the corner" from the giant branch to the main sequence. It required the rather vague supposition that the giants subsist on one kind of matter, and that when this is exhausted, they "tap" another source of energy and start down the main sequence.

A new turn was given to the ideas of stellar development in 1936, when an actual nuclear process capable of supplying stellar energy was simultaneously discovered by Hans Bethe and Carl Friedrich von Weizsäcker: the "carbon cycle" (p. 348). If this process of building hydrogen into helium is operating, a star in the course of its development would consume only a very small fraction (less than 1%) of its mass, even if it used all its hydrogen. And in the process it would not become fainter, but rather brighter, because a star poor in hydrogen is brighter, for its mass, than a hydrogen-rich star. Therefore stars could no longer be thought of as sliding down the main sequence, but rather as climbing up it, if the carbon cycle were the source of energy. The proton-proton reaction, now surmised to be as potent as the carbon cycle in many stars, would have much the same effect, since it also subsists on hydrogen.

The net result of drawing stellar energy from the carbon cycle and proton-proton reaction is that a star can no longer be supposed to be able to traverse the whole spectrum-luminosity domain (the Hertzsprung-Russell diagram); its peregrinations are restricted and its mass can change but little from consumption of its store of atoms, which probably means that its luminosity is restricted also. In other words, the "sequences" in the spectrum-luminosity diagram are no longer *paths* taken by developing stars; they are places where stars "accumulate." Now we may picture a developing star as moving almost horizontally across the H-R diagram, loitering long in the neighborhood of the main sequence.

Even though the carbon cycle is capable of delivering internal energy over very long periods, the lifetime of a main-sequence star is limited by the amount of its available hydrogen, that is to say, by its mass. If energy output (i.e., luminosity) were *proportional* to the mass of a star, the lifetimes of all stars on the main sequence would be the same. But we have seen (p. 354) that the luminosity goes up as some power of the mass—the approximate theory quoted leads to the fifth power, while the observed mass-luminosity curve suggests more nearly the cube, of the mass. Therefore the more massive stars may be expected to have shorter lifetimes.

When the available hydrogen is exhausted, the carbon cycle must cease to supply energy, but the star may continue to shine if it releases gravitational energy by contraction. This necessarily causes a rise in central temperature. When the central temperature has risen to perhaps 100 million degrees, it seems possible that a new source of energy will be tapped—helium nuclei will begin to undergo exothermic reactions. The consequence will be an increase in the size of the star, which now moves away from the main sequence, and may become the kind of giant star that is found in the color-luminosity array of a globular cluster—a giant star of Population II. The characteristic relation between brightness and color for the brighter members of globular clusters (Figures 15.10, 15.11) have been convincingly predicted by Sandage and Schwarzschild in such terms as these.

We possess convincing evidence of the existence of a group of stars that have exhausted all possible sources of nuclear energy, and must shine entirely by dint of contraction—the white dwarfs (p. 357), which have densities between 10^5 and 10^8 grams/cm^3. Their interiors are in the "degenerate state" in which all atoms are ionized, and the particles packed together as closely as possible. The interiors of white dwarfs may be regarded as equivalent to gigantic molecules in the lowest possible energy state. The absence of available energy is illustrated by the statement that these stars cannot even solidify. They seem to have attained a physical impasse.

The theory of a degenerate gas leads to the conclusion that only stars with less than about 1.4 times the solar mass can become white dwarfs. Most stars actually have masses below this limit. Those of higher mass could become white dwarfs only if they lost the excess. Possibly a supernova is a star that is reducing, by a violent outburst, to a mass within the white dwarf limit; at least one supernova (p. 392) is known to have survived in the form of the Crab Nebula, a white dwarf and a massive, turbulent envelope that represents the debris of the explosion.

The white dwarfs are very numerous among the stellar population of our own vicinity. Only the faint stars of the main sequence outnumber them. Therefore they represent a frequent (and certainly a lengthy) epoch in the stellar lifetime. But their understanding is beset with difficulty. From what has been said it would be inferred that they must be very old stars. Yet their motions do not seem to display the high-velocity characteristics of Population II, and a few have been found in galactic clusters, which are generally regarded as relatively youthful groups. Thus, although the physical condition of white dwarfs is probably more completely understood than that of most stars—far more completely than that of the giants and supergiants—their place in the evolutionary scheme is still a mystery. We do not know by what process a pair of stars as dissimilar as Sirius and its companion can have come into being and developed.

Such is the rather hazy picture that comes from considering definite nuclear processes for the supply of stellar energy. But there are other possible effects on stellar development. Stars may lose matter (and therefore energy) by explosion or spinning, or they may pick it up from space. But only the very fast-spinning stars can spill matter in important quantities, and a star can pilfer from space only if it almost stands still within the dust to do it. And so, even though the two processes of shedding and pilfering may be important, they are limited to a minority of the stars.

We have carried the argument as far as a consideration of energy-sources can carry it. A broader view takes in not only the stars' nature, but also their antecedents. They are members of a community, and their history is that of their community. Stellar populations are the basic clue to stellar development. Their differences of membership, motion, and distribution probably contain the essentials of the solution.

Populations I and II, for example, are probably the limiting possible stellar communities (the extreme left and right, so to speak, in stellar politics). They are well defined, and if we pick out conspicuous members of either party, we obtain a fairly narrow band; but many individuals may lie between, even though almost all may have tendencies toward one party or the other.

Whether the individual sequences are really narrow and distinct is a basic problem. If they really are—and the present information is conflicting and opinion divided—ideas of stellar development are put to a severe strain, and we must practically think of whole sequences of stars remaining, so to speak, rigidly connected and gradually developing into other sequences.

The interpenetration of the two populations in a galaxy like our own is significant. Population II forms the substratum; it is deficient in supergiants, dust, gas, classical Cepheids (a special case of supergiants), and in the swift rotation that characterizes Population I. It is, in fact, notable chiefly for what it lacks. Population I forms the arms; it contains the dust, gas, and luminous stars, all of which form a system in complex, rapid rotation (wheel-like within, "planetary" farther out). It also contains, apparently, specimens of everything that is found in Population II—though this may be an accident of position and motion. How are we to account for the coexistence of the two different communities in spiral galaxies?

The dust and gas are the important difference between the two environments. It is clear from the Andromeda spiral that the rule is: no dust, no supergiants and no arms. There are two possibilities. Either the supergiants feed upon the dust, and keep the lineaments, if not the fact, of youth, or they are continually born from the dust. How can we decide between these two possibilities?

The time scale comes to the rescue here. Luminous supergiants, like Rigel, or brilliant main-sequence stars like Zeta Puppis, cannot (on their own resources) be more than 10 million years old. Either they were born within that interval, or are pilfering from space. And yet we know that fainter stars (the sun, for instance) are at least a hundred times older. It is hardly likely that the luminous stars, even though they lie in the dust-layer, can have picked up matter on so large a scale—it would amount to grand larceny rather than pilfering. More probably they are stellar infants.

Another line of argument leads to the same conclusion. The shearing rotation of a spiral galaxy like our own must in time distort and finally destroy spiral arms. We know the rate of rotation and the amount of shear, and it is evident that spiral arms have a lifetime of between 10 million and 100 million years—otherwise the arms would be wound up like a ball of twine, whereas in fact few galaxies show more than one or two coils. Spiral arms, indeed, are of about the same age as supergiants, and much younger than we must suppose the main bodies, or substrata, of the galaxies to be.

Although it seems fantastic to suppose that some of the brightest stars in the sky are actually younger than life on earth (even than vertebrate life), recent theories, developed by Whipple and by Spitzer, have shown that stars can be born as long as the raw material—interstellar dust clouds—is avail-

able. They are therefore not only being born all the time, but will continue
to be born until the supply of raw material is exhausted or removed. Collisions between galaxies might remove the material. The newborn stars
(not at all the same as "new stars," of course) are probably produced by
the concentration of a considerable aggregate of dark material. Many dark
flecks are known that are capable of becoming stars; and the Coal Sack, for
instance, has the makings of a whole star cluster such as the Pleiades. The
concentration would be sudden, and the development of a star from a dark
cloud would probably be a very violent collapse, which would carry it
rapidly from the right to the left of the spectrum-luminosity array, so that
it would appear to make its debut as a main-sequence star. Very large
masses with slow rotation might turn into supergiants; smaller masses would
take their own places farther down the sequence; enormous masses would
probably break up into clusters. Very likely small masses would appear
more often than large ones; hence the preference of the main sequence
for small, faint stars.

Possibly we are actually witnessing the process of the growth of stars
in the peculiar variable stars within dark nebulae; most of them are fairly
faint main-sequence stars in the making—or perhaps they are simply main
sequence stars that are feeding on the nebular grains.

Very likely the problem of stellar development will be solved as a consequence of the solution of the problem of the development of galaxies.
This may seem like "the blind leading the blind"; but actually the development of galaxies (though by no means solved) is the less difficult problem.
It depends on the known principles of dynamics, rather than on the more
difficult and more diversified problems of nuclear energy-sources, and the
very complex and variegated possibilities as to the stars' internal structure.
When the development of galaxies (and in particular the differences in
history of substratum and arms) is understood, the development of the
stars that compose them will at least be greatly elucidated.

The history of a galaxy may be somewhat as follows. An irregular
galaxy is first formed from a cloud of gas and dust. As it develops, it acquires an axis of rotation, and the dust and gas that have not become stars
accumulate gradually in a plane perpendicular to that axis. The original
stars constitute the substratum, and Population II is their ultimate loitering-
place. The dust and gas continually give birth to fresh stars, which are situated in the same plane as, and share the current rotation of, matter from
which the stars arose. The Type I stars are those most recently formed; as
time goes on they join Population II in properties (perhaps renewing their
youth somewhat, at the expense of the nonstellar material in which they
are traveling). The supply of young stars is replenished until the dust and
gas are in some way removed. When a galaxy arrives at this point, or soon
after, the substratum is all that is left, and the end-point of the process is
an elliptical galaxy. A small system has, perhaps, never been a spiral; a

large system like our own galaxy may finally turn into something similar to Messier 87, which has not a much lower luminosity than the substratum of our own system.

Probably the original substrata are nearly spherical; but the stars that attain Population II come, with advancing time, from more and more flattened regions—there seems to be no way in which the orbits described by the stars around the center can be systematically changed (in inclination and eccentricity) *once they have become stars*. Therefore the final stage of a stellar system that has been a spiral galaxy is probably a highly flattened, nucleated elliptical system. All this is, of course, highly speculative, and another turn of the kaleidoscope may alter the picture completely.

The detailed development of particular stars has merely been mentioned. Here the main role is probably played by stellar rotation. Most of the hot, luminous main-sequence stars are rotating rapidly. Struve has suggested the possible life history of such an object: it begins as a single spinning star and may increase its rotation as it contracts (and if it does, it must spin faster and faster if angular momentum is to be conserved), and it may then either begin to spill at the edges or split in two. That a large fraction of stars are double suggests the second step as very likely. Even when a star has divided, the rapid spinning of the resulting pair will continue to shed material, so that the stars will steadily lose mass and luminosity and will move down the main sequence. (Admittedly, the *observed* rate of loss by spinning pairs is not large enough to allow the components to move very far). Finally, Struve suggests, such a star may reach the status of the sun, and develop a "solar system," thus getting rid of angular momentum; most of the angular momentum of our solar system is in the planets and the sun has very little. This idea looks superficially rather unlike those that were discussed in Chapter VIII for the genesis of the solar system; but Kuiper's theory envisages the formation of planets as a special case of double-star formation.

A few of the bright main-sequence stars seem to be rotating slowly. These, perhaps, develop into the "giant stars." They gradually build up a huge envelope which gives them a low superficial temperature; however, they are not spinning fast enough to shed it. Giant stars, according to this idea, are constructed quite differently from main-sequence stars; their inner structure has long been a puzzle, and current opinion suggests that if their energy source is similar to that of the main-sequence stars, they must be more internally condensed, so that their cores are equally hot. They are essentially main-sequence stars with huge envelopes. The giant stars, of course, are very rare compared to the main-sequence stars; it is quite likely that they represent the uncommon accident of a star originally formed with a small rotation.

The picture of stellar development just sketched is fairly coherent; but it is incomplete, and it may be quite incorrect. Time will tell.

I well consider all that ye have said,
And find that all things steadfastness do hate
And changed be: yet being rightly weighed
They are not changed from their first estate;
But by their change their being do dilate;
And turning to themselves at length again,
Do work their own perfection so by fate:
Then over them change doth not rule and reign,
But they reign over change and do their states maintain.

EDMUND SPENSER
Faery Queen, VII, Canto vii

PHYSICAL CONSTANTS
AND UNITS EMPLOYED

A great many units are needed to express the physical quantities measured in astronomy. These units are not all, however, quite independent of one another. If we exclude electric and magnetic properties, all the units that we employ are compounded of the three basic units: *mass, length,* and *time.* Speed, or velocity, for example, may be expressed in kilometers per second. The kilometer is a unit of length, and the second, of time. It is conventional to say that velocity *has the dimensions* of length divided by time, and to write the relation:

$$[\text{velocity}] = [L/T] = [LT^{-1}].$$

The *numerical value* of a particular velocity depends, of course, on the actual sizes of the adopted units of length and time. The velocity may be expressed in kilometers per second (the basic units being the kilometer and the second) or in miles per hour (the basic units being the mile and the hour). Thus a velocity of 60 miles per hour is a velocity of 0.0268 kilometers per second. Evidently the basic units in which a quantity is expressed must be stated.

The units that are employed in this book, and their dimensions, as defined above, are summarized below.

Units of Mass [M]. The mass of a body is related to the quantity of matter that it contains. At any one place, the weights of bodies are proportional to their masses, weight being the force exerted on a mass by gravitational attraction. The same mass will have different weights even at different parts of the earth's surface, because of the differences in gravity (g) at different points.

1 gram (g) = the basic metric unit of mass (it is a small quantity: a half dollar has a mass of 12½ g).
1 kilogram (kg; 1000 g) = nearly 2.2 lb avoirdupois.
1 solar mass = 1.992×10^{33} g.

Units of Length [L]. The metric unit of length, the meter, was originally defined with reference to the supposed circumference of the earth;

later measures showed that it does not conform exactly to this definition. The meter is now defined empirically by the length of a carefully preserved standard bar.

> 1 meter (m) = 3.281 ft.
> 1 centimeter (cm) = 10^{-2} m.
> 1 Angstrom unit (A) = 10^{-8} cm = 10^{-10} m.
> 1 kilometer (km; 1000 m) = 0.621 miles.
> 1 astronomical unit (A.U.) = 1.495×10^8 km.
> 1 light year (l.y.) = 9.4627×10^{12} km.
> 1 parsec (ps) = 3.084×10^{13} km.
> 1 kiloparsec (kps) = 1000 ps.
> 1 megaparsec (mps) = 10^6 ps.

Units of Time [*T*]. The basic unit of time used in astronomy is the mean solar second (see p. 55). The larger units of time, the day, month, and year, must be defined according to the system to which they apply.

> MEAN SOLAR TIME
>
> 1 second (s; sec).
> 1 hour (h; hr) = 3.600×10^3 s.
> 1 day (d) = 8.600×10^4 s = 24 h.
> 1 synodic month = 29 d 12 h 44 m 2.8 s
>   = 29.530588 d.
> 1 sidereal month = 27 d 7 h 43 m 11.5 s
>   = 27.321661 d.
> 1 nodical month = 27 d 5 h 5 m 35.8 s
>   = 27.212220 d.
> 1 tropical year = 365 d 5 h 48 m 46.0 s
>   = 365.24220 d = 3.15569×10^7 s.
> 1 sidereal year = 365 d 6 h 9 m 9.5 s = 3.15581×10^7 s
>   = 365.25636 mean solar days
>   = 366.25636 sidereal days.
> 1 anomalistic year = 365 d 6 h 13 m 53.01 s
>   = 365.25964 d.
> 1 sidereal day = 23 h 56 m 4.091 s of mean solar time.

The so-called *cosmic year* is the time taken by the sun to make one complete revolution around the center of our galaxy. It equals about 2×10^8 years. It is, however, not an accurately determined quantity and is used only in rough and general statements.

Units of Velocity [*LT*$^{-1}$].

> kilometers per second (km/sec).
> parsecs per million years (1 ps/10^6 y) = nearly 1 km/sec.

Radial velocity, *v*, is deduced from the Doppler shift by the formula $d\lambda/\lambda = v/c$, where λ is the unaltered wave length, $d\lambda$ the change in wave length, and *c*, the velocity of light (2.997×10^{10} cm/sec = 2.997×10^5 km/sec).

Units of Acceleration $[LT^{-2}]$. Acceleration is the rate of change of velocity; it may be expressed in the following terms.

centimeters per second per second (cm/sec^2).
gravitational acceleration at earth's surface (standard) $= 980.665$ cm/sec^2.

Units of Force $[MLT^{-2}]$. 1 Dyne is the (unopposed) force that will, if acting continuously, impart a uniform acceleration of 1 cm/sec^2 to a mass of 1 g.

1 gram weight $= 980.665$ dynes.

Units of Work, Energy, and Heat $[ML^2T^{-2}]$. 1 Erg is the work done by a force of 1 dyne if it acts through a distance of 1 cm in its own direction.

1 joule (j) $= 10^7$ ergs.
1 gram-calorie $=$ heat required to raise the temperature of 1 g of water through 1°C (varies slightly with temperature).
1 mean gram-calorie $= 4.186$ j.
1 electron volt (ev) $= 1.6020 \times 10^{-12}$ ergs.
1 million electron volts (Mev) $= 1.6020 \times 10^{-6}$ ergs.

Units of Power $[ML^2T^{-3}]$.

1 watt (absolute) (w) $= 10^7$ erg/sec $= 1$ j/sec.
1 horsepower (hp) $= 746$ w.

Unit of Magnetic Field $[M^{\frac{1}{2}}L^{-\frac{1}{2}}T^{-1}\mu^{-\frac{1}{2}}]$.

1 gauss is the strength of a magnetic field such that unit magnetic pole situated in it experiences a force of 1 dyne.

Unit magnetic pole is a pole that repels an equal and similar pole, at a distance of 1 cm in air, with a force of 1 dyne.

Note that magnetic field is the only quantity that we have defined that is not expressible solely in terms of mass, length and time. It requires the additional quantity μ, known as the *magnetic permeability*.

THE CONSTELLATIONS

Andromeda	And	Andromeda
Antlia	Ant	The Air Pump
Apus	Aps	The Bird of Paradise
Aquarius	Aqr	The Water-Carrier
Aquila	Aql	The Eagle
Ara	Ara	The Altar
Aries	Ari	The Ram
Auriga	Aur	The Charioteer
Boötes	Boo	The Herdsman
Caelum	Cae	The Graving Tool
Camelopardalis	Cam	The Giraffe
Cancer	Cnc	The Crab
Canes Venatici	CVn	The Hunting Dogs
Canis Major	CMa	The Great Dog
Canis Minor	CMi	The Little Dog
Capricornus	Cap	The Goat
Carina	Car	The Keel
Cassiopeia	Cas	Cassiopeia
Centaurus	Cen	The Centaur
Cepheus	Cep	Cepheus
Cetus	Cet	The Whale
Chamaeleon	Cha	The Chameleon
Circinus	Cir	The Pair of Compasses
Columba	Col	The Dove
Coma Berenices	Com	Berenice's Hair
Corona Australis	CrA	The Southern Crown
Corona Borealis	CrB	The Northern Crown
Corvus	Crv	The Crow
Crater	Crt	The Goblet
Crux	Cru	The Cross
Cygnus	Cyg	The Swan
Delphinus	Del	The Dolphin
Dorado	Dor	The Swordfish
Draco	Dra	The Dragon
Equuleus	Equ	The Little Horse
Eridanus	Eri	The River Eridanus
Fornax	For	The Furnace
Gemini	Gem	The Twins
Grus	Gru	The Crane
Hercules	Her	Hercules

Horologium	Hor	The Clock
Hydra	Hya	The Sea-Serpent
Hydrus	Hyi	The Water-Snake
Indus	Ind	The Indian
Lacerta	Lac	The Lizard
Leo	Leo	The Lion
Leo Minor	LMi	The Little Lion
Lepus	Lep	The Hare
Libra	Lib	The Scales
Lupus	Lup	The Wolf
Lynx	Lyn	The Lynx
Lyra	Lyr	The Lyre
Mensa	Men	The Table [mountain]
Microscopium	Mic	The Microscope
Monoceros	Mon	The Unicorn
Musca	Mus	The Fly
Norma	Nor	The Ruler
Octans	Oct	The Octant
Ophiuchus	Oph	The Serpent-Bearer
Orion	Ori	Orion
Pavo	Pav	The Peacock
Pegasus	Peg	Pegasus
Perseus	Per	Perseus
Phoenix	Phe	The Phoenix
Pictor	Pic	The Easel
Pisces	Psc	The Fishes
Piscis Austrinus	PsA	The Southern Fish
Puppis	Pup	The Prow
Pyxis	Pyx	The Mariner's Compass
Reticulum	Ret	The Net
Sagitta	Sge	The Arrow
Sagittarius	Sgr	The Archer
Scorpio	Sco	The Scorpion
Sculptor	Scl	The Sculptor's Tools
Scutum	Sct	The Shield
Serpens	Ser	The Serpent
Sextans	Sex	The Sextant
Taurus	Tau	The Bull
Telescopium	Tel	The Telescope
Triangulum	Tri	The Triangle
Triangulum Australe	TrA	The Southern Triangle
Tucana	Tuc	The Toucan
Ursa Major	UMa	The Great Bear (Big Dipper)
Ursa Minor	UMi	The Little Bear (Little Dipper)
Vela	Vel	The Sails
Virgo	Vir	The Virgin
Volans	Vol	The Flying Fish
Vulpecula	Vul	The Fox

PROBLEMS

O dear Ophelia, I am ill at these numbers.
SHAKESPEARE
Hamlet, Act II, Scene ii

CHAPTER I

1. Discuss the contrasting points of view of the two quotations at the head of the chapter.

2. If Odysseus followed Calypso's advice, in which direction did he sail?

3. Discuss, on the basis of the information furnished by Aratus, whether the Greeks or the Phoenicians were the better navigators.

CHAPTER II

1. Discuss the relationship between astronomy and astrology.

2. Look for as many of the naked-eye planets as are currently visible, and discuss the appropriateness of the names given them by the Greek astronomers.

Sections 1–2:

1. Observe the setting sun. Describe the changes that it seems to undergo in color and in shape, and try to account for them.

Sections 3–5:

1. Look for the earthlight near the time of new moon; compare what you can see of the surface with Figure 5.4.

2. Draw a diagram to demonstrate that if Eratosthenes had made observations from three places at the same longitude, he could have shown directly that the sun is very distant, instead of having been obliged to assume it.

3. Determine by measurement the oblateness of a pumpkin, and the prolateness of a football.

Sections 6–10:

1. Why does the sun rise at a different point on the horizon, as seen from a given place, at midsummer and at midwinter? Notice the point at

which the sun rises at weekly intervals. At what times of year will it rise furthest north and furthest south?

2. Construct ellipses of eccentricity 0.8, 0.3 by means of a string (length equal to the major axis) attached at its ends to two pins (stuck into a board at the positions of the two foci).

3. What features of our present calendar do you consider the most objectionable? How would you suggest that they should be corrected?

CHAPTER III

Sections 1–2:

1. As a ship travels from America to Europe, should the clocks be advanced or retarded? Why?

2. How would you design a sundial for use in the tropics? At the equator?

3. In which direction should one cross the international date line in order to go back to the previous day?

Section 3:

1. What is the frequency of radio waves of wave length 2000 meters? Of microwaves of wave length 1 centimeter? Of ultraviolet light of wave length 1000 Angstroms?

2. Assume the temperature of the surface of the sun to be 6000°K. What is the temperature of the surface of a star that radiates 16 times as much light per square centimeter? That radiates 1/81 times as much light per square centimeter?

3. The sun (assuming temperature 6000°K) has maximum energy in the yellow-green wave length, 5000 Angstroms. What is the temperature of a star whose maximum energy is at 2500 Angstroms? At 10,000 Angstroms?

Section 4:

1. What properties are required of a telescope for the study of (a) planetary surfaces, (b) very faint stars, and (c) very faint nebulosities?

2. Discuss the change in the astronomical picture made by (a) the invention of the telescope, and (b) the photographic plate.

CHAPTER IV

Sections 1–3:

1. If the sun were three times as distant, how large would it appear to us?

2. If the sun were at its present distance, but of twice the diameter and half the surface temperature, by what factor would its apparent brightness be altered?

3. Suppose an area of the sun's surface is twice as bright as the neighboring regions, and that the difference of brightness is a consequence of

difference in temperature. If the temperature of the neighboring regions is 6000°K, what is the temperature of the brighter region?

Sections 5–7:

1. Suppose two points of the sun's surface to be at the same (solar) longitude, one at (solar) latitude 20°, the other at 60°. How long a time must elapse before they are again at the same solar longitude? Which one will then have gained one lap on the other?

2. Study the spectra, Figures 4.14 and 4.17, and pick out lines that show: the lower temperature of the sunspot; the high magnetic field of the sunspot.

CHAPTER V

1. How much would a body that weighs 10 lb on the earth weigh on the moon?

2. If the moon were twice as distant, by what factor would its brightness be altered?

3. If the moon had an albedo of 0.50, by what factor would its brightness be altered?

4. Simulate the lunar rays by smacking the back of the bowl of a spoon sharply onto a small mound of flour or facepowder.

5. What kinds of observation do you suppose that Hipparchus must have made in order to discover the evection and the variation?

CHAPTER VI

Section 1:

1. If the moon's orbit were twice as large as it is, how would eclipses of the moon differ from those now observed?

2. Under the same conditions, how would eclipses of the sun differ?

3. Imagine yourself on the moon at a time of total eclipse of the sun on earth. What would you expect to observe?

4. What would you expect to observe from the moon at the time of a total lunar eclipse on the earth?

Section 2:

1. In what parts of the earth would you expect the diurnal inequality to be (a) largest, (b) least, and why?

2. At what intervals does perigee occur? At what intervals do new and full moons occur? What, therefore, will be the interval between the especially high spring tides that characterize the coincidence of these events?

CHAPTER VII

Section 1:

1. Can Mercury be visible during a total eclipse of the moon? Can Jupiter? Why?

2. Under what circumstances can Venus be visible at local midnight?

3. From what part of the solar system could a transit of Jupiter be observed?

4. If a planet had a synodic period of fifty-two years, what would its sidereal period be?

5. What phases would the earth display, as seen by an observer on Mars? On Mercury?

Section 2:

1. From what feature of the narrative of Herodotus could it have been inferred that the earth is not flat?

2. If Aristarchus' estimate of the diameters of the sun and moon were correct, what would be their actual distances from the earth? Assume the diameter of the earth to be known to have its modern value.

3. If Heraclitus' estimate that the sun is 1 foot in diameter were correct, how far from the surface of the earth would the sun be?

4. What observations, made without instruments, give evidence of the obliquity of the ecliptic?

Sections 3–5:

1. What do you think would have been the effects on human history if the heliocentric system had been accepted from the time of Aristarchus?

2. Use Kepler's third law to determine: (a) the period (in years) of a planet twice as distant from the sun as the earth; (b) the distance from the sun (in astronomical units) of a planet with a period twice the earth's; (c) the period of a "planet" that exactly grazes the sun's surface; (d) the period (in days) of a terrestrial satellite that moved in an orbit 500 kilometers from the earth's surface at the equator; (e) the distance from the earth's surface of a satellite that moved around above the equator with a period of seven solar days.

Sections 6–7:

1. Make a table of Galileo's principal observations, and estimate the scientific importance of each.

2. Identify the principal features of the moon's motion in Newton's summary.

Section 8:

1. Explain why favorable oppositions of Mars occur only about once in sixteen years.

2. If future observations were to increase the accepted mean value of the solar parallax by 1%, what would that value be? What would the size

of the astronomical unit be? What would be the effect on our ideas of the scale of the solar system?

CHAPTER VIII

Sections 1–2:

1. Could an observer on Mercury observe Venus in the crescent phase? Why?

2. How much would a body that weighs 10 pounds on earth weigh on Mercury?

3. Discuss what kinds of information a (hypothetical) inhabitant of Venus could obtain about the rest of the solar system.

Sections 3–4:

1. Tabulate (a) the features in which Mars resembles the earth, and (b) those in which the two planets differ.

2. Consider Mars as a possible site for life of our kind on the basis of the answer to Question 1 above.

3. By what factor would the average diameter of the earth, as seen from Mars, differ from the average diameter of Mars, as seen from the earth?

4. If our moon was the size of Phobos, and otherwise unchanged, by what factor would its observed brightness differ?

5. If our moon had an orbit of the same size as that of Deimos, what would be its period? Why would this period not be the same as that of Deimos? By what factor would its apparent diameter and brightness differ from that which we observe?

Section 5:

1. Tabulate the features in which Jupiter (a) resembles, and (b) differs from the earth.

2. If Io, Jupiter 6, and Jupiter 9 were satellites of the earth with their present periods, what would be their mean distances?

3. If Jupiter 5, Callisto, and Jupiter 11 were satellites of the earth with their present mean distances, what would be their periods?

Section 6:

1. How many times have Saturn's rings been presented to the earth edgewise since Galileo first saw them in this orientation?

2. By what factor would the mean diameter of Saturn differ, as observed from Mimas and from Phoebe?

3. By what factor would the average brightness of Saturn's ball differ, as observed from Enceladus and from Hyperion?

4. If Mimas were a satellite of Mars, with its present mean distance, what would be its period? Why would this period differ from the present period of Mimas?

Sections 7–9:

1. Compare the mean apparent diameter and mean brightness of the sun, as seen from Mercury and from Pluto.

2. What is the interval between the closest approaches of Neptune and Pluto to one another?

CHAPTER IX

Sections 1–2:

1. The mean distances of Ceres and Eros from the sun are 414 and 218 million kilometers respectively; the eccentricities of their orbits are 0.076 and 0.223 respectively. Calculate the greatest and least distances of these asteroids from the sun.

2. Make a tabular comparison of Pallas, Vesta, and our own moon; if these two asteroids were placed at the distance of the moon, by what factor would their brightness differ from hers? Assume that Pallas has an albedo of 0.15, Vesta, of 0.50.

Sections 3–4:

1. Calculate the semimajor axes of the orbits of the following comets, and their aphelion distances.

Comet	*Eccentricity*	*Perihelion Distance* (A.U.)
Halley	0.967	0.587
Encke	0.850	0.332
Giacobini-Zinner	0.716	1.000
Schwassmann-Wachmann II	0.395	2.09

[Note: $e = (s - p)/s$, and $a = 2s - p$, where s is the semimajor axis, p the perihelion distance, a the aphelion distance, and e the eccentricity of the orbit (see Fig. 2.12).]

2. The orbit of Lexell's Comet changed twice in the latter part of the eighteenth century. Use the data given to calculate the eccentricity and aphelion distance on each occasion.

Date	*Semimajor Axis* (A.U.)	*Perihelion Distance* (A.U.)	*Period* (yr)
Before 1767	5.06	2.96	11.4
1770	3.15	0.67	5.6
After 1779	6.37	3.33	16.2

Verify Kepler's third law; why is the numerical agreement not exact?

CHAPTER XI

1. Discuss how far the study of the solar system prepares us for a study of the stars. In what respects is it inadequate?

CHAPTER XII

Sections 1–2:

1. By what factor do the brightnesses of the sun and of the full moon exceed that of Sirius?

2. If the components of Spica were of equal brightness, what would be the apparent magnitude of each?

3. Calculate the distance of Sirius, Procyon, and Kapteyn's star in parsecs; in light years.

Sections 3–4:

1. Calculate the radii of Sirius, Procyon, Kapteyn's star from the data given in Table 12.2 and Table 12.3.

2. Compare the diameters of Procyon and Canopus (spectrum F0) and of Betelgeuse and Kapteyn's star.

3. Which is the largest star of Table 12.2, and which is the smallest? Compare their diameters.

Section 5:

1. Calculate the tangential velocities of Sirius, Barnard's star, Kapteyn's star, and Alpha Centauri (Table 12.2).

2. Which star of Table 12.2 has the largest tangential velocity? The smallest? What would be the tangential velocity of a star whose annual proper motion was numerically equal to its parallax?

Section 6:

1. Calculate the total velocities of Sirius, Barnard's star, Kapteyn's star, Alpha Centauri, Washington 5583 (Table 12.5).

2. List the stars that are common to the tables of the brightest stars and the nearest stars. Calculate their total velocities.

3. List the stars that are common to the lists of the nearest stars and the stars of largest proper motion. Calculate their total velocities.

4. List the stars that are common to the tables of the brightest stars and the stars of the largest proper motion.

5. Which star in the three tables mentioned has the largest total velocity; the smallest total velocity?

Section 7:

1. Calculate the greatest and least distances between the components of Sirius in astronomical units.

2. Calculate the semimajor axes of the orbits of the components of Procyon about their common center of gravity.

3. In Table 12.7, which pair of stars approach most closely? Which has the greatest maximum separation?

4. Which of the principal planets have orbits larger than the orbit of the less massive component of Capella about the more massive one?

Section 8:

1. Suppose a totally eclipsing system with spherical components to consist of stars with surface temperatures of 5000° and 12,000°. Calculate the ratio of the depths of the two eclipses, and the difference in depth in magnitudes.

2. In a totally eclipsing system with spherical components, one eclipse is exactly three times as deep as the other. What is the ratio of surface temperatures of the two stars?

CHAPTER XIII

Sections 1–2:

1. Compare Figure 13.5 and Figure 4.15. Pick out features that indicate that Arcturus is at a lower temperature than the sun.

Sections 3–4:

1. Compare the spectra of the Orion Nebula (Figure 13.17), of NGC 7662 (Figure 13.20) and of NGC 6572, 7027 (Figure 13.21). List the similarities and differences that you can detect.

Section 5:

1. Calculate in mass units and in Mev, the amount of energy absorbed by the reaction of Figure 13.22 using 1 gram of $_7N^{14}$.

2. If a star converts 10^{23} grams of hydrogen into helium by means of the carbon cycle, calculate the energy liberated in mass units.

3. A star has liberated 10^{29} grams of energy as a result of the proton-proton reaction. What mass of hydrogen has been consumed, and what mass of helium produced?

Section 8:

1. Calculate the mean molecular weight of a completely ionized mass of (a) hydrogen, (b) lithium, (c) iron, and (d) barium. The approximate nuclear masses are respectively 1, 7, 56, and 137; the atomic numbers are 1, 3, 26, and 56.

2. Compare the central temperatures of the stars of Table 12.4 on the assumption that all are built on the same model and have the same mean molecular weight.

CHAPTER XIV

Sections 1–2:

1. From Figure 14.4 deduce the approximate absolute photographic magnitudes of the Cepheids of Table 14.1.

2. Use the "giant" color indexes of Table 12.3 to obtain the absolute visual magnitudes of the Cepheids of Table 14.1 for the *average* spectrum (it will be necessary to interpolate in Table 12.3).

3. From average color index and absolute visual magnitude (see Question 2) calculate the radii of the Cepheids by the formula of p. 281.

4. What distance for the Magellanic Clouds is implied by the vertical scales of Figure 14.4?

Sections 3–4:

1. The average apparent photographic magnitude of the RR Lyrae stars in the globular cluster ω Centauri (Figure 15.8) is 14.37. Calculate the distance of the cluster. In this and similar problems obscuration is supposed to be absent unless the contrary is stated.

2. If RR Lyrae stars of the twentieth apparent photographic magnitude are just observable, up to what distance can they be detected?

3. The radial velocity of an RR Lyrae star of apparent magnitude 7 is —100 km/sec, and its proper motion is 0″.012 per year. Calculate its total velocity.

Section 5:

1. A long-period variable is 1.5 magnitudes brighter at maximum (temperature 2700°) than at minimum. Assume that the whole change is caused by change of temperature, and calculate the temperature at minimum.

2. Assume that Mira Ceti has a mass of ten suns, and take its diameter from p. 284; calculate its mean density (a) in solar densities, and (b) in grams per cubic centimeter.

Section 6:

1. A nova reaches apparent magnitude +6 at maximum. Assume its absolute magnitude at maximum to be —7, and calculate its distance.

2. If the star of Question 1 were a supernova of absolute magnitude —17 at maximum, what would its distance be?

3. Suppose stars of the twentieth apparent magnitude to be detectable. How distant would the most distant detectable nova be? The most distant detectable supernova?

4. Suppose the sun to become a nova, the surface rising at 1500 km/sec. What would be the volume after twenty-four hours? If the surface brightness per square centimeter remained constant during the expansion, how much would the "nova" increase in brightness in twenty-four hours?

5. A nova (absolute magnitude —7) appears at a distance of 8 kiloparsecs. What is its apparent magnitude at maximum?

6. If the star of Question 5 were a supernova, absolute magnitude —17, how bright would it appear to us?

Section 7:

1. Assume that U Geminorum stars have absolute magnitude +4 at maximum, and that the stars of the twentieth magnitude are just detectable. Up to what distance can U Geminorum stars be observed?

2. A U Geminorum star is of apparent magnitude +8 at maximum. Calculate its distance.

3. A flare on the sun has 1% of the sun's total brightness. How much would a similar flare on Proxima Centauri change the apparent magnitude of that star?

CHAPTER XV

Sections 1–3:
1. The Hyades contain both A0 and K0 stars of absolute visual magnitude 0. The brightest stars in Pleiades are of Class B5 and absolute visual magnitude −2.5. The double cluster in Perseus contains both B0 and M0 stars of absolute visual magnitude −6. Compare the sizes of the stars mentioned (the data are approximate).

Section 4:
1. Our galaxy contains about 10^{11} stars and 100,000 RR Lyrae stars. The globular cluster ω Centauri contains about 130 RR Lyrae stars and its total population is estimated at 100,000 stars. Compare the RR Lyrae populations of the galaxy and the cluster.

2. The nova T Scorpii (adopt magnitude +7 at maximum) appeared in 1860 near the globular cluster Messier 80. If its absolute magnitude at maximum was −7, and it was in the cluster, calculate the distance of the cluster. If the apparent magnitude of the cluster is +6.8, calculate its absolute magnitude.

3. The apparent magnitude of the globular cluster Messier 3 is +4.5; the average apparent magnitude of its RR Lyrae stars is 15.50; calculate its distance and absolute magnitude.

4. Compare the absolute magnitudes obtained in Questions 2 and 3; discuss the possible reasons for their difference.

CHAPTER XVI

Sections 1–2:
1. The nova RY Doradus (12.4 at maximum) appeared in 1926 in the Large Magellanic Cloud; if it was a typical nova, what is the distance of the Large Cloud?

2. The R Coronae variable W Mensae is in the Large Magellanic Cloud. From its apparent magnitude, +13.8 at maximum, deduce the absolute magnitude of an R Coronae star, using the result of Question 1.

3. If the zero point of the period-luminosity curve were revised, and all absolute magnitudes made brighter by 1.5 magnitudes, by what factor would our value for the distance of the Magellanic Clouds be changed? What would the revised distance be?

Sections 3–4:
1. In 1885, a 6.5 magnitude supernova appeared in Messier 31. If its absolute magnitude was −17, calculate the distance of Messier 31.

2. If the distance of Messier 31 is as given in Question 1, what will be the apparent magnitude of its RR Lyrae stars? How bright would a star like the sun appear at this distance?

3. Compare the dimensions of Messier 31 and Messier 33.

Section 5:

1. If there were no obscuration within the galaxy, how bright would an RR Lyrae star appear at the near and far edges of the system? (Assume the diameter of the galaxy to be 24 kps, and that we are 8 kps from the center.)

2. If we can see nineteenth magnitude RR Lyrae stars toward the galactic center, 8 kps distant, calculate the absorption per kps in the galactic plane. On the basis of this calculation, how faint would the RR Lyrae stars on the extreme far side of our system appear to be?

3. If a supernova has absolute magnitude —15, how bright would a supernova on the far edge of the galactic plane appear to be? Can we assume that all galactic supernovae in the past 2000 years have been recorded? In the past 200 years?

Section 6:

1. In 1919 a supernova appeared in Messier 87 (Figure 16.15); if it had apparent magnitude +12.3 and absolute magnitude —17, calculate the distance of Messier 87. If the apparent magnitude of Messier 87 is +10.1, what is its absolute magnitude?

2. Calculate the percentages of spiral, irregular, and elliptical galaxies in Table 16.1.

CHAPTER XVII

1. Calculate the percentages of spiral, irregular, and elliptical galaxies in Table 17.1.

2. By what factor does the largest galaxy in Table 17.1 exceed the smallest?

3. By what factor does the brightest galaxy in Table 17.1 exceed the faintest?

4. Calculate the mean diameter and mean absolute magnitude of the Sb, Sc, irregular and elliptical galaxies in Table 17.1.

INDEX